Bicycle Transportation 2018

Transportation Research Record: Journal of the Transportation Research Board Volume 2672, Issue 36

ISBN: 978-1-5108-8244-7

Printed from e-media with permission by:

Curran Associates, Inc.
57 Morehouse Lane
Red Hook, NY 12571

Some format issues inherent in the e-media version may also appear in this print version.

Published by Sage Publications USA

ISBN (Print) 978-1-5108-8244-7
ISBN 2018 Printed Set (All Issues) 978-1-5108-7735-1

Printed by Curran Associates, Inc. (2019)

Additional copies of this publication are available from:

Curran Associates, Inc.
57 Morehouse Lane
Red Hook, NY 12571 USA
Phone: 845-758-0400
Fax: 845-758-2633
Email: curran@proceedings.com
Web: www.proceedings.com

Volume 2672 Number 36 December 2018

TRR
JOURNAL OF THE TRANSPORTATION RESEARCH BOARD

Contents

Bicycle Transportation 2018

Articles

Perceived Safety and Separated Bike Lanes in the Midwest: Results from a Roadway Design Survey in Michigan 1
Rebecca L. Sanders and Belinda Judelman

Effects of Building a Stock of Bicycling Experience in Youth 12
Calvin G. Thigpen and Susan L. Handy

Equity in Spatial Access to Bicycling Infrastructure in Mid-Sized Canadian Cities 24
Meghan Winters, Jaimy Fischer, Trisalyn Nelson, Daniel Fuller, and David G. T. Whitehurst

Lowering Bicycle Stress One Link at a Time: Where Should We Invest in Infrastructure? 33
Sarah K. Moran, William Tsay, Sean Lawrence, and Gregory R. Krykewycz

Who Are Public Bicycle Share Programs Serving? An Evaluation of the Equity of Spatial Access to Bicycle Share Service Areas in Canadian Cities 42
Kate Hosford and Meghan Winters

Empirical Analysis of the Macroscopic Characteristics of Bicycle Flow during the Queue Discharge Process at a Signalized Intersection 51
Bernat Goñi-Ros, Yufei Yuan, Winnie Daamen, and Serge P. Hoogendoorn

The Keys to Connectivity: The District of Columbia's Innovative Approach to Unlocking Low-Stress Bicycle Networks 63
Conor Semler, Meredyth Sanders, Darren Buck, Stephanie Dock, Burak Cesme, and Shi Wang

Park 'n' Roll: Identifying and Prioritizing Locations for New Bicycle Parking in Québec City, Canada 73
Marie-Pier Veillette, Emily Grisé, and Ahmed El-Geneidy

Generation of "Biking Schedules" for Bicycle Travel Analysis 83
Amr Mohamed and Alexander Y. Bigazzi

Use of Deep Learning to Predict Daily Usage of Bike Sharing Systems 92
Hong Yang, Kun Xie, Kaan Ozbay, Yifang Ma, and Zhenyu Wang

Introducing a Design Framework for a Multi-Modal Public Transportation System, Focusing on Mixed-Fleet Bike-Sharing Systems 103
Mehrnaz Ghamami and MohammadHossein (Sam) Shojaei

The Relationship between Experienced and Imagined Bicycling Comfort and Safety 116
Dillon T. Fitch and Susan L. Handy

Analytical Model of Passing Events for One-Way Heterogeneous Bicycle Traffic Flows 125
Liang Xu, Meiqi Liu, Xiang Song, and Sheng Jin

Operational Evaluation of Central Sharrows and Dooring Zone Treatment on Road User Behavior in Ottawa, Canada 136
Ali Kassim, Karim Ismail, and Shawn McGuire

Quality of Bicycle Traffic Management at Urban Road Links and Signalized Intersections Operating under Mixed Traffic Conditions 145
Sambit Kumar Beura and Prasanta Kumar Bhuyan

Design and Evaluation of K-Pass: A Bicycle-Friendly Modification of Speed Bumps 157
Vinod Vasudevan, Aniruddha Rajurkar, Rahul Soni, and Akhil Tiwari

Transportation Officials, Washington, D.C.
K. Jane Williams, *Acting Administrator, Federal Transit Administration, U.S. Department of Transportation*
Paul Zukunft *(Admiral, U.S. Coast Guard), Commandant, U.S. Coast Guard, Washington, D.C.*

---

* Membership as of February 2019.

TRB

TRANSPORTATION RESEARCH BOARD

CORE SPONSORS

Transportation Departments of the 50 States and the District of Columbia

Federal Government

U.S. Department of Transportation
- Federal Aviation Administration (FAA)
- Federal Highway Administration
- Federal Motor Carrier Safety Administration (FMCSA)
- Federal Railroad Administration (FRA)
- Federal Transit Administration
- National Highway Traffic Safety Administration (NHTSA)
- Office of the Assistant Secretary of Transportation for Research and Technology (OST-R)

U.S. Air Force Civil Engineer Center (AFCEC)
U.S. Coast Guard (USCG)
U.S. Department of Energy
U.S. Department of the Interior (DOI)
U.S. Environmental Protection Agency (EPA)

Nongovernmental Organizations

American Public Transportation Association (APTA)
Association of American Railroads (AAR)
California Air Resources Board

The National Academies of
SCIENCES · ENGINEERING · MEDICINE

The **National Academy of Sciences** was established in 1863 by an Act of Congress, signed by President Lincoln, as a private, nongovernmental institution to advise the nation on issues related to science and technology. Members are elected by their peers for outstanding contributions to research. Dr. Marcia McNutt is president.

The **National Academy of Engineering** was established in 1964 under the charter of the National Academy of Sciences to bring the practices of engineering to advising the nation. Members are elected by their peers for extraordinary contributions to engineering. Dr. C. D. Mote, Jr., is president.

The **National Academy of Medicine** (formerly the Institute of Medicine) was established in 1970 under the charter of the National Academy of Sciences to advise the nation on medical and health issues. Members are elected by their peers for distinguished contributions to medicine and health. Dr. Victor J. Dzau is president.

The three Academies work together as the National Academies of Sciences, Engineering, and Medicine to provide independent, objective analysis and advice to the nation and conduct other activities to solve complex problems and inform public policy decisions. The Academies also encourage education and research, recognize outstanding contributions to knowledge, and increase public understanding in matters of science, engineering, and medicine.

Learn more about the National Academies of Sciences, Engineering, and Medicine at **www.national-academies.org.**

The **Transportation Research Board** is one of seven major programs of the National Academies of Sciences, Engineering, and Medicine. The mission of the Transportation Research Board is to increase the benefits that transportation contributes to society by providing leadership in transportation innovation and progress through research and information exchange, conducted within a setting that is objective, interdisciplinary, and multimodal. The Board's varied activities annually engage about 7,000 engineers, scientists, and other transportation researchers and practitioners from the public and private sectors and academia, all of whom contribute their expertise in the public interest. The program is supported by state transportation departments, federal agencies including the component administrations of the U.S. Department of Transportation, and other organizations and individuals interested in the development of transportation.

Learn more about the Transportation Research Board at **www.TRB.org**.

Article

TRR
JOURNAL OF THE TRANSPORTATION RESEARCH BOARD

Transportation Research Record
2018, Vol. 2672(36) 1–11

DOI: 10.1177/0361198118758395
journals.sagepub.com/home/trr
SAGE

Perceived Safety and Separated Bike Lanes in the Midwest: Results from a Roadway Design Survey in Michigan

Rebecca L. Sanders[1] and Belinda Judelman[1]

Abstract

This article presents the results of an address-based sample survey (n = 351) conducted in the fall of 2016 for the Michigan Department of Transportation (MDOT) as part of an effort to provide guidance for building sidepaths. The survey investigated attitudes toward bicycling among drivers and bicyclists, bicycling habits, barriers to bicycling, and roadway design preferences regarding bicycle infrastructure in Michigan. In particular, this survey explored design preferences while bicycling with children, bicycling by oneself, and driving. Safety emerged as a key barrier to bicycling, as did distance, weather, and the difficulty of carrying things or traveling with others. Roadway design preferences were clearly weighted toward greater separation when sharing the roadway whether as a bicyclist or a driver, and this trend was most pronounced ($p < 0.001$) when considering bicycling with children. In all cases, ratings for one-way separated bike lanes were similar to those for sidepaths, suggesting that separated bike lanes could be a key part of addressing the safety and comfort concerns of more cautious riders. Preferences for separation were strongly associated with perceived safety as a barrier. These results were even stronger for non-transport-cyclists, although all groups, regardless of frequency or type of bicycling, preferred more separation. These results corroborate past research and add compelling evidence for separated facilities as a key part of expanding the potential for bicycling trips in general, and particularly with children. The survey findings will inform guidance about sidepath design for MDOT.

The last two decades have seen tremendous growth in bicycling in the U.S. In the last decade alone, major U.S. cities have experienced an average 47% increase in bike commuting (*1*). In some cities, such as Lexington, Kentucky, and Portland, Oregon, the percentage of bike commuters has grown over 300% since 2000, and many other U.S. cities have seen growth in excess of 100–200% (*2*). This trend toward increased bicycling has led to a need to better understand how to accommodate bicyclists into streets historically designed for the car. Additionally, cities attempting to encourage bicycling are exploring ways to make cycling more attractive for the "interested but concerned" cyclists, which many see as key to increasing bicycling mode share (*3*).

This paper presents findings from research examining roadway design preferences when bicycling with children, bicycling alone, and driving on multi-lane, commercial streets. The findings are based on results from a recent survey exploring attitudes toward driving and bicycling, bicycling habits, barriers to bicycling, and roadway design preferences among Michigan residents. The results corroborate past research findings on bicyclists' and drivers' roadway design preferences, and contribute a new understanding of design preferences for bicyclists specifically traveling with children. These findings also clearly tie perceived safety to preferences, and support an alignment between roadway users for design preferences, with both drivers and bicyclists—regardless of bicycling frequency or type of bicycling—preferring greater separation on the roadway.

Literature Review

Barriers to Bicycling

Research on barriers to bicycling in North America has documented strong desires for additional bicycle facilities. For example, in their 2009 online survey of 1,605 Texas bicyclists, Sener, Uluru, and Bhat found that nearly 80% of respondents characterized the overall quality of bicycle facilities in their communities as "inadequate" or "very inadequate" (*4*). Similarly, Dill and Voros conducted a telephone survey of 566 Portland residents, and found that 37% of those who wanted to bicycle more reported that there were not enough bike lanes or trails near where they wanted

[1]Toole Design Group, LLC, Portland, OR

Corresponding Author:
Address correspondence to Rebecca L. Sanders: rsanders@tooledesign.com

to go (*5*). In their adaptive stated preference survey of 167 University of Minnesota employees, Tilahun, Levinson, and Krizek found that the presence of a bicycle lane had a much greater impact on the odds of choosing a facility than did the elimination of on-street-parking or the presence of an off-road-facility (*6*). This pattern was magnified among women, and was not significantly tied to bicycling experience.

Bicyclists' Roadway Design Preferences

More recent research has focused specifically on roadway design preferences. In their survey of 1,402 Vancouver bicyclists (all experience levels, including potential cyclists), Winters and Teschke found that off-street paths were the most preferred by all cyclist types, whereas major city streets without bicycle facilities were the least preferred. Barrier-separated cycle paths next to major streets were the fourth-ranked option for potential and occasional cyclists, and ranked second and third, respectively, for regular and frequent cyclists (*7*). A telephone survey of 908 residents from Portland, Oregon, found similarly strong preferences—particularly among potential cyclists and women—for separated bicycle facilities (*8*). These results were corroborated by a more recent survey of nearly 2,300 residents in five major U.S. cities (*9*). A separate national study of 3,000 people in 50 major U.S. cities found that the "interested but concerned" group of cyclists strongly preferred separation in the form of a path or trail over a major street with a striped bike lane (*3*).

Research on route choice supports these preferences. Winters et al. examined the routes of 74 participants to determine the distance people detour from the shortest path to use a bicycle facility (*10*). They found that bike trips were significantly more likely to occur along routes with enhanced bicycle facilities including traffic calming, stencils, and signage; although only 21% of trips would be along designated bike routes in a shortest path scenario, on average, 49% of actual trip distance took place along a bike route. Broach, Dill, and Gliebe used GPS monitors to gather data on the routes of 164 bicyclists over several days in Portland, Oregon. After modeling the results to account for tradeoffs in topography, traffic volumes, and street network characteristics, they found that bicyclists travel out of their way to reach bicycle infrastructure, and particularly bicycle boulevards (*11*).

Gap in the Literature

Knowing bicyclists' roadway design preferences is critical to encouraging more bicycling, but research has not always examined these preferences in a nuanced way. For example, a person may prefer the most direct route when commuting, but may prioritize roads with low-stress bicycle facilities when riding with children. Furthermore, given the needs of practitioners to plan for all modes, they also need to know how motorists who share the road with bicyclists view various design options. Few studies have examined drivers' design preferences, although Sanders found strong preferences for separation among both drivers and bicyclists in the San Francisco Bay Area (*12*). The research presented in this paper strengthens the findings about drivers' roadway design preferences and contributes to a nuanced understanding of bicyclists' design preferences, including how they differ when bicycling by oneself versus with children.

Methodology

Survey Construction and Recruitment

The findings presented in this paper are based on research conducted as part of a larger project to develop design guidance for sidepaths for the Michigan Department of Transportation (MDOT). The survey was developed based on past research on attitudes, behavior, and roadway design preferences (*3, 12*) and was designed to take 10–15 minutes to complete. The survey focused on drivers and bicyclists to capture as wide an audience as possible, and because conflicts on sidepaths often involve motor vehicles. An effort was made to collaborate with related national research (NCHRP 08-102: Bicyclist Facility Preferences and Effects on Increasing Bicycle Trips), and the Michigan survey modeled the NHCRP survey where feasible. In particular, the survey graphics were almost all developed by the NCHRP team (*13*). This collaboration will allow for certain responses to be compared with a larger sample from around the country, which may ultimately enhance the reach of the end user tool. More detail about the entire project, including a copy of the survey, can be found in the related MDOT report (*14*).

In October of 2016, approximately 5,000 letters printed in both English and Spanish were mailed to a random selection of Michigan residents who were in the telephone directory, asking them to take the survey. To increase the likelihood that recipients would open the letters, the research team hand-addressed each envelope and used first class stamps. The letter explained the purpose of the survey and directed the recipients to the website www.michdrivebike.org, where there was a link to take the survey in either English or Spanish; recipients could also request a paper copy of the survey. The first 250 respondents were offered a US$5 "gourmet coffee" gift card upon completion of the survey; the remaining respondents were entered into a raffle for one of twenty US$5 gift cards. Respondents were given 2 weeks to respond to the survey, and two reminder postcards were sent out, approximately 6 days apart. To mitigate potential response bias, the letter requested that only the person in the household whose birthday was most recent and who was at least age 18 take the survey.

Seven pairs of duplicates were detected, after which one of each pair was randomly removed from the sample. The final cleaned sample contained 351 respondents, resulting in a 4.9% response rate. Guidance on surveys suggests that this percentage is well within the range of what one can expect from a mail survey (*15*).

Respondent Cyclist Typology

To facilitate analysis, respondents were categorized according to how often they bicycled for "work/school", "transportation other than to work or school (e.g., errands)", and recreation or exercise. The categories equated to the following:

Frequent cyclist (*n* = 116): Respondent who reported bicycling at least once a week for any purpose (transportation, recreation, or exercise), and not being "absolutely limited" by not having a bike or not knowing how to ride a bike.

Occasional cyclist (*n* = 83): Respondent who reported bicycling at least once a month (but less than once a week) for any purpose, or reported biking at least once a week but also reported being "absolutely limited" by not having a bike or not knowing how to bike.

Rare cyclist (*n* = 93): Respondent who reported bicycling occasionally, but less than once a month for any purpose.

Never cyclist (*n* = 54): Respondent who reported never bicycling for any purpose; or who did not indicate how often they bike for a particular purpose, but is either "absolutely limited" by not knowing how to ride a bike or not owning a bike, or indicated that they "cannot bike at all."

Five respondents were unable to be classified because they did not indicate: 1) how often they biked for transportation, recreation, or exercise purposes, 2) whether they could ride a bike at all, or 3) whether they were limited by not having a bike or not knowing how to bike.

The findings in this paper focus on roadway design preferences. Respondents were asked to indicate their agreement or disagreement with comfort for 1) bicycling by oneself, 2) bicycling with children, and 3) driving on a series of seven multi-lane, commercial roadway designs (with one 2-lane exception). Agreement was measured on a 5-point Likert scale, with a neutral option and the modifier "completely" for the ends of the scale. The photos, displayed together in Figure 4, were manipulated via photo-editing software to show a variation on the original roadway design. Respondents were presented with the photos one-by-one, and the photos were randomized within each survey to control for ordering effects. Respondents did not know which designs the choice set contained.

The results were analyzed using Microsoft Excel and the statistical package STATA. Kruskal –Wallis and Wilcoxon–Mann–Whitney tests, both of which are non-parametric versions of ANOVA appropriate for ordinal variables, were used to determine if differences in responses between different types of cyclists were statistically significant. These instances are noted where applicable in the text.

Limitations

Due to the racial homogeneity of the sample, care should be taken when extending these conclusions to races and ethnicities other than White/Caucasian. However, recent research indicates that traffic safety is a significant concern among Black and Latino bicyclists as well—albeit not the only safety concern (*16*). As with all surveys, there may be some bias because people are more likely to respond when interested in a subject. In addition, the seven roadway designs did not represent a complete set of designs available; there may be roadway designs that respondents would have preferred more or less than the ones discussed in this paper. It also cannot be guaranteed that all respondents held the same definition of "comfort" when answering the questions.

Findings

Demographic Characteristics of Survey Participants

The 351 respondents live in more than 20 different cities and towns across Michigan, ranging in size from less than 3,000 to over 500,000 residents. Respondents provided their race or ethnicity, age, gender, annual household income, household composition, and bicycle and vehicle ownership, as shown in Table 1. The majority of survey respondents identified as White or Caucasian (only); the racial distribution across the different cyclist types was relatively similar. None of the respondents took the survey in Spanish; all but two respondents took the survey online.

The median age range of respondents was 45–54. There were more respondents over age 65 than between the ages of 18 and 24, which may be due to the representation of those age groups in a telephone-directory-based sample. Seniors were overrepresented in the "never cyclist" category (33%, compared with less than 20% of the other categories).

Forty-one percent of respondents identified as female. The gender split was less balanced among frequent and occasional cyclists than among rare and never cyclists, fitting with past research on bicycling demographics (*3, 9*). Of the approximately 80% of the sample who provided their annual household income, the responses indicate a significantly higher percentage earning US$50,001–US$75,000 compared with all other income groups, except those whose households earn more than US$125,000 ($p < 0.02$). The highest income respondents were more common among the frequent and occasional cyclists than the rare and never cyclists.

Nearly 72% of respondents reported having no children under age 16 in their household. Never cyclists were the least likely to report having children under age 16, likely related to the higher percentage of senior citizens in this group. Forty percent of occasional cyclists have at least one child in their household, suggesting the potential to make bicycling easier for children may be welcomed in this group. In terms of vehicle and bicycle ownership, only 2% of respondents represented zero-vehicle households, 24% had one vehicle, and 74% had two or more vehicles. Similarly, 14% of respondents had no bicycles in their households, 19% had one bicycle, and 65% had two or more bicycles.

Respondents were also asked about their commute habits and how often they bicycle for "transportation other than to work or school", fun/recreation/leisure, and exercise/fitness. Nearly 90% of the sample drives alone to work or school at least once a week. However, the sample also includes a fair amount of walking and bicycling, with approximately 26% walking to work or school at least once a week, and approximately 14% bicycling at least once a week for the same purpose. These numbers may indicate a bias toward walking and bicycling within the sample population, given that the American Community Survey indicates that only 0.5% of Michiganders bicycle to work on average, but they may also reflect the multi-modal nature of commuting that is difficult to capture via the American Community Survey (*17*).

Nearly one-quarter of respondents bicycle for transportation other than to work or school at least once a month, whereas approximately half of respondents bicycle primarily for fun/recreation/leisure or exercise/fitness purposes at least once a month. These findings reveal a fair amount of bicycling within the sample, and support the need to understand bicycling desires and patterns via research like this.

Views on Transportation

To gain a deeper understanding of respondents' attitudes toward bicycling, they were asked to choose how strongly they agreed or disagreed (on a 5-point Likert scale) with several statements about their lifestyles and travel habits. Figure 1 presents the distribution of responses from all respondents for statements about attitudes toward driving, bicycling, walking, and exercise. Although the general trends between the four groups are relatively similar, there are a few statistically significant differences, as indicated by asterisks in the chart.

The majority of survey respondents hold generally positive attitudes toward driving, but feel less uniformly positive toward active transportation. A large majority of each group agreed or strongly agreed that "owning a car is an important sign of (their) freedom" and that they "like traveling by car." Despite this clear appreciation for cars, a much smaller percentage of all groups agreed or strongly agreed that helping cars to get around better should be a top transportation concern. As might be expected, frequent and occasional cyclists were significantly less likely ($p<0.05$) to agree with the latter two statements than rare or never cyclists.

Figure 1 also shows the significant variation ($p < 0.001$) in the degree to which the four types of cyclists feel positively about bicycling, walking, and exercise. The majority of frequent and occasional cyclists agreed or strongly agreed with all four statements about bicycling, walking, and exercise, whereas the majority of never cyclists agreed or strongly agreed with only the two statements that did not focus on bicycling. Note the similarity in responses between enjoyment of bicycling and wanting to "see more people bicycling in [their] city."

General Barriers to Bicycling

Respondents were also asked about general barriers to bicycling. Forty-three percent of respondents agreed or strongly agreed that they would bicycle more if their friends or family came with them; this seemed to be particularly salient for occasional cyclists. Only 38% of respondents agreed or strongly agreed that they are often accompanied by children or older adults when they travel, but more occasional cyclists (46%) expressed agreement with this statement than other groups. Even fewer respondents (34%) agreed or strongly agreed that they live too far away from things to walk or bicycle, although never cyclists were significantly more likely ($p < 0.001$) to agree with this statement. Over 70% of respondents disagreed with the statement that "adults who bicycle for transportation are viewed as odd or too poor to own a car."

Respondents were also asked about concerns about bicycling, particularly with regard to safety. Less than 15% of each cyclist type agreed or strongly agreed that they would discourage others from bicycling to work or school, although never cyclists were significantly more likely ($p<0.02$) to agree with this statement than frequent cyclists. However, over 40% of the sample agreed or strongly agreed with the statement that "many bicyclists appear to have little regard for their personal safety," suggesting that behavioral trends may need to be further explored in Michigan. Agreement differed by bicycling frequency (45% of never cyclists compared with 29% of frequent cyclists). Moreover, the large majority (73%) of the sample agreed or strongly agreed with the idea that "many drivers don't seem to notice bicyclists," suggesting that people may not feel safe bicycling due to driver behavior. Frequent cyclists (65% agreement) were the least likely to agree with this statement.

Barriers to Bicycling to Work or School

Participants were also asked how much various factors limited their ability to bicycle to work or school (options were "absolutely limits", "limits quite a bit", "limits somewhat", or "does not limit"). Figure 2 shows the percentage of each cyclist group that is limited "quite a bit" or "absolutely" from biking to work or school by each barrier. The data indicate that safety, distance, and weather concerns topped the list for occasional, rare, and never cyclists, whereas frequent cyclists seemed more focused on safety and roadway/infrastructure conditions. In all instances, frequent cyclists were the least likely to indicate being limited by the barriers, whereas never cyclists were generally the most likely to indicate a higher level of limitation. However, other barriers are as limiting to occasional or rare cyclists as they are to never cyclists, again suggesting that addressing these barriers (e.g., building more high-quality

Table 1. Demographic Characteristics of Survey Participants by Cyclist Type

Demographic characteristics	Never cyclists (n=54)	Rare cyclists (n=93)	Occasional cyclists (n=83)	Frequent cyclists (n=116)	Total (n=351)[1]	Michigan[2]
Age						
18-24	4%	3%	0%	1%	2%	10%
25-34	7%	22%	17%	14%	15%	12%
35-44	11%	12%	24%	19%	17%	12%
45-54	20%	23%	23%	22%	23%	14%
55-64	22%	22%	23%	24%	23%	14%
65+	33%	17%	12%	19%	18%	15%
Unknown	2%	2%	1%	2%	2%	-
			Kruskal Wallis significant (p < 0.05)			
Sex						
Female	46%	52%	37%	31%	41%	51%
Male	52%	46%	60%	68%	58%	49%
Unknown	2%	2%	2%	1%	1%	-
			Kruskal Wallis significant (p < 0.05)			
Race/ethnicity						
White or Caucasian (only)	85%	81%	83%	83%	83%	76%
Hispanic or Latino	2%	2%	4%	1%	2%	5%
Black or African American	4%	1%	5%	3%	3%	14%
Asian, Native Hawaiian, or Pacific Islander	0%	3%	4%	2%	2%	3%
American Indian or Alaskan Native	2%	0%	0%	0%	0%	0%
Multiracial	0%	3%	0%	2%	1%	2%
Unknown	7%	10%	5%	10%	9%	-
			Kruskal Wallis not significant			
Annual household income						
$30,000 or less	7%	10%	8%	3%	9%	29%
$30,001 to $50,000	13%	13%	7%	15%	12%	20%
$50,000 to $75,000	24%	20%	13%	20%	19%	18%
$75,001 to $100,000	7%	15%	18%	9%	12%	12%
$100,001 to $125,000	13%	14%	10%	12%	12%	8%
More than $125,000	9%	8%	24%	22%	17%	13%
Unknown	26%	20%	19%	20%	21%	-
			Kruskal Wallis significant (p < 0.05)			
Children ≤ age 16 in household[3]						
None	85%	73%	59%	73%	71%	70%
At least one	15%	27%	40%	23%	27%	30%
Unknown	0%	0%	1%	3%	2%	-
			Kruskal Wallis significant (p < 0.01)			

Note: - = data not applicable.
[1]Total includes five respondents unable to be classified in the cyclist typology.
[2]American Community Survey 2011–2016, Five Year Estimates.
[3]Data for Michigan is based on the number of children in a household under age 18, not age 16.

infrastructure, teaching people how to dress for and be safe while bicycling in inclement weather) may encourage bicycling within certain groups to different degrees. Note that more cyclists of each type indicated that safety concerns were a greater barrier than a lack of bike lanes or trails, although the difference between the two diminishes with increasing bicycling frequency. These findings suggest that the provision of bike lanes and trails should be accompanied by more general traffic-calming strategies where possible. For example, bicycle boulevards—traffic-calmed neighborhood streets that prioritize bicycle traffic—offer a potential solution to simultaneously address both barriers.

For the less limiting barriers in Figure 2, responses between groups varied significantly except regarding security concerns about crime. Occasional cyclists were the most likely to indicate a strong limitation in the case of negative effect on appearance and difficulty traveling with others, but occasional and rare cyclists tend to feel similarly; a one-way ANOVA confirmed that there were no significant differences between the responses of occasional and rare cyclists for any of the barriers listed.

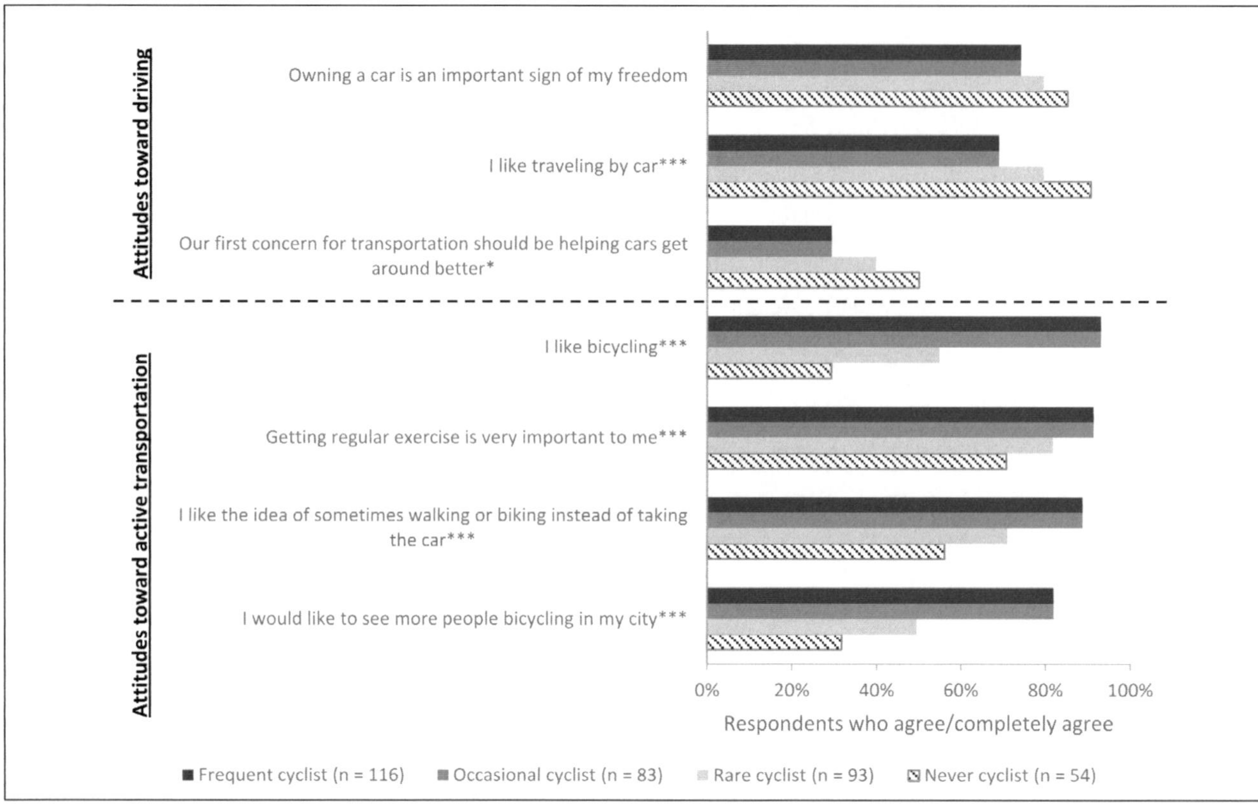

Figure 1. Attitudes toward driving, cycling, walking, and exercise, by cyclist type.
* $p < 0.05$, ** $p < 0.01$, and *** $p < 0.001$.

Note that the barrier of not having a bike/not knowing how to ride a bike was cited by some respondents of each group, but overwhelmingly so by never cyclists. It is impossible to tell which part of the question respondents were referring to due to question wording. However, only 20% of never cyclists indicated via an earlier question that they cannot ride a bike, suggesting that this barrier refers to not having a bicycle for the majority of never cyclists.

Factors That Would Encourage More Bicycling

Participants were also asked to choose three changes that would encourage them to bicycle more in their community (Figure 3). Seventy-five percent of all respondents indicated that the installation of separated bicycle facilities would encourage them to bicycle more. Rare cyclists were more likely to choose this response than occasional cyclists, possibly corroborating past research findings that a lack of perceived bicycling comfort and safety is a critical barrier to increasing bicycling (*5, 12, 18*). The fact that about 72% of frequent cyclists also chose separated bike facilities suggests that this lack of comfort and safety is experienced even by those who currently bicycle. This connection between increased bicycling and increased perceived safety and comfort is underscored by the clear difference in the percentage of people who chose separated bike facilities compared with more bike facilities. This choice seems to be particularly salient for never and rare cyclists, with almost twice as many choosing separated bike facilities over more facilities in general. These findings reflect clear preferences for bike facilities that provide safety and comfort (see Figures 5 and 6).

In comparison, about half of respondents indicated that adding more bicycle facilities or a complete bicycle network would encourage them to bicycle more, whereas a minority suggested that improved law enforcement of motorist and bicyclist behavior would encourage them to bicycle more. Increased law enforcement, easier access to education, and lower speed limits were unpopular generally. As might be expected, never cyclists were more likely to choose increased law enforcement of bicyclists over that of motorists, whereas frequent cyclists were much more likely to choose the opposite. Overall, these findings indicate a clear preference for infrastructure in the form of separated bike facilities and a complete bicycle network to encourage the sample population to bicycle more often.

Confidence While Bicycling

Respondents were also asked how confident they felt riding a bicycle. The majority of frequent cyclists and occasional

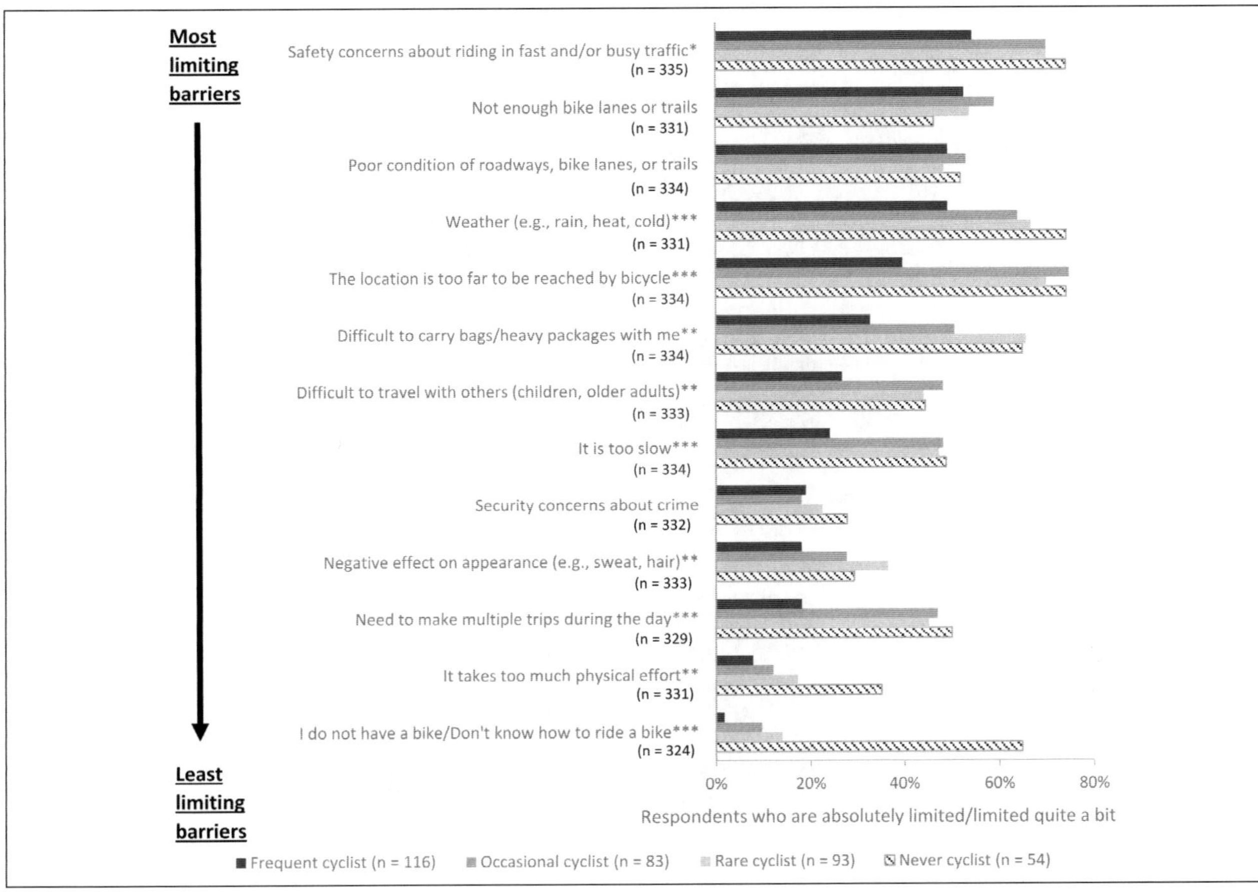

Figure 2. Strength of barriers to biking to work/school, by cyclist type.
* $p < 0.05$, ** $p < 0.01$, and *** $p < 0.001$.

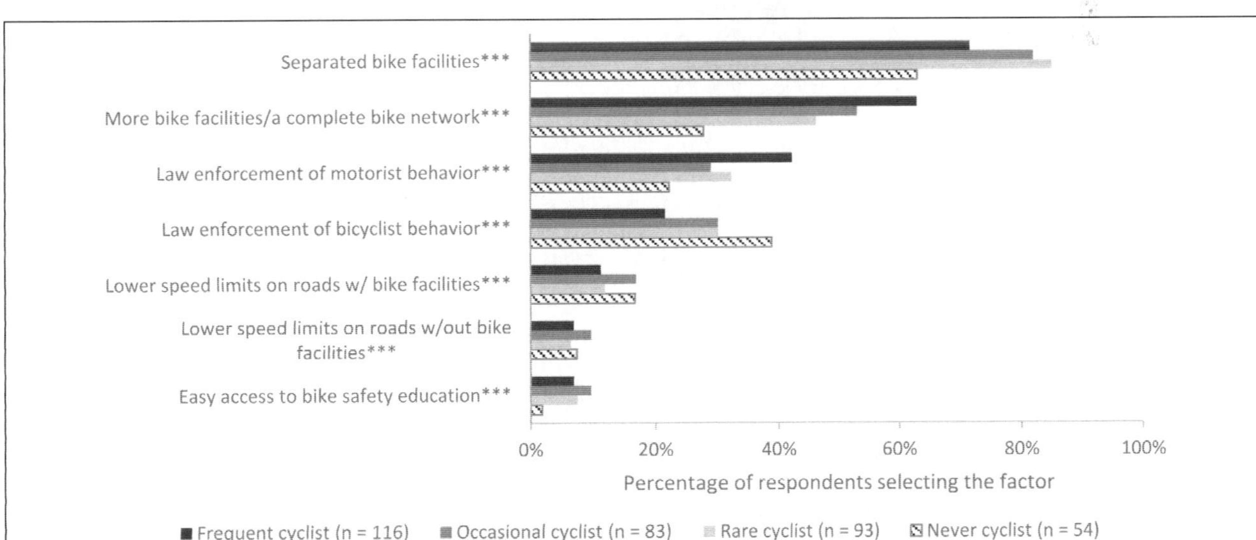

Figure 3. Factors that would encourage you to bicycle more often.
***$p < 0.001$.
Note: Separated bicycle facilities include those that provide increased separation from traffic (e.g., median, landscape buffer, etc.). Percentages do not add up to 100 because respondents could choose up to three factors.

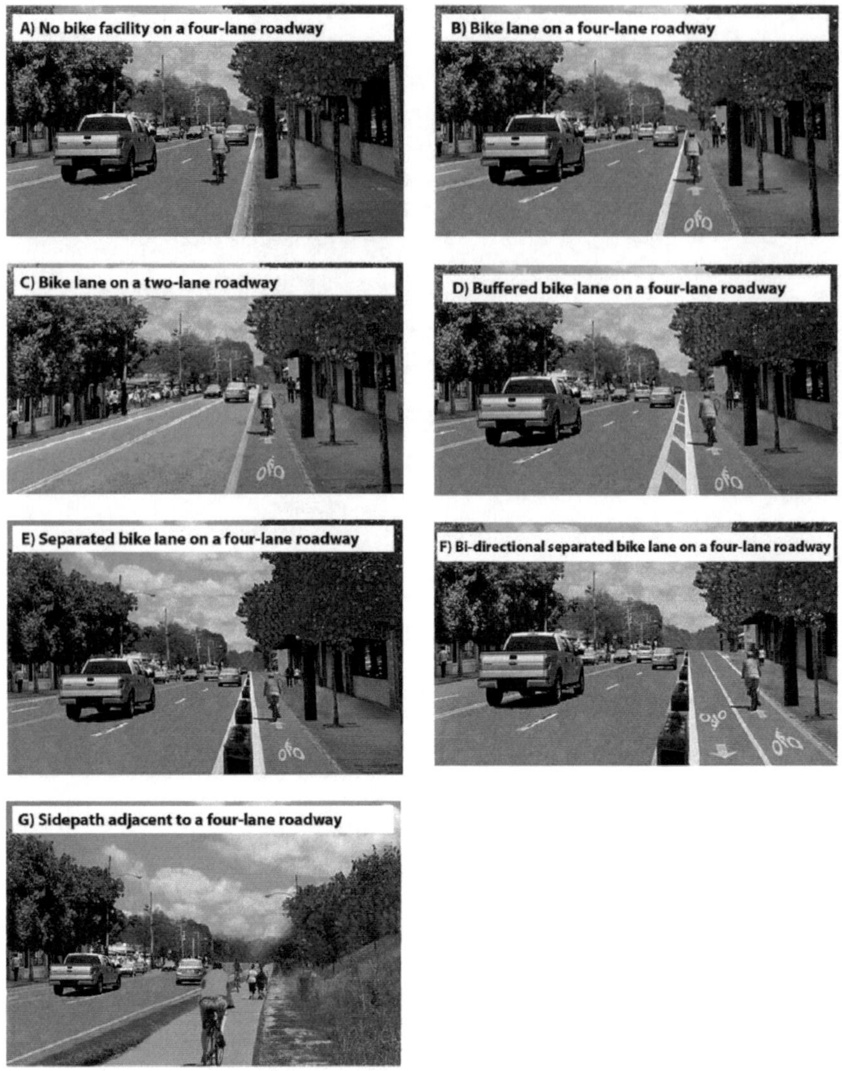

Figure 4. Roadway design options. Figures are as follows: A) No bike facility on a four-lane roadway; B) Bike lane on a four-lane roadway; C) Bike lane on a two-lane roadway; D) Buffered bike lane on a four-lane roadway; E) Separated bike lane on a four-lane roadway; F) Bi-directional separated bike lane on a four-lane roadway; G) Sidepath adjacent to a four-lane roadway.

cyclists feel "very confident" riding a bicycle. Nearly 90% of rare cyclists feel at least "somewhat confident" riding a bicycle, as do 59% of never cyclists, although the split between "very" and "somewhat" confident was much more even for these two groups. Never cyclists were also much more likely to report not feeling confident than the other cyclist types (41% not confident or cannot ride a bike at all, compared with only 11% of rare cyclists not feeling confident). While bicycling frequency is not solely explained by the ability to ride a bike, there is clearly a connection between the two ($p < 0.001$). Additionally, being "somewhat confident" may be a barrier to more riding in itself, or it may be the result of not being able to bicycle frequently due to other barriers.

Roadway Design Preferences

Respondents were then asked a series of questions about their level of comfort and experience riding and driving on seven different roadway designs (see Figure 4). The results indicate several key trends in respondents' preferences for driving near bicyclists and bicycling near drivers.

Figure 5 shows how comfortable respondents reported feeling when considering riding by themselves, riding with children, and driving on each of the seven roadway designs. The data show that respondents would be more comfortable with both bicycling by oneself and driving with increased separation, which research has found to be associated with perceived

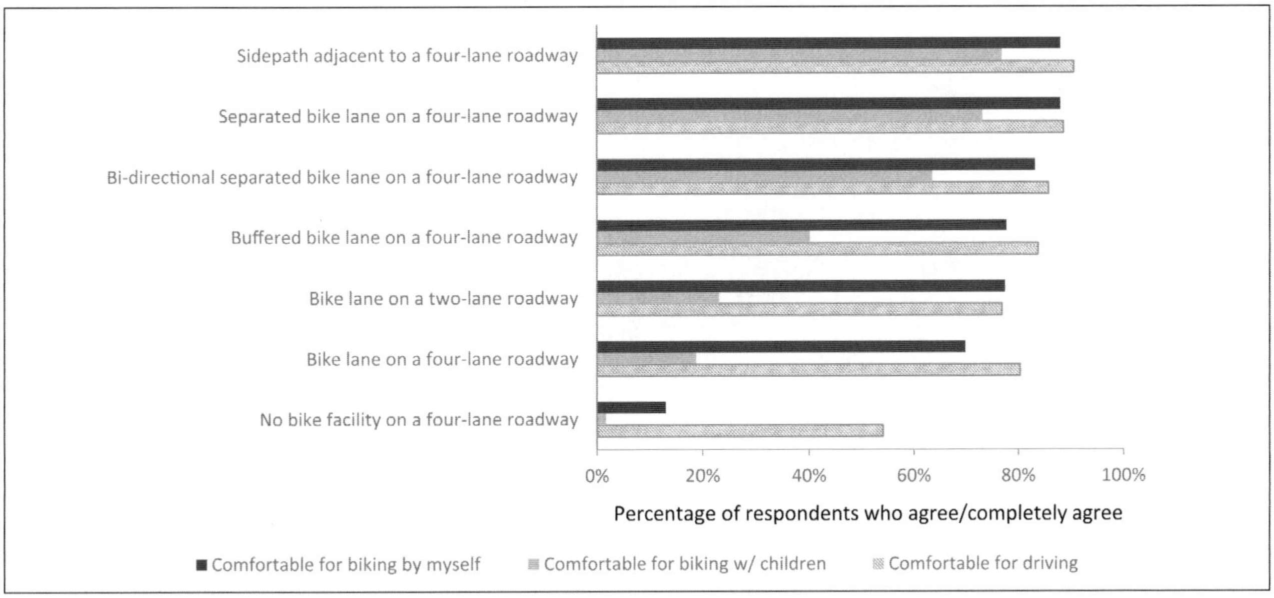

Figure 5. Roadway design preferences by percentage of respondents who would feel comfortable bicycling in various scenarios.

predictability on the roadway (*12*), although the majority of respondents would feel comfortable in most scenarios with some type of bicycle facility. A greater percentage of respondents would feel comfortable driving on a four-lane roadway with a bike lane than on a two-lane roadway with a bike lane—likely related to opportunities for drivers to further separate themselves from bicyclists on the multi-lane roadway.

When considering bicycling with children, however, comfort declines rapidly without separation from cars: less than 50% of the sample would be comfortable in a buffered bicycle lane; less than 20% would be comfortable in a regular bicycle lane on a four-lane roadway; and less than 2% would be comfortable on a four-lane roadway with no bicycle facility. These findings underscore the increased perceived vulnerability of bicycling with children and the commensurate need for clear and strong separation from traffic for comfort. Note that sidepaths and separated bicycle lanes are approximately evenly comfortable for each purpose, with just a slight dip for bicycling with children. Given that paths have tended to be the gold standard for comfort, this finding is encouraging for cities aiming to use separated bike lanes to encourage more bicycling.

Comfort was also examined by type of bicycling (non-transport-bicycling compared with all-purpose bicycling). The same preference order was found for both groups, although non-transport cyclists were significantly less likely to feel comfortable bicycling alone or with children than all-purpose cyclists—with the exception of the sidepath and separated bike lane designs.

These data corroborate past research on roadway design preferences, indicating that most respondents felt considerably less comfortable riding on a roadway without a bicycle facility than riding on a roadway with any type of bike facility (*3, 12*). They also support findings presented earlier in the paper regarding bicyclists' concerns about safety and their desire for more separated bike lanes.

The survey also asked about respondents' willingness to try bicycling on the various facilities. Nearly 90% of respondents would be willing to try bicycling on a separated facility, while nearly 70% would be willing to try even just a bicycle lane. In contrast, only 24% indicated a willingness to try bicycling on a four-lane roadway without a bicycle facility.

Roadway Design Preferences by Cyclist Type. Comfort levels for each roadway design were also examined by cyclist type (data not shown; general patterns fit with Figure 5). In each case, never cyclists were the least comfortable with the bicycle facilities, whether bicycling alone or with children. Comfort levels among frequent, occasional, and rare cyclists were similar for all of the separated facilities, with rare cyclists slightly lower in general, and the difference becoming more pronounced for non-separated facilities. The percentage of respondents who indicated comfort bicycling by oneself without a bicycle facility on a four-lane roadway was low for all groups (less than 20%), but particularly so for never cyclists.

When considering riding with children, all bicyclists felt less comfortable as separation decreased, and dramatically so as the separation decreased to just a line (i.e., bike lane) and then no facility at all. Comfort for driving was pretty consistently high, although less so for the bike lane on the two-lane roadway and the no facility option, similar to Figure 5. For both of these design options, comfort increased with bicycling frequency.

Finally, most respondents indicated a willingness to try bicycling on the various street designs, although the willingness steadily decreased as separation decreased. Never cyclists were consistently the least likely to indicate a

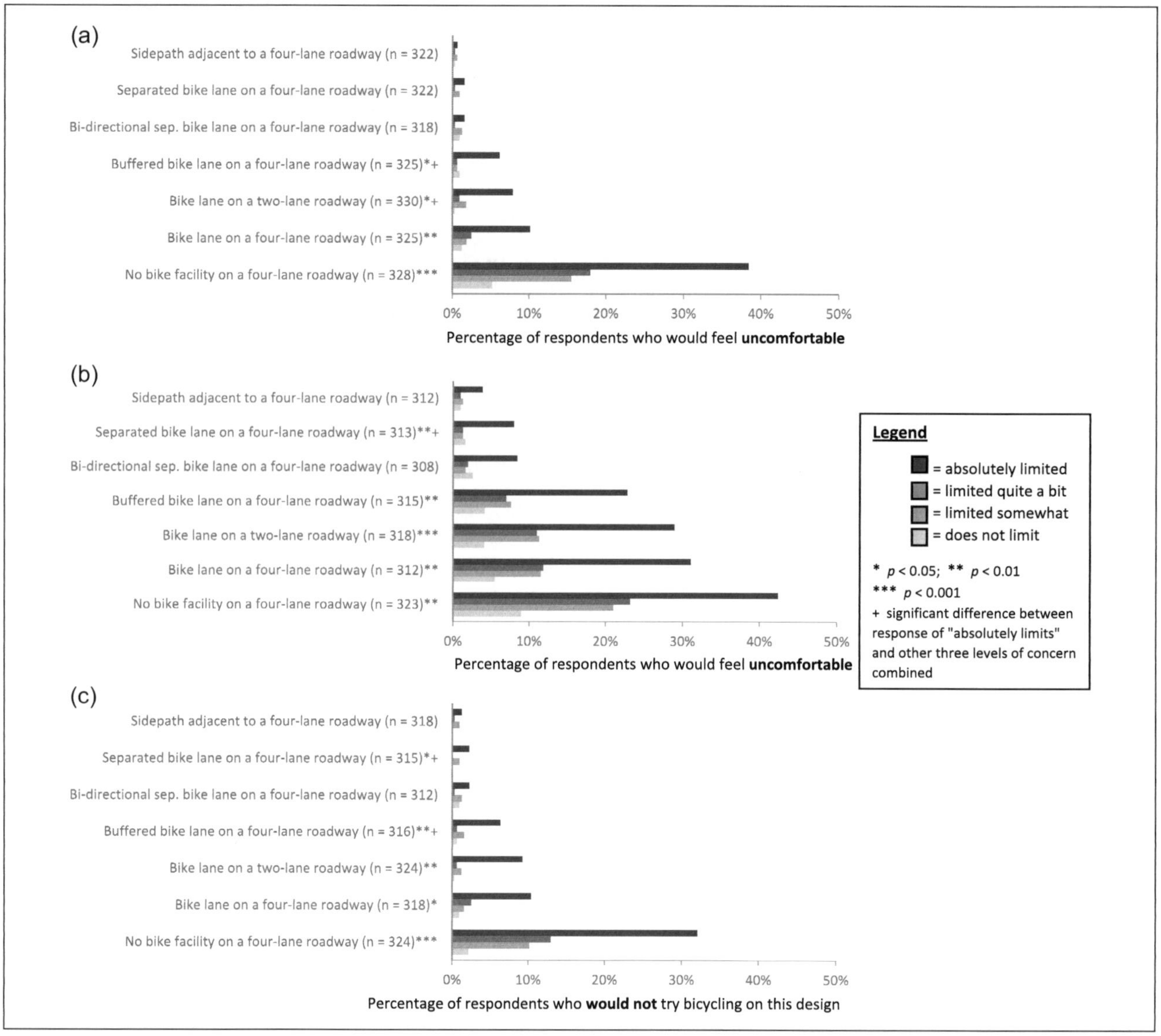

Figure 6. Comfort level for bicycling by level of concern for safety, (*a*) bicycling by oneself, (*b*) bicycling with children, (*c*)"Something I'd try bicycling."

willingness, although over 70% of them indicated a willingness to try bicycling on a separated bike lane or a sidepath. The no bike facility option remained unpopular across all groups.

Roadway Design Preferences by Respondents Concerned about Safety. Additional exploration was conducted to determine whether reported comfort levels on various roadway designs differed based on respondents' concerns about safety. Figure 6 shows the percentage of respondents who disagreed or strongly disagreed that they would feel comfortable bicycling or driving on the various roadway designs, by concerns about safety. Fitting with Figure 5, a majority of respondents indicated that they would not feel uncomfortable on any facility that had some type of bicycle lane. This suggests a population fairly comfortable with bicycling as long as infrastructure is present, although it should be noted that a significantly higher percentage of respondents who are very concerned about safety reported that they would feel uncomfortable riding by themselves on a roadway with no bike facilities compared with the rest of the respondents ($p < 0.001$). Physical separation seems to be key for this group: when it is present, very concerned cyclists feel a similar level of comfort to the rest of the respondents.

The differences in the responses of those who are very concerned and those who are not concerned about safety are even more pronounced when considering bicycling with children, although all groups indicate less comfort on the various roadways when considering bicycling with children. These findings lend support to the findings presented in Figure 3: overall, respondents of all types, whether bicycling alone or with children, prefer physically-separated bike facilities over striped bike lanes, and striped bike lanes over no lanes at all.

Conclusions

This survey captured a cross-section of Michigan residents to gain a more nuanced understanding of their attitudes toward bicycling and driving, bicycling habits, barriers to bicycling, and roadway design preferences. The findings overwhelmingly suggest a preference for more bicycle accommodations, and more separated facilities in particular. Seventy-five percent of all respondents indicated that the installation of separated bicycle facilities would encourage them to bicycle more, with almost twice as many rare cyclists choosing separated bike facilities over more facilities in general.

Relatedly, safety concerns, distance, and weather appeared to be the most limiting barriers for all cyclist types. Nearly 89% of respondents reported that safety concerns about riding in fast and/or busy traffic at least somewhat limited their ability to bike to work or school, with 68% saying that safety concerns limited them "quite a lot" or "absolutely." Frequent cyclists were less likely to indicate barriers than other cyclists.

The presence of bicycle facilities increases respondents' comfort and willingness to try bicycling on a roadway. Most respondents felt considerably more comfortable bicycling on a roadway with any type of bike facility over one lacking a bicycle facility, and this preference was even stronger when the facility was separated from drivers by a physical barrier. Separation was even more important when considering bicycling with children, with comfort levels declining rapidly without separation from cars when children are considered. Respondents were also more likely to indicate comfort while driving with greater separation from bicyclists.

These findings corroborate past research, clarify the nuance of bicycling considering trip companions, and show a clear pathway from safety as a barrier to bicycling to something that can be addressed through infrastructure.

Acknowledgments

Thanks to Josh DeBruyn and the Michigan Department of Transportation for supporting a rigorous attempt to understand design preferences among Michigan residents. We also appreciate the help of Hannah Pritchard, Connor Cox, and our Wayne State partners for helping deploy the survey.

References

1. Flusche, D. *Bicycle Commuting Data*. 2012. http://www.bikeleague.org/news/acs2010.php. Accessed June 12, 2013.
2. Flusche, D. *Highlights from the 2009 National Household Travel Survey*. League of American Bicyclists. Washington, D.C., 2010.
3. Dill, J., and N. McNeil. *Revisiting the Four Types of Cyclists: Findings from a National Survey*. Transportation Research Board, Washington, D.C., 2016.
4. Sener, I. N., N. Eluru, and C. Bhat. Who are Bicyclists? Why and How Much Are They Bicycling? *Transportation Research Record: Journal of the Transportation Research Board*, 2009. 2134: 63–72.
5. Dill, J., and K. Voros. Factors Affecting Bicycling Demand: Initial Survey Findings from the Portland, Oregon, Region. *Transportation Research Record: Journal of the Transportation Research Board*, 2007. 2031: 9–17.
6. Tilahun, N. Y., D. M. Levinson, and K. J. Krizek. Trails, Lanes, or Traffic: Valuing Bicycle Facilities with an Adaptive Stated Preference Survey. *Transportation Research Part A*, Vol. 41, 2007, pp. 287–301.
7. Winters, M., and K. Teschke. Route Preferences among Adults in the Near Market for Bicycling: Findings of the Cycling in Cities Study. *American Journal of Health Promotion*, Vol. 25, No. 1, 2010, pp. 40–47.
8. Dill, J., and N. McNeil. Four Types of Cyclists? Examination of Typology for Better Understanding of Bicycling Behavior and Potential. *Transportation Research Record: Journal of the Transportation Research Board*, 2012. 2387: 129–138.
9. Goddard, T., C. Monsere, and N. McNeil. *Can Protected Bike Lanes Help Close the Gender Gap in Cycling? Lessons from Five Cities*. Presented at *Urban Studies and Planning Faculty Publications and Presentations*, 2014, p. 123.
10. Winters, M., K. Teschke, M. Grant, E. Setton, and M. Brauer. How Far out of the Way Will We Travel? Built Environment Influences on Route Selection for Bicycle and Car Travel. *Transportation Research Record: Journal of the Transportation Research Board*, 2010. 2190: 1–10.
11. Broach, J., J. Dill, and J. Gliebe. Where Do Cyclists Ride? A Route Choice Model Developed with Revealed Preference GPS Data. *Transportation Research Part A*, Vol. 46, 2012, pp. 1730–1740.
12. Sanders, R. L. We Can All Get Along: The Alignment of Driver and Bicyclist Roadway Design Preferences in the San Francisco Bay Area. *Transportation Research Part A*, Vol. 91, 2016, pp. 120–133.
13. Watkins, K., et al., NCHRP 08-102. In progress.
14. Toole Design Group. *The Importance of Space: Bicyclists, Drivers, and Bicycle Lanes*. Michigan Department of Transportation, 2017.
15. Aday, L. A., and L. J. Cornelius. *Designing and Conducting Health Surveys*. John Wiley & Sons, Inc., San Francisco, Calif., 2006.
16. Brown, C., and S. Blickstein. *Bicycling Among Black and Latino Women: Focus Group Summary Report*. Rutgers University, New Brunswick, 2016.
17. American Community Survey (2011–2015). *Sex of Workers by Means of Transportation to Work*. American Community Survey. Accessed July 15, 2017.
18. Manaugh, K., G. Boisjoly, and A. El-Geneidy. Overcoming Barriers to Cycling: Understanding Frequency of Cycling in a University Setting and the Factors Preventing Commuters from Cycling on a Regular Basis. *Transportation*, Vol. 44, No. 4, 2017, pp. 871–884.

The Standing Committee on Bicycle Transportation peer-reviewed this paper (18-06369).

TRR
JOURNAL OF THE TRANSPORTATION RESEARCH BOARD

Article

Transportation Research Record
2018, Vol. 2672(36) 12–23

DOI: 10.1177/0361198118796001
journals.sagepub.com/home/trr
SAGE

Effects of Building a Stock of Bicycling Experience in Youth

Calvin G. Thigpen[1,2] and Susan L. Handy[1]

Abstract

Travel is difficult without the supportive attitudes, norms, and skills that make up "motility," the capacity for travel. Travel experiences are critical to developing motility. This study identifies factors that influence the development of bicycling motility through analysis of a prospective panel of 19 children, interviewed at ages 9, 12, and 15. This study is set in Davis, California, USA, where bicycling infrastructure is comprehensive, allowing a focus on the role of bicycle experiences. We analyze the interviews using structural and longitudinal coding techniques and find that bicycling experiences are associated with the types of attitudes held by the participants, especially at age 15, when the children come to value the independence and convenience of bicycling. Parental behavior and rules had important associations with the children's bicycling behavior, particularly at younger ages. Friends reinforced bicycling behavior directly through the logistical challenges of traveling as a group at age 15. Gender was not associated with bicycling behavior in this setting. This study provides confirmatory evidence for the importance of motility and its behavioral determinants.

Driver's training is considered a rite of passage for most US teenagers. Through online courses and in-vehicle training, they acquire the knowledge and skills needed to earn their driver's license and become safe drivers. Depending on where they live, they may also informally acquire skills and knowledge necessary to make use of other modes of transportation, such as riding a bicycle or taking the bus. The competence to make use of transportation modes is called "motility," a construct comprised of several dimensions: transportation access, skills and knowledge, and attitudes and norms (*1*).

Motility is necessary for travel. Those lacking formal (e.g., a driver's license) and informal (e.g., years of driving experience) elements of motility face difficulties traveling, with implications for safety, accessibility, and equity. Yet travel experiences are necessary to build motility. An adult who never learned to bicycle as a child is unlikely to pick up the habit later in life, due to a lack of bicycling skills and knowledge, and perhaps negative perceptions as well. Despite its importance, the process of motility development has not been well studied, as most studies of travel behavior investigate the influence of attitudes, norms, and skills on behavior rather than the reverse, and usually at only one point in time (*2*).

We focus in this study on the development of bicycling motility, motivated by the environmental and health benefits of this mode. We conducted a prospective panel of interviews with parents and their children at ages 9, 12, and 15 in Davis, California, USA, to investigate how children build motility. Davis, with its extensive network of bicycle facilities, offers the opportunity to examine motility's psychological constructs and causal influences in a context where infrastructure is not a limiting factor. We explore the questions: (a) How do bicycling behavior, attitudes, and norms change through childhood? (b) What role do bicycling experiences play in developing bicycling attitudes and norms? We consider as well, the influence of parents, peers, friends, and personal characteristics on behavior and motility. The results show a strong connection between bicycling motility and bicycling behavior.

Literature Review and Conceptual Model

Many studies have demonstrated the influence of the physical environment (e.g., distance to school, pedestrian and bicycle infrastructure) on active school travel (see Stewart et al. (*3*) for a review). Evidence exists for the

[1]Institute of Transportation Studies, University of California at Davis, Davis, CA
[2]School of Geographical Sciences and Urban Planning, Arizona State University, Tempe, AZ

Corresponding Author:
Address correspondence to Calvin G. Thigpen: cgthigpen@ucdavis.edu

influence of children's personal characteristics and social spheres, as well. Though these factors are not the primary focus of this study, we include them as part of a larger socio-ecological framework (*4*) in examining motility development. The literature on motility and on the influence of personal and social-environmental characteristics informed our conceptual model.

Motility

The term "motility" has a well-established meaning in the fields of biology and medicine, referring to the ability to move (*1*). But in the early 2000s, the sociologist Vincent Kaufmann co-opted the term, applying it more abstractly to the skill and knowledge needed for traveling (*1*), such as understanding a timetable or bus route map. Kaufmann went further in postulating that motility also includes an individual's "representations" of modes of travel, such as their attitudes or norms, as well as the constraints imposed by context, including physical or monetary barriers. Importantly for this study, individuals can increase their motility over time through their travel experiences, he suggested.

To more clearly operationalize the concept of motility, which at times is hazily described in the literature, we borrow the elements of the theory of planned behavior (TPB) (*5*)—attitudes, subjective norms, and perceived behavioral control—as elements of motility. In contrast to most travel behavior studies, which use psychological constructs to explain or predict travel behavior, we focus on motility as the outcome variable of interest and behavior as a key determinant.

The theory of cognitive dissonance is particularly relevant to the motility framework's positioning of behavior as a determinant of attitude. Cognitive dissonance theory originally posited that individuals are likely to work to re-align dissonant (i.e., conflicting) attitudes and behaviors, with attitudes more likely to change to be consonant (i.e., in agreement) with behaviors than the reverse (*6*). One might expect, given the authority dynamic between children and their parents, that children who are encouraged, instructed, or otherwise told to bicycle would bring their bicycling attitudes in line with their behavior. But studies show this tendency holds true among adults as well. Kroesen et al. (2017) found that dissonant Dutch travelers (i.e., those whose travel behavior and attitudes did not match) in a panel survey were more likely to shift their attitudes to a consonant state than to shift their behavior (*7*).

Researchers have also demonstrated that norms, that is, unwritten rules of proper behavior, have a strong relationship with behavior. Psychologists often distinguish between different types of norms, such as personal, descriptive, and injunctive. Personal norms are strongly tied to self-concept, in that violating personal norms tends to result in guilt while complying yields feelings of pride (*8*). Descriptive norms relate to what is perceived as typical behavior, and injunctive norms describe what behaviors are considered acceptable (*9*). Though the motility framework does not mention norms explicitly, it describes the importance of "appropriations" and "project[ions]," which refer to a bundle of psychological elements, including perceptions and personal meanings of different travel patterns (*10*). Research into bicycling (*11*) and other modes of travel confirms that social norms have strong influences on behavior.

Several studies have demonstrated that bicycling skills are influenced by maturation as well as experience (*12*), which is consistent with the power-law rule of practice: additional time spent learning a skill is (log-log linearly) associated with improved ability (*13*). In addition to the work of bicycle safety researchers, travel behavior scholars have also examined the influence of bicycling skill as described by perceived behavioral control or the related construct of "self-efficacy" (*14*), that is, what individuals think that they are able to do (e.g., ride a bicycle, read a transit map). Studies demonstrate that these constructs are associated with greater bicycling frequency (*15*).

Influence of Personal Characteristics and the Social Environment

Personal characteristics, especially gender and age, directly or indirectly influence children's travel behavior and motility, according to prior studies. Girls are less likely to bike to school than boys (*16*), and young women tend to hold more negative attitudes toward bicycling, especially in adolescence (*17*). These gender differences often extend into adulthood as well (*18*).

Parents play a particularly strong role in early childhood development of bicycling skills and attitudes through the decisions they make (*19*) and the rules and examples they set (*20*). In contrast, friends' behavior and attitudes may increase in importance as the child enters their teenage years (*21*). Researchers have not examined the role of friends and peers as closely as they have the influence of parents, but those studies that have analyzed friend and peer influence have demonstrated its importance (*22*). This effect appears to be particularly strong for active travel modes among girls, while boys appear to be indifferent to whether they feel that their peers think bicycling is cool (*17*).

Conceptual Model

Our conceptual model reverses the TPB to resemble the motility framework: bicycle behavior is the independent variable of interest, and the multi-dimensional construct of bicycling motility is the principal dependent variable. This study primarily attends to attitudes and norms as the underlying dimensions of motility, as the early rounds of interviews did not systematically assess bicycling skills and knowledge. This study also includes independent variables from the personal, social, and environmental spheres, consistent with the socio-ecological approach (*4*). All four independent factors (personal, parental, environmental, and friends' influences) are conceived as influencing bicycling behavior directly, while friends, parents, and personal characteristics also influence bicycling motility. Personal characteristics, particularly age, are expected to moderate the influence of the other three independent variables on behavior and to moderate the influence of parents and friends on motility. We include a causal link from motility to behavior, reflecting the idea that attitudes and norms also influence behavior, in addition to being shaped by behavior, in a causal loop.

Methodology

Setting

This study uses a prospective panel of interviews in the city of Davis, CA (*23*). Davis, known as the bicycling capital of the United States, is virtually flat, has a Mediterranean climate, and is home to roughly 65,000 residents. Davis resembles, to a notable degree, a Dutch city: with over 50 miles (80 kilometers) of on-street bicycle lanes and 50 miles (80 kilometers) of off-street shared use paths in its 10 square miles (16 square kilometers) (*23*), Davis residents can reach many destinations within a short distance on safe facilities. In a striking similarity to Dutch statistics, 30%–50% of the city's children aged 5 to 18 bicycle to school (*24*), 50% of the city's university students bicycle to campus (*25*), and approximately 20% of staff and faculty at the University of California, Davis, the largest employer in the city, commute by bicycle (*25*).

Despite these conditions, a substantial proportion of children are driven to school and other activities, suggesting that bicycling infrastructure is a necessary but insufficient condition for bicycling (*26*). Davis thus enables us to study how personal and social factors rather than infrastructure influence children's bicycling motility and behavior.

Sampling, Interviews, and Analysis

In the first round of interviews in 2009, Maiss and Handy recruited participants for the panel via flyers, convenience sampling, and newspaper ads (*27*). The resulting sample of 20 children, aged 8 and 9, was spatially distributed around Davis and balanced between boys and girls. Driller and Handy followed up with the panel in 2012 (*28*), refreshing and augmenting the panel as needed to build a new sample of 25 children, who at this stage were about 12 years old. In 2015, we conducted a third round of interviews with the participants, now aged 15. We added five new participants to account for dropouts, for a sample size of 24. Eleven of the original 20 children in 2009 participated in all three rounds of interviews, and an additional eight children participated in both 2012 and 2015. These 19 children form the basis for this study's analysis. Since most of the children in the sample were age 9 in 2009, 12 in 2012, and 15 in 2015, the participants in these three rounds are collectively referred to as 9, 12, and 15-year-olds, respectively, in the results section.

Each round of semi-structured interviews used a script from which the interviewer could deviate slightly as appropriate to clarify meaning or elaborate (*29*). Though the focus of each round shifted slightly (e.g., Maiss and Handy studied mental maps (*27*)), we asked a core set of interview questions regarding bicycling attitudes and behavior in all three rounds. In the 2015 interviews, we supplemented this core with questions about parental attitudes and behavior, friends' attitudes and behavior, and subjective norms. We also issued a short survey to supplement the interviews and to better gauge the representativeness of the sample compared to the Davis population (using 2011–15 American Community Survey 5-Year estimates (*30*)). The sample was approximately representative of the overall Davis population by measures of their families' income, education, race, and gender (Table 1).

Each interview was audio-recorded, and a professional transcription service converted the interview recordings into text. For the first cycle of analysis, we used a structural coding approach, in which we assigned a code representing one of our variables or other aspects of our conceptual model to specific statements by the participants (*31*). We conducted structural coding for all three rounds of the panel. We then applied a longitudinal coding scheme to the structural codes in a second cycle of analysis, allowing us to examine how the children's bicycling frequency was associated with their bicycle use and motility (attitudes and norms), as identified in the structural codes. We present the longitudinal coding results in both descriptive and explanatory fashions in the results section.

Limitations

The small sample of this study reduces its generalizability, but the qualitative nature of the interviews and the prospective panel design contributes to good internal validity. The interviews more readily permit identification of causal mechanisms and important contextual

Table 1. Sociodemographic Characteristics of the Sample and the Davis Population

Variable	Characteristics	Sample[a]		Population[b]
Household income (USD)	Less than 100,000/year	24%	(4)	46%
	100,000 to 200,000/year	53%	(9)	35%
	More than 200,000/year	24%	(4)	20%
Parent educational attainment	Less than 4-year college degree	11%	(2)	28%
	4-year college degree	37%	(7)	29%
	Graduate degree	53%	(10)	43%
Parent race	Caucasian	78%	(14)	56%
	Asian–American	11%	(2)	22%
	Hispanic	11%	(2)	14%
	Other	0%	(0)	8%
Child's gender	Female	63%	(12)	46%
	Male	37%	(7)	54%

[a]Sample sizes within each variable row may not sum to 19 households due to missing responses. Percentages within each variable row may not sum to 100% due to rounding error.
[b]Data for the city of Davis' characteristics were obtained from tables B01001, B03002, B15003, and S1901 using the 2011–15 American Community Survey 5-Year estimates (*30*). Percentages within each variable row may not sum to 100% due to rounding error.

factors than surveys alone, while the panel design allows us to assess the influence of characteristics in one time period on attributes in a later period.

The study has a number of potential biases. The types of families who move to Davis may be more pro-bicycling than families in other cities, resulting in self-selection bias. The sample did include families with a range of attitudes towards bicycling, however. The drop-out rate of panel participants would represent a source of bias if the probability of dropping out systematically varied with important characteristics of the participants. However, the children who dropped out were evenly split by gender and were somewhat more likely to bicycle frequently, resulting in more varied mode use among the remaining group of panel participants.

The generalizability of the specific results obtained in this study to other communities is uncertain, as Davis possesses unique attributes that encourage bicycling. However, the more universal relationships identified in this study with respect to travel behavior and motility are likely to generalize to other modes of travel and other contexts.

Results

Descriptive Analysis

Descriptive analysis of the temporal change in bicycling behavior, attitudes, and norms provides a starting point for understanding motility development in this study. In the discussion that follows, we provide the number of participants reporting a particular characteristic in parentheses (e.g., "[three]" to indicate three participants). The inset, italicized quotations include a letter in parentheses indicating the unique ID for that participant (e.g., *"'It's fun to bike.' [G]"* refers to a quote from participant "G").

Behavior. At age 9, most participants (six) rode their bicycle frequently (once a week or several times a week), while a smaller number rode daily (three) or infrequently (two; less than once a week; see Table 2). Three years later (after the addition of eight new panel members), this panel was composed of more frequent riders: eight rode daily, seven rode frequently, and four rode infrequently. The slight increase was due to increased bicycling frequency among the panel participants who participated in all three rounds, not due to bias in the panel participants that joined in the second round. Approximately the same was true at age 15, as nine rode daily, eight rode frequently, and two rode infrequently. Overall, the panel increased how often they rode their bikes, though in each round of interviews at least two participants rode their bicycle infrequently. This is the reverse of the age-related trend in bicycling seen elsewhere in the United States, where bicycling frequency is typically at its highest among young children and declines as individuals age (*32*).

At age 9, the participants provided a wide variety of reasons for bicycling, though biking for enjoyment took precedence. The most common response (nine) was "for fun"—the intrinsic enjoyment of biking. On a related note, seven of the participants said they biked in order to get to the park, where they presumably had fun. Five of the respondents biked to school, and four each biked to meet with friends (typically at their house) or to go shopping. Three of the respondents mentioned biking

Table 2. Children's Bicycling Behavior Over Time

	Age 9 (2009)		Age 12 (2012)		Age 15 (2015)	
ID	Frequency	Purposes	Frequency	Purposes	Frequency	Purposes
A	Frequently	Fun, shopping, friends' houses, school, summer camp	Daily	School, restaurants, friends' houses	Daily	School, friends, other activities
B	Frequently	Fun, park, triathlons	Frequently	School, swimming pool, Target	Frequently	School, downtown
C	–	–	Daily	School, exploring, fun	Daily	School, downtown, restaurants, friends, shopping
D	Frequently	Fun, sports	Frequently	School, sports, friends' houses, restaurants, exploring	Frequently	Downtown, friends, sports, movie theater
E	Frequently	Downtown, farmer's market, fun, park, school	Infrequently	Downtown, summer camp, UC Davis	Infrequently	School, friends, library, downtown
F	–	–	Infrequently	Fun, friends, park, restaurants	Frequently	Downtown, friends, pool, park
G	Frequently	Fun, park, school, store	Daily	School, sports	Daily	School, sports, friends, downtown, swimming pool
H	–	–	Frequently	Downtown, friends, shopping	Daily	Downtown, friends, restaurants, school
I	Daily	School, doctor, downtown, friends' houses, grocery store, library, park, sports, restaurants, shop, movies	Daily	School, park, shop, grocery store, friends' houses, downtown	Daily	School, sports, downtown, friends, parks, anywhere
J	Daily	Fun, downtown, friends' houses, park, swimming pool	Daily	Sports, shop	Frequently	Park, friends, library, job, other activities
K	Infrequently	Fun, summer camp	Frequently	School	Frequently	Extracurricular activities, swimming pool, downtown, other activities
L	Daily	Friends' houses, park, school, shop	Daily	School, shop, friends' houses	Daily	School, downtown, friends, restaurants, shop
M	Frequently	Fun	Frequently	School, park	Daily	School, downtown, friends, swimming pool, other activities
N	–	–	Daily	School, shopping, grocery store, friends	Daily	School, coffee, downtown
O	–	–	Daily	School	Daily	School, grocery store
P	–	–	Frequently	Fun	Frequently	Shopping, park
Q	–	–	Infrequently	Fun	Frequently	School
R	–	–	Infrequently	Shopping	Frequently	School, friends, downtown
S	Infrequently	Fun, park	Frequently	School, shop, farmer's market, downtown	Infrequently	School, downtown, friends

Note: "Daily" = every day; "Frequently" = once a week or a few times a week; "Infrequently" = less than once a week; – = the question was not asked of the participant because they were not yet a part of the study.

Table 3. Children's Bicycling Attitudes Over Time

	Age 9 (2009)			Age 12 (2012)			Age 15 (2015)		
ID	Like?	Likes	Dislikes	Like?	Likes	Dislikes	Like?	Likes	Dislikes
A	Yes	Fun, environment, exercise	Traffic	Yes	Environment, fast	Slower bicyclists, hill, turns	Yes	Time to think, exercise, environment, independent, flexibility	Slowness, extreme rain
B	Yes	Fun, no seatbelt	No dislikes	Yes	New bike	No dislikes	Neutral	With friends	Slow, hot
C	–	–	–	Yes	Fast, can sit down	Uncomfortable seat, hot	No	Practical, easy	Hills, tiring, helmet, too much biking
D	Yes	Fun, friends and family, ice cream, restaurants	Hot	–	–	–	Yes	Independent, convenience, only option	Doesn't ride in rain
E	Yes	Fun, fast, exercise	Slow, tiring	Yes	Feels good, environment, exercise	Traffic, cold, fear of falling off	Yes	Fast, can avoid strangers, environment, exercise	Cold, risk of crashes
F	–	–	–	Yes	Enjoyable, fast, easy, exploring, fun, in control, with friends	Occasional mechanical issues, cold, traffic/cars	Yes	Fast, independent, convenient, with friends	Hills
G	Yes	Getting outside, exercise	Hot	Yes	Fast	Carrying musical instrument	Yes	Independent, flexible, fast	Hot, distance/slow
H	–	–	–	–	–	–	Yes	Fast, flexible, easy to carry things, fresh air, independent, with friends, refreshing	Rain, hot, helmet hair
I	Neutral	Scenery	Hot, tiring	Yes	Fast, freedom	Biking in morning, rain, tiring	Yes	Easy to get around, flexible, independent, listen to music, biking slowly, with friends	Weather, slow
J	Yes	No likes	Fear of falling off	Yes	Freedom, moving, exploring	Traffic/cars	Yes	Simple and easy, flexible, environment	Slow
K	Yes	Fast	Hot, helmet	Yes	Fast, wind in face, scenery, fun	Uncomfortable seat, tiring	Yes	Flexible, independent, with friends, wind in face, notice scenery, down time	Weather, inconvenient
L	Yes	Fun	Tiring	Yes	Nice	Slow, tiring	Yes	Flexible, independent, low stress, fast, with friends	Helmet, tiring
M	Yes	Feels good, exercise, wind in face	Tiring	Yes	Fun	Fear of falling off	Yes	Flexible, independent, social, in control, easy	Tiring
N	–	–	–	Yes	Can multitask	Cold, annoying school bike partner	Yes	Fast, listen to music, relaxing, easy, with friends	Hot, morning, tiring, windy
O	–	–	–	Yes	In control, easy	No dislikes	Yes	Easy, convenient, no need for gas	Traffic danger

(continued)

Table 3. (continued)

ID	Age 9 (2009)			Age 12 (2012)			Age 15 (2015)		
	Like?	Likes	Dislikes	Like?	Likes	Dislikes	Like?	Likes	Dislikes
P	–	–	–	Yes	Fast, tricks	Locking it up	Yes	Fast, easy, convenient, better than alternatives	Locking it up
Q	–	–	–	Yes	Fun, with friends, fast	Helmet hair	No	For fun or recreation	Traffic danger, being seen by peers, sweaty, constrains clothing and hair choices
R	–	–	–	Yes	Fast, fun	Hills	Neutral	Fast, nice when hot because breeze	Locking it up, mechanical issues
S	Neutral	Mobility/ freedom	Fear of crash	Yes	Exercise, environment, different types of bikes	Slow, traffic/cars, fear of falling off	Yes	Cheap, environment, efficient, fast	Parent rules

Note: – = the question was not asked of the participant, either due to omission or because they were not yet a part of the study.

downtown. The minimum number of purposes provided was one (by one respondent), the maximum was eleven (five excluding this apparent outlier), and the median was four purposes.

Three years later, respondents indicated fewer purposes for bicycling and the most common purposes had shifted. Twelve participants biked to school, eight rode their bicycle to go shopping, while six biked to meet with friends and five each for fun or to go downtown. The maximum number of purposes at age 12 dropped to six and the median dropped to three, suggesting that the overall variety of trips being taken by the 12-year-olds decreased slightly.

At age 15, the most common purposes shifted yet again, to biking to meet with friends (thirteen) and to reach downtown destinations (fourteen), suggesting that social outings were particularly important to this age group. Bicycling to school remained one of the most common purposes (fourteen), while participants mentioned other purposes substantially less often. At age 15, the minimum number of purposes remained at one, the median increased to four, and the maximum remained at six, a slight uptick in the overall variety of trips taken by the teenagers.

Overall, bicycling frequency was associated with the number of bicycling purposes: those who bicycled more often were more likely to report a larger number of purposes. Over time, the participants increased how often they rode their bicycles and simultaneously shifted the types of trips they made by bike. While 9-year-olds overwhelmingly rode for the intrinsic enjoyment of bicycling, by 12 the same children were more likely to be bicycling to get to school and at 15 most of the panel cohort had added biking for social purposes to their repertoire.

Attitudes. Across all three rounds of interviews, the participants almost unanimously reported liking to bike, overall (Table 3). The real nuance in understanding the children's attitudes thus lies in the characteristics they cited as being reasons they like or dislike bicycling. As the children aged, the characteristics of bicycling that they like changed. 9- and 12-year-old children were more likely to enjoy intrinsic aspects (e.g., fun, movement, speed) of bicycling as well as the fun and ability to explore that they derived from bicycling. Yet these same children at age 15 appreciated the functional characteristics of bicycling as a way to travel quickly, independently, and flexibly, both with friends and to meet friends, and as a way to relax. In all three rounds, the most common characteristics of bicycling that participants disliked were the weather and effort associated with bicycling. Concerns about traffic and falling were somewhat prevalent at age 9 and 12, while concerns about appearance arose among some participants by age 15.

At age 9, among the 11-person panel, the most frequently cited aspects of bicycling that the participants liked were related to the intrinsic enjoyment of biking as "fun" (six). Though their descriptions at this age were plain, they were clear:

> It's fun to bike. (G)

Four also said they enjoyed it for the exercise; three said they liked biking for the physical, embodied experience (its movement and freedom); two reported enjoying being outside, seeing the scenery, and not being confined to a car; and two liked the speed of bicycling.

After the panel was augmented by an additional eight participants at age 12 (bringing the panel size to 19), many of the same characteristics were mentioned again: speed (nine), fun (seven), the physicality of bicycling (five), and seeing scenery/exploring (three):

> There's wind in your face and stuff, and it's nice and you can look around more than in a car because there's only a window. You can look around and you see all sorts of things. (K)

One change was an increase in the number of children (three) noting the environmental benefits of bicycling as a reason they liked it:

> Well, it's healthy for you and it's healthy for the environment and stuff. (S)

At age 15, the 19 participants were much more likely to cite functional aspects of bicycling as reasons they liked to ride a bicycle. Flexibility, convenience, or practicality (fifteen), independence (nine), and speed (especially relative to walking; nine) were the most frequently-mentioned characteristics at this age. These children appreciated being able travel efficiently and quickly on their own (with parental permission) without needing to schedule a car ride with their parents in advance. Many of these teenagers liked the convenience of leaving when they wanted, rather than potentially being restricted to leave at a certain time in order to match their parents' schedules:

> I don't have to wait for my parents to be able to take me places so I can just go when I'm ready. (G)

They also attributed their enjoyment of bicycling at this age to socializing (eight) and relaxation (six):

> Being with your friends is kind of more fun. And I guess in general when you're with your friends more, and it wouldn't really matter where you're going because you're with your friends and you can like talk and stuff. (B)

A small number of participants labeled the physicality of bicycling (e.g., being in control, moving; four) and the environment (four) as reasons for liking to bike.

Though most children reported a substantially larger number of qualities they liked than disliked about bicycling, almost all had at least one quality of bicycling that they said they disliked in each round of the interviews. At age 9, unpleasant weather (four) and the effort and slow speed of bicycling (four) were the most commonly cited reasons for disliking to bike:

> If you have to bike really far sometimes it's kind of tiring and you have to stop at lights and stuff. (E)

Fear of traffic and cars were mentioned by two respondents and concerns with falling were mentioned by one participant.

At age 12 (keeping in mind that the panel increased in size from 11 to 19 children), the effort and slow speed of bicycling (seven) and unpleasant weather (five) remained common responses. The children became more likely to report traffic/cars (four) and fear of falling (three) than they were at age 9:

> I'm scared of crossing the street. [And] I just get scared that I'm going to fall off sometimes. That's why I don't like riding my bike alone. (E)

Three years later at age 15, the 19-child panel provided unpleasant weather (nine) as well as the effort and slow speed of bicycling (eleven) as the primary reasons for not liking to bicycle. Fear of falling was not mentioned, while fear of traffic remained present but uncommon (three). Concerns about appearance became prevalent (four), primarily related to helmets and their impact on hair style choices, as well as the need to choose certain types of clothing if riding a bike.

At ages 9 and 12, several of the children had concerns regarding falling off their bicycle, suggesting that these children may not have been very confident in their bicycle handling skills, particularly on city streets. Though skills and knowledge were not directly assessed in the interviews, these attitudinal responses point to low bicycling skill (i.e., perceived behavioral control or self-efficacy). By age 15, in contrast, the slow speed of bicycling and the effort required were the main characteristics of bicycling that these children disliked, and fear of falling had completely disappeared, indicating bicycle handling proficiency.

Norms. We assessed injunctive norms (i.e., unwritten rules about how you should behave) at age 15 through questions about peers' personal attitudes toward bicycling and the image of bicycling at school (Table 4). Participants viewed their peers as holding neutral or slightly positive attitudes toward bicycling: nine said that

Table 4. Children's Perceptions of Peer Norms at Age 15

ID	Age 15 (2015)
	Peers
A	Neutral. Image of bicycling as normal.
B	Some like it, otherwise neutral. Half bike to school, half are driven. Image of bicycling is neutral.
C	Neutral, no image of bicycling.
D	At least neutral, considered a part of life.
E	Most like to bike for independence. Considered normal because so prevalent.
F	Most like to bike, because it is easy and independent. No image of bicycling because it is common.
G	At least neutral, for independence.
H	At least neutral. No image of bicycling because it is common.
I	Most are neutral, just the way it is in Davis.
J	Neutral, mode to school doesn't matter.
K	Neutral.
L	Neutral.
M	Most like to bike, accepted as normal.
N	At least neutral, not enthusiastic. No image of bicycling but bike fashion.
O	Most like to bike, but no image of bicycling.
P	Neutral, because it is convenient, best way to get around Davis. No image because it is the norm.
Q	At least neutral, no image.
R	Neutral, image of bicycling as uncool (compared to driving).
S	Neutral, part of lifestyle.

the average attitude toward bicycling was neutral, six said that their peers were at least neutral toward it, while four said that most of their peers liked to bike:

> I'd say at least the majority of people bike to school [and] they enjoy it. … It's just sort of normal, it's like eating. (A)

Several participants justified this perspective by suggesting that bicycling is part of the Davis lifestyle:

> I think biking is just kind of part of life in Davis. (D)

Explanatory Analysis

Our goal for this study was to answer the question: *What role do bicycling experiences play in developing bicycling attitudes and norms?* To do so, we begin by dividing participants according to their typical bicycling frequency over the three rounds of interviews. We define three groups: infrequent bicyclists (E, F, Q, R, and S), frequent bicyclists (B, D, H, K, M, and P), and daily bicyclists (A, C, G, I, J, L, N, and O). Each individual child's bicycling frequency was remarkably stable, shifting by no more than one category throughout the six years of this panel study. We assess systematic differences in attitudes and norms between these three groups before examining the additional influence of parents, friends, and personal characteristics.

Attitudes and Attitude Change. All three groups had similar bicycling attitudes at age 9 and 12 (Table 5). The biggest divergence seems to occur at age 15: just two of the five infrequent bicyclists came to appreciate the independence and flexibility afforded by bicycling, while all but one of their peers who bicycled frequently or daily came to like that aspect of bicycling. This result is intuitive, as infrequent bicyclists are unlikely to experience the independence and flexibility provided by a bicycle if they do not ride it often enough. But it has important ramifications, as failing to gain an awareness of how bicycling can fit into a (young) adult life could possibly hinder later adoption of bicycling.

Dislikes of bicycling were also associated with bicycling behavior. Fears of falling or crashing were more likely to be reported by the infrequent bicyclists, suggesting that less bicycling experience reduced these participants' bicycling skills, another important component of motility. At age 12, three of the five infrequent bicyclists mentioned fear of car traffic as a reason they disliked bicycling, whereas their peers were unlikely to list these fears at all. Similarly, two of the infrequent bicyclists mentioned fears of falling off their bike, while only one frequent bicyclist and one daily bicyclist identified this as a dislike. Although concerns about bicycle handling disappeared by age 15, two infrequent bicyclists and one daily bicyclist still cited fear of crashing as a characteristic of bicycling that they disliked.

Norms and Norm Change. Norms, which we operationalize as perceived peer attitudes toward bicycling, were negatively associated with personal bicycling behavior. Infrequent bicyclists were proportionally more likely to feel that their peers liked to bike (two of five infrequent

Table 5. Differences in Bicycling Attitudes and Norms at Age 15 by Categories of Bicycling Frequency

	Age 9			Age 12			Age 15		
	Infrequent	Frequent	Daily	Infrequent	Frequent	Daily	Infrequent	Frequent	Daily
Bicycling attitudes: Likes									
Fun	1	2	2	4	2	0	1	0	0
Exercise	1	1	2	2	0	0	1	0	1
Exploring or being outside	0	0	2	1	1	2	0	2	0
Movement	1	1	0	1	2	2	1	2	0
Speed	1	1	0	3	2	4	4	2	3
Environmentalism	0	0	1	2	0	1	2	0	2
Flexibility or convenience	0	0	0	0	0	0	2	5	8
Independence	0	0	0	0	0	0	1	4	4
Social	0	1	0	2	0	0	1	4	3
Relaxing	0	0	0	0	0	0	0	2	4
Bicycling attitudes: Dislikes									
Fear of car traffic or crashes	1	0	1	3	0	1	2	0	1
Weather	0	2	2	2	0	3	1	4	4
Slow	1	0	0	1	0	1	0	1	4
Tiring	1	1	2	1	1	3	2	1	3
Fear of falling	0	0	1	2	1	0	0	0	0
Helmet	0	1	0	1	0	0	1	1	2
Norms—Perceived peer attitudes									
Neutral	–	–	–	–	–	–	2	2	5
Neutral or like	–	–	–	–	–	–	1	3	2
Like to bike	–	–	–	–	–	–	2	1	1
Column totals	2	4	5	5	6	8	5	6	8

Note: – = the question was not asked of the participants because it was not part of the interview script in that year.

bicyclists) compared to their frequent and daily bicyclist peers (one of six and one of eight, respectively). Though this association is tenuous, this pattern suggests that the infrequent bicyclists see their peers as liking to bike more than they do, as a way to explain to themselves why their peers are bicycling at higher rates.

Influences of Parents, Friends, and Personal Characteristics. Parents' bicycling behavior was associated with their children's bicycling behavior and attitudes, though not with their children's injunctive norms about what is acceptable. Parents' bicycling appears to serve as a model to the children, and the parents' bicycling experiences could inform rules requiring their child to bike as well as aid them in helping their child overcome common bicycling barriers. Some parents imposed rules to limit the distance or location of independent bicycling at age 9 and 12. For example, some parents identified key locations that caused them worry, such as a bicycle underpass beneath a railroad track with low visibility. However, these rules declined in strictness as the child aged, in line with this study's conceptual model.

We could not explore the influence of friends' behavior, attitudes, and norms as easily as those of parents, who were directly involved in the interviews. The evidence we do have suggests that the influence of friends is similarly potent, particularly at older ages. At age 15, the social element of bicycling rose to the fore. Through the interviews, it became apparent that friends likely reinforced positive bicycling attitudes and norms indirectly through the relatively mundane act of coordinating social gatherings with friends. For example, if several friends were meeting downtown, and one friend was driven by a parent, the teenagers would complain that they were hindered from moving around downtown easily by bicycle because they would need to walk with their bicycle-less friend. They therefore much preferred when all members of the group had bicycled to get downtown so that they could spontaneously create or change plans.

Gender differences in bicycling seen elsewhere in the United States do not seem to hold in Davis. Both sexes are well represented in each bicycling behavior group: infrequent bicyclists (1 male, 3 female), frequent bicyclist (3 males, 3 females), and daily bicyclists (3 males, 5 females). However, at age 15, young women were the only panel participants to report neutral (two) or negative (two) attitudes toward bicycling. This may be related to external, societal standards of appearance for young women, reflected in their reported concerns over the impacts of bicycling on perspiration, hair style, and clothing choice.

Discussion of Results

This study finds that children's bicycling frequency is associated with their bicycling attitudes and norms, in support of the motility framework and this study's conceptual model. At age 15, frequent and daily bicyclists almost universally voiced their appreciation of the independence and convenience provided by the mode. Infrequent bicyclists were less likely to note these characteristics as reasons for liking to bike, likely because they did not use it often enough or in sufficient settings for this benefit to become apparent. This research identifies behavioral antecedents of these attitudes: because these individuals were bicycling at high levels before they acquired their liking of the independence and convenience of bicycling, we can conclude that their bicycle behavior in part caused the development of these attitudes.

Whether positive bicycling attitudes have formed by age 15 is especially relevant from a practical standpoint, since at this age the teenagers' ineligibility to hold a full driver's license restricted their mobility. Holding positive attitudes toward bicycling at this life stage is therefore likely to facilitate independent travel (and vice versa) rather than a dependence on chauffeured trips by parents.

Bicycling behavior was also associated with the perception of norms held by peers and with the bicycling behavior of friends. Infrequent bicyclists were more likely to perceive peers as liking to bike, perhaps as a way for them to explain to themselves why others are bicycling more frequently. The social influence of friends increased bicycling and contributed to more positive attitudes about bicycling, creating positive feedback loops to reinforce bicycling behavior.

Previous research at both aggregate (*32*) and disaggregate (*21*) scales suggests that bicycling rates decline when American children reach their teenage years. The results of this study demonstrate the opposite, showing that children in Davis, CA, are as, if not more, likely to ride their bicycles as young teenagers. Though the influence of the built environment is not the primary focus of this paper, these results inevitably point to the value of the comprehensive, safe bicycle network that Davis planners and elected officials have cultivated over the past decades (*23*) in not just promoting bicycling but also building bicycling motility. Another round of interviews with this panel will enable us to assess the degree to which bicycling motility changes as these teenagers become eligible for driver's licenses.

Acknowledgments

The authors thank the University of California Transportation Center, National Center for Sustainable Transportation Dissertation Fellowship, and Dwight D. Eisenhower Transportation Fellowship for funding interview administration and analysis. They also thank Rachel Maiss and Brigitte Driller for their excellent stewardship of the first two rounds of interviews. Lastly, they wish to thank the Davis, California, families for taking the time to share their travel experiences and attitudes over multiple rounds of interviews.

Author Contributions

The authors confirm contribution to the paper as follows: study conception and design: S. Handy, C. Thigpen; data collection: S. Handy, C. Thigpen; analysis and interpretation of results: S. Handy, C. Thigpen; draft manuscript preparation: S. Handy, C. Thigpen. All authors reviewed the results and approved the final version of the manuscript.

References

1. Flamm, M., and V. Kaufmann. Operationalising the Concept of Motility: A Qualitative Study. *Mobilities*, Vol. 1, No. 2, 2006, pp. 167–189. https://doi.org/10.1080/17450100600726563.
2. Handy, S., B. van Wee, and M. Kroesen. Promoting Cycling for Transport: Research Needs and Challenges. *Transport Reviews*, Vol. 34, No. 1, 2014, pp. 4–24. https://doi.org/10.1080/01441647.2013.860204.
3. Stewart, O., A. V. Moudon, and C. Claybrooke. Common Ground: Eight Factors That Influence Walking and Biking to School. *Transport Policy*, Vol. 24, 2012, pp. 240–248. https://doi.org/10.1016/j.tranpol.2012.06.016.
4. Sallis, J. F., N. Owen, and E. B. Fisher. Ecological Models of Health Behavior. In *Health Behavior and Health Education: Theory, Research, and Practice* (Glanz, K., B. K. Rimer, and K. Viswanath, eds.), Jossey-Bass, San Francisco, CA, 2008, pp. 465–485.
5. Ajzen, I. The Theory of Planned Behavior. *Organizational Behavior and Human Decision Processes*, Vol. 50, 1991, pp. 179–211. https://doi.org/10.1016/0749-5978(91)90020-T.
6. Festinger, L., and J. M. Carlsmith. Cognitive Consequences of Forced Compliance. *Journal of Abnormal and Social Psychology*, Vol. 58, 1959, pp. 203–210. https://doi.org/10.1146/annurev.psych.48.1.609.
7. Kroesen, M., S. Handy, and C. Chorus. Do Attitudes Cause Behavior or Vice Versa? An Alternative Conceptualization of the Attitude-Behavior Relationship in Travel Behavior Modeling. *Transportation Research Part A: Policy and Practice*, Vol. 101, 2017, pp. 190–202. https://doi.org/10.1016/j.tra.2017.05.013.
8. Schwartz, S. H. Normative Explanations of Helping Behavior: A Critique, Proposal, and Empirical Test. *Journal of Experimental Social Psychology*, Vol. 9, No. 4, 1973, pp. 349–364. https://doi.org/10.1016/0022-1031(73)90071-1.
9. Cialdini, R. B., C. A. Kallgren, and R. R. Reno. A Focus Theory of Normative Conduct: A Theoretical Refinement and Reevaluation of the Role of Norms in Human Behavior. *Advances in Ex*, Vol. 24, 1991, pp. 201–234.
10. Kaufmann, V., G. Viry, and E. D. Widmer. Motility. In *Mobile Living across Europe II. Causes and Consequences of Job-Related Spatial Mobility in Cross-National Comparison* (Schneider, N., and B. Collet, eds.), Barbara Budrich Publishers, Opladen & Farmington Hills, MI, 2010, pp. 95–111.

11. De Bruijn, G. J., S. P. J. Kremers, H. Schaalma, W. Van Mechelen, and J. Brug. Determinants of Adolescent Bicycle Use for Transportation and Snacking Behavior. *Preventive Medicine*, Vol. 40, No. 6, 2005, pp. 658–667. https://doi.org/10.1016/j.ypmed.2004.09.003.
12. Zeuwts, L., P. Vansteenkiste, G. Cardon, and M. Lenoir. Development of Cycling Skills in 7- to 12-Year-Old Children. *Traffic Injury Prevention*, Vol. 17, No. 7, 2016, pp. 736–742. https://doi.org/10.1080/15389588.2016.1143553.
13. Newell, A., and P. S. Rosenbloom. Mechanisms of Skill Acquisition and the Law of Practice. In *Cognitive Skills and their Acquisition* (Anderson, J. R., ed.), Erlbaum, Hillsdale, NJ, 1981, pp. 1–55.
14. Bandura, A. Self-Efficacy Mechanism in Human Agency. *American Psychologist*, Vol. 37, No. 2, 1982, pp. 122–147. https://doi.org/10.1037/0003-066X.37.2.122.
15. Dill, J., C. Mohr, and L. Ma. How Can Psychological Theory Help Cities Increase Walking and Bicycling? *Journal of the American Planning Association*, Vol. 80, No. 1, 2014, pp. 36–51. https://doi.org/10.1080/01944363.2014.934651.
16. McDonald, N. C. Is There a Gender Gap in School Travel? An Examination of US Children and Adolescents. *Journal of Transport Geography*, Vol. 20, No. 1, 2012, pp. 80–86. https://doi.org/10.1016/j.jtrangeo.2011.07.005.
17. Goddard, T., and J. Dill. Gender Differences in Adolescent Attitudes about Active Travel. Presented at 93rd Annual Meeting of the Transportation Research Board, Washington, D.C., 2014.
18. Emond, C. R., W. Tang, and S. L. Handy. Explaining Gender Difference in Bicycling Behavior. *Transportation Research Record: Journal of the Transportation Research Board*, 2009. 2125: 16–25. https://doi.org/10.3141/2125-03.
19. Susilo, Y. O., and C. Liu. The Influence of Parents' Travel Patterns, Perceptions and Residential Self-Selectivity to Their Children Travel Mode Shares. *Transportation*, Vol. 43, No. 2, 2016, pp. 357–378. https://doi.org/10.1007/s11116-015-9579-0.
20. Broach, J., and J. Dill. Modeling Children's Independent Walk and Bike Travel to Parks and School. Presented at 92nd Annual Meeting of the Transportation Research Board, Washington, D.C., 2013.
21. Underwood, S. K., S. L. Handy, D. A. Paterniti, and A. E. Lee. Why Do Teens Abandon Bicycling? A Retrospective Look at Attitudes and Behaviors. *Journal of Transport & Health*, Vol. 1, 2014, pp. 17–24.
22. Baslington, H. Travel Socialization: A Social Theory of Travel Mode Behavior. *International Journal of Sustainable Transportation*, Vol. 2, No. 2, 2008, pp. 91–114. https://doi.org/10.1080/15568310601187193.
23. Buehler, T., and S. Handy. Fifty Years of Bicycle Policy in Davis, California. *Transportation Research Record: Journal of the Transportation Research Board*, 2008. 2074: 52–57.
24. Fitch, D. T., C. G. Thigpen, and S. L. Handy. Traffic Stress and Bicycling to Elementary and Junior High School: Evidence from Davis, California. *Journal of Transport and Health*, Vol. 3, No. 4, 2016, pp. 457–466. https://doi.org/10.1016/j.jth.2016.01.007.
25. Thigpen, C. *Results of the 2014-15 Campus Travel Survey*. Publication UCD-ITS-RR-15-15. Davis, CA, 2015.
26. Ahlport, K. N., L. Linnan, A. Vaughn, K. R. Evenson, and D. S. Ward. Barriers to and Facilitators of Walking and Bicycling to School: Formative Results From the Non-Motorized Travel Study. *Health Education & Behavior*, Vol. 35, No. 2, 2008, pp. 221–244. https://doi.org/10.1177/1090198106288794.
27. Maiss, R., and S. Handy. Bicycling and Spatial Knowledge in Children: An Exploratory Study in Davis, California. *Children, Youth and Environments*, Vol. 21, No. 2, 2011, pp. 100–117.
28. Driller, B. K. *Children's Attitudes Toward Transportation Modes: A Qualitative Interview Study in Davis, California*. University of California, Davis, CA, 2013.
29. Fylan, F. Semi-Structured Interviewing. In *A Handbook of Research Methods for Clinical and Health Psychology* (Miles, J., and P. Gilbert, eds.), Oxford University Press, New York, NY, 2005.
30. U.S. Census Bureau. American FactFinder. factfinder.census.gov. Accessed July 5, 2016.
31. Saldana, J. *The Coding Manual for Qualitative Researchers*. SAGE Publications, Thousand Oaks, CA, 2009.
32. Pucher, J., and R. Buehler. Making Cycling Irresistible: Lessons from The Netherlands, Denmark and Germany. *Transport Reviews*, Vol. 28, No. 4, 2008, pp. 495–528. https://doi.org/10.1080/01441640701806612.

The Standing Committee on Bicycle Transportation (ANF20) peer-reviewed this paper (18-00147).

Article

TRR
JOURNAL OF THE TRANSPORTATION RESEARCH BOARD

Transportation Research Record
2018, Vol. 2672(36) 24–32

DOI: 10.1177/0361198118791630
journals.sagepub.com/home/trr
SAGE

Equity in Spatial Access to Bicycling Infrastructure in Mid-Sized Canadian Cities

Meghan Winters[1], Jaimy Fischer[1], Trisalyn Nelson[2], Daniel Fuller[3], and David G. T. Whitehurst[1,4]

Abstract

The impacts of active transportation planning on equity are often overlooked, potentially leading to disparities in who receives benefits of infrastructure investment. This study examined income inequalities in spatial access to bicycling infrastructure in three mid-sized Canadian cities: Victoria and Kelowna (British Columbia), and Halifax (Nova Scotia), using non-spatial and spatial methods. We compiled municipal bicycling infrastructure data and calculated access to bicycling infrastructure (m/km^2) for dissemination areas (DAs) within each city. We analyzed trends in access across median household income quintiles, and characterized spatial patterns using a local measure of spatial autocorrelation. DAs in Kelowna ($n = 168$) had the greatest access to infrastructure (median infrastructure = 2,915 m/km^2), followed by Victoria ($n = 386$ DAs; median = 2,157 m/km^2), and Halifax ($n = 312$ DAs; median = 0 m/km^2). Lower income areas in Victoria and Kelowna had greater access to infrastructure compared with higher income areas. The majority of DAs in Halifax had no infrastructure (59%), consistent across income quintiles. Spatial pattern analysis identified clusters of low income areas with poor access in each city, which may be targets for strategic, equitable investment. Although in many cities bicycling infrastructure planning is not driven by equity considerations, there is increasing political pressure to ensure equitable access to safe bicycling. Measuring and mapping trends in access to transportation resources from an equity perspective are requisite steps in the pathway toward healthy, sustainable cities for all.

Access to high-quality bicycling infrastructure can increase overall comfort, safety, and uptake of bicycling (*1*) regardless of sociodemographic background (*2*). The distribution of bicycling infrastructure correlates with the uptake of bicycling, and thus who receives benefits (*3–5*). In North America, the physical and social benefits of bicycles have gained increasing attention in transportation, urban planning, and public health research. Despite this, the provision of bicycling infrastructure remains sparse and equity is often not a primary consideration in its implementation (*6, 7*). Equity issues related to active transportation pertain to both the social (*who is* walking and bicycling) and spatial (*who has access to* walking and bicycling infrastructure), and affect how benefits and risks are distributed across population groups (*5*).

Past research on equity in access to active transportation has had variable findings. For example, a U.S. study of bike-share programs in New York, Boston, Washington DC, Arlington, Chicago, Denver, and Seattle found that bike share stations were disproportionately located in neighborhoods with greater proportions of White people and of higher socioeconomic status (*8*). In contrast, other studies have found that neighborhoods with lower socioeconomic status have more supportive settings for active transportation. For instance, socially disadvantaged groups tended to have better access to public transit and bike share in Montreal, Canada (*9, 10*), and to bicycling infrastructure in Melbourne, Australia (*11*). A study across 63,000 U.S. census tracts found that disadvantaged neighborhoods were more walkable than higher income neighborhoods,

[1]Faculty of Health Sciences, Simon Fraser University, Burnaby, BC, Canada
[2]School of Geographical Sciences and Urban Planning, Arizona State University, Tempe, AZ
[3]School of Human Kinetics and Recreation, Memorial University of Newfoundland, St. John's, NL, Canada
[4]Centre for Clinical Epidemiology and Evaluation, Vancouver Coastal Health Research Institute, Vancouver, BC, Canada

Corresponding Author:
Address correspondence to Meghan Winters: mwinters@sfu.ca

a finding framed as the "disadvantaged advantage" (*12*). A number of studies have expressly looked at changes in both demographics and infrastructure over time. Two such studies (one in Chicago and Portland, and another of Birmingham, Chicago, Minneapolis, and Oakland) found that marginalized (or formerly marginalized) neighborhoods experienced increased investment in bicycling infrastructure concurrent with an increase of more privileged populations (*13, 14*), rather than impacts on equity in access. On the whole, these contrasting results support that the relationship between urban form and neighborhood socioeconomic composition are complex, and that context is important (*5, 12*).

Much of the past research has been in large U.S. cities where racial and socioeconomic segregation across neighborhoods may be more pronounced than in Canada (*15*). In Canada, many cities are mid-sized (populations between 50,000 and 500,000), with shorter travel distances and different land use and transportation networks relative to larger cities. This study examined income inequalities in access to bicycling infrastructure in three mid-sized Canadian cities using non-spatial and spatial analysis. We analyze access to bicycling infrastructure by quintiles of median household income. Additionally, we advance this research area by incorporating spatial pattern analysis, which can reveal spatial patterns and identify localized opportunities for investment in bicycling infrastructure that could increase equity in access (*6, 12*). To date, spatial considerations have often been overlooked despite the fact that infrastructure and population demographics are spatially patterned.

Materials and Methods

Study Area

This study focuses on three Canadian mid-sized cities: Victoria (British Columbia), Kelowna (British Columbia), and Halifax (Nova Scotia). These were selected as study cities in the larger "Impacts of Bicycling Infrastructure in Mid-Sized Cities" (IBIMS) project, a natural experiment study of an investment in a citywide bicycling network in the City of Victoria (described elsewhere [*16*]). Kelowna and Halifax are control cities in the IBIMS study and were identified based on size, urban layout, and climate.

The geographic boundaries for the study areas were based on regional travel patterns, and determined together with local municipal partners. In Victoria, the study area includes the city proper, together with the adjacent municipalities (Esquimalt, Oak Bay, and Saanich). We use the census subdivision (city limits) in Kelowna. Halifax is a regional municipality, including both urban and rural environments. For the IBIMS study we include the metropolitan core that includes the peninsula, Mainland Halifax and Dartmouth. Population, bicycling mode share, infrastructure types, and density for each city are presented in Table 1.

We use dissemination areas (DAs) as our mapping unit (population of 400–700 persons [*17*]), which are the smallest standard geographic unit for which census data are available. DAs with suppressed income data were excluded.

Access to Bicycling Infrastructure

We assess spatial access to bicycling infrastructure in the first year of the IBIMS study. We obtained spatial data for 2016 from regional transportation authorities in Victoria and Halifax, and from the Open Data Catalogue for Kelowna (*18*). We standardized classification of bicycling specific infrastructure (*1, 19, 20*) using the following categories: separated cycle tracks, on-street bicycle lanes, off-street paths (either bicycle only or multi-use), and local street bikeways. We excluded sharrows and shared lanes, which are shared sections of road along major or minor arterial roads. Here, local street bikeways were defined as routes along streets with lower speeds and motor vehicle traffic volumes, signs, or pavement markings to indicate bicyclists (*19*), and which may or may not have traffic calming measures or bicyclist activated crossings. Route-type classifications were cross-validated using Google Earth and in consultation with city partners. Local street bikeways were the most inconsistently classified routes in the municipal data.

We operationalized spatial access as the density of bicycling infrastructure, measured in meters of infrastructure/km^2 (m/km^2) of land area. Within each DA we calculated total length (in meters) of bicycling infrastructure and normalized by area to account for differences in the size of DAs. As bicycling infrastructure often corresponds to edges of DAs along road centerlines (*17*), we buffered DAs to include all infrastructure with 10 m of boundaries.

Income Inequalities

To assess inequalities, we used median household income, derived from the 2011 National Household Survey (*21*). For the non-spatial analysis we created city-specific quintiles of median household income, comparable with past studies (*22–24*). Quintile 1 (Q1) is the highest income and Quintile 5 (Q5) is the lowest income. For spatial pattern analysis, we used median household income as a continuous measure, as is appropriate for studies of spatial autocorrelation using Moran's I measures (*25*).

Table 1. Population, Mode Share, and Infrastructure Characteristics of Study Cities in the Impacts of Bicycling Infrastructure in Mid-Sized Cities Study

				Bicycling infrastructure (km)				
City	Population	Population density (people/km^2)[b]	Journey to work bicycle mode share[b]	Cycle track	On-street bicycle lane	Off-street path (bicycle only or multi-use)	Local street bikeway	Total
Victoria	85,792 (235,689[a])	4,388 (1,673[a])	9%[a]	3.1[a]	91.8[a]	72.6[a]	26.6[a]	193.8[a]
Kelowna	128,669	602	4%	4.5	157.3	81.2	0	243
Halifax[c]	204,927	1,674	2%	0.6	35.9	44.7	1.2	82.4

[a]Study area population including adjacent municipalities of Esquimalt, Oak Bay, and Saanich.
[b]Statistics Canada. Census Dictionary, 2011. http://www12.statcan.gc.ca/census-recensement/2011/ref/dict/index-eng.cfm. Accessed Jul. 27, 2017.
[c]Includes the metropolitan core of the downtown peninsula, mainland populous area and Dartmouth.

Analysis

We analyzed associations between income and access to bicycling infrastructure using non-spatial and spatial analysis. In non-spatial analysis, we assessed trends in access to bicycling infrastructure across quintiles of income, and used linear regression to assess significant differences (using $p < 0.05$ as a threshold for test of statistical significance) in access to bicycling infrastructure across quintiles, with Q1 (highest income) as the reference category.

The spatial pattern analysis characterizes the distribution of income and infrastructure across the cities to highlight spatial patterns and identify localized opportunities for investment in bicycling infrastructure. We used a local measure of spatial autocorrelation (Local Moran's I_i), which quantifies spatial autocorrelation and characterizes similarity and dissimilarity in nearby locations (*26*). Specifically, we used a variant of Local Moran's I_i designed for bivariate data. Bivariate Local Moran's I_i quantifies the local relationship between two variables (e.g., income and infrastructure) against the overall global pattern. It results in an index value for each DA, which can be used to map where spatial patterns are different than would be expected if the spatial pattern was random, identifying where both variables are extreme highs, where both variables are extreme lows, and where one variable is high and the other low (or vice versa) (*25*). Statistical significance was determined using Monte Carlo Randomization (MCR), a random permutation procedure that compares the pattern of observed spatial data to patterns generated through random processes. Observations are shuffled (permuted) within the study area and a local statistic is calculated for each permutation, creating a reference distribution of what patterns could be if the processes that generated them were random. For each observation, if the observed value of the local statistic falls in the upper tail of the reference distribution it is considered part of a cluster (*27*). The permutation process generates a pseudo p-value that is used to test whether spatial autocorrelation is significant. We used 999 permutations which approximates a pseudo significance of $p = 0.001$ (*27*).

We used maps of Bivariate Local Moran's I_i results to visualize clusters ($p \leq 0.05$) of DAs with high median income and high access to bicycling infrastructure; high median income and low access to bicycling infrastructure; low median income and low access to bicycling infrastructure; and low median income and high access to bicycling infrastructure. Data processing and analysis were completed using ArcGIS 10.4.4 (*28*), R Statistics 3.3.2 (*29*) and GeoDa 1.6.7 (*30*)

Results

Non-Spatial Analysis

There were 386 DAs in the Victoria study area, 167 in Kelowna, and 317 in Halifax. Overall, areas in Kelowna had the greatest access to bicycling infrastructure (median = 2,915 m/km^2, 25th percentile [p25] = 1177, 75th percentile [p75] = 5256), followed by Victoria (median = 2157, p25 = 213, p75 = 3994), and then Halifax (median = 0, p25 = 0, p75 = 1417). In Halifax, 59% of DAs had no access to bicycling infrastructure at all (188/317); in Victoria 18% (70/386) of DAs had no infrastructure, whereas in Kelowna all DAs had access to at least some bicycling infrastructure. In Victoria, 55% of bicycling infrastructure was on-street bicycle lanes, 73% in Kelowna, and 53% in Halifax.

Figure 1 shows trends in access to bicycling infrastructure by income quintile for each city. The boxplots illustrate that in Victoria and Kelowna there is greater access to bicycling infrastructure in lower income quintiles. In Halifax, the dominant message is that there is low access to infrastructure across all income quintiles.

We assessed statistical significance of associations in city-specific linear regression models (Table 2). For

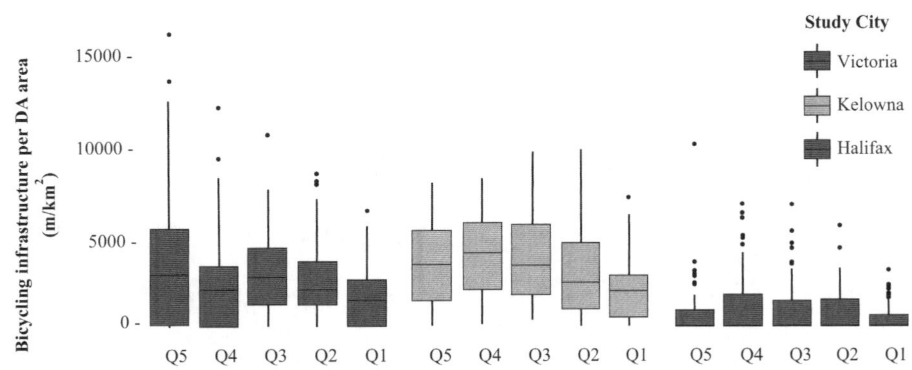

Figure 1. Access to bicycling infrastructure by quintiles of area-level income. Boxplots show that in Victoria and Kelowna there is greater access to bicycling infrastructure in lower income quintiles, and in Halifax, the main trend is that there is low access to infrastructure across all income quintiles.

Table 2. Associations between Quintiles of Median Income and Mean Total Bicycling Infrastructure per DA Area (m/km^2)

	Facilities/DA area (m/km^2)		
Income quintile	Victoria	Kelowna	Halifax
Q1 (high income)	Reference	Reference	Reference
Q2	860.0* (23.3, 1,696.7)	904.9 (−287.04, 2,096.73)	393.4 (−128.3, 915.1)
Q3	1,336.2* (499.5, 2,172.9)	2,191.8* (999.9, 3,383.6)	429.1 (−92.6, 950.8)
Q4	862.58* (25.9, 1,699.3)	1,974.4* (791.3, 3,157.5)	636.4* (116.8, 1,156)
Q5 (low income)	2,076.9* (1,242.9, 2,911.0)	1,403.8* (220.7, 2,586.9)	182.4 (−337.2, 702.1)
Adjusted r^2	0.05	0.07	0.01

Note: * = significant difference ($p \leq 0.05$) from highest median household income quintile (Q1).

Victoria, all other quintiles (Q2–Q5) had significantly greater access to infrastructure relative to those in Q1 (highest income) areas. In Kelowna, Q3–Q5 (mid to lowest income quintiles) had significantly more access than the highest income quintile (Q1). In Halifax, only Q4 had significantly more access to bicycling infrastructure than Q1.

Spatial Pattern Analysis

Figure 2 shows maps of clusters of DAs with high median income and high access to bicycling infrastructure; high median income and low access to bicycling infrastructure; low median income and low access to bicycling infrastructure; and low median income and high access to bicycling infrastructure as identified using Bivariate Local Moran's I_i. In Kelowna and Victoria, the lower income clusters are in the city cores (shown in blues), and higher income areas are outside the city core (in reds). In Halifax, lower income communities tend to be outside the city core, in lower density areas. Overall, 16% of people in Victoria live in DAs with high spatial access to bicycling infrastructure (either low income or high income areas), 18% of people in Kelowna, and only 7% of the population in Halifax.

These maps suggest insights for bicycling infrastructure planning. The low income–low infrastructure clusters (dark blue) are clear opportunities for investment, if the goal is to reduce income inequalities in access to bicycling infrastructure. In Victoria and Kelowna, many of these are within the city core. Of note, they are often adjacent to low income–high infrastructure clusters, suggesting there is bicycling infrastructure nearby, just not in the immediate area. There are also areas with high income–low infrastructure (light red), which may benefit from infrastructure investment. Investment here may increase access, but would not reduce inequities in access.

Finally, the maps also suggest connectivity issues. Areas where high infrastructure and low infrastructure clusters are adjacent to each other (e.g., light blue areas

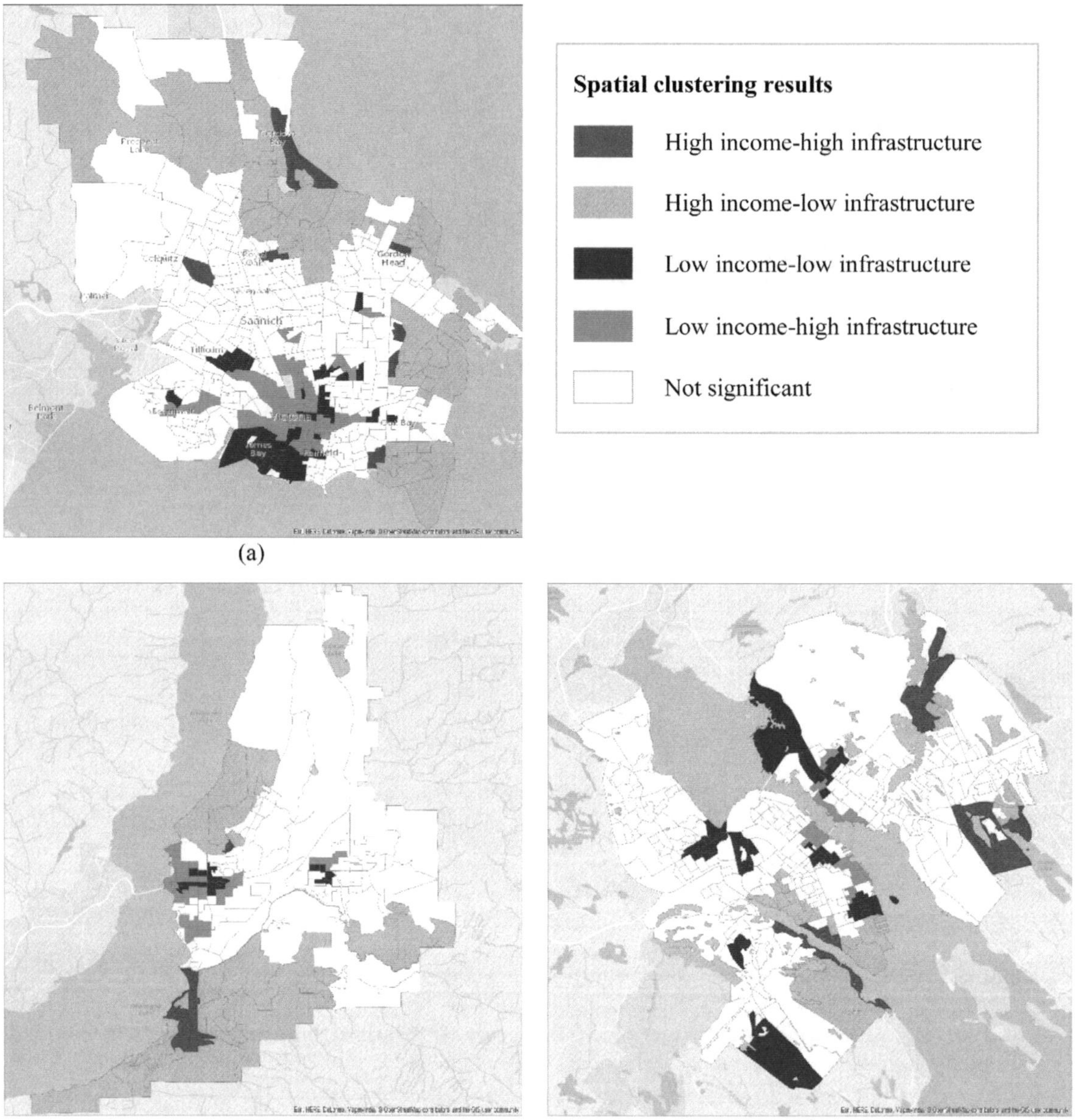

(a)

(b) (b)

Figure 2. Bivariate Local Moran's I_i results for spatial patterns between access to bicycling infrastructure and area-level income in study cities (*a*) Victoria, (*b*) Kelowna, and (*c*) Halifax. In Victoria and Kelowna, clusters of lower income are in the city cores (blues), and clusters of higher income areas are outside the city core (reds); in Halifax, lower income clusters tend to be on the periphery. Low income–low infrastructure clusters (dark blue) are potential areas for investment, and high infrastructure and low infrastructure clusters beside each other (light blue areas adjacent to dark blue areas) may highlight network connectivity gaps.

adjacent to dark blue areas) may highlight gaps in infrastructure provision, and highlight areas were investment is needed to improve connectivity across the region.

Discussion

There is increasing discussion on equity in active transportation research and practice, considering both spatial and social equity (*5, 6*), as well as horizontal equity (equal spatial distribution across a region) and vertical equity (greater access to vulnerable populations) (*31–33*). This field needs to consider accessibility for those with the greatest needs, and ultimately assess their uptake of bicycling infrastructure. Social inequities may be exacerbated if infrastructure is not targeted in lower income or disadvantaged neighborhoods. In this study, we analyzed equity in access to bicycling infrastructure in three Canadian mid-sized cities. We found that in Victoria and Kelowna, there was greater access to bicycling infrastructure in low income areas, whereas Halifax was a city with much lower access across all areas. Our inclusion of spatial methods is a novel contribution to this field, and enables the identification of areas in need of investment so as to improve equity and connectivity.

Although past research has typically been conducted in large North American cities, we focused on mid-sized cities. Using the definition of "a center with a population of 50,000 to 500,000," there are just over 50 mid-sized cities in Canada, and they are home to about 8.1 million of the 35 million people living in Canada (*34–36*). Mid-sized cities bring a different context for active transportation planning and investment, and many of these cities are poised to develop their active transportation networks, albeit from a different baseline than large urban centers. In Halifax, a city with lower bicycling mode share, the majority of DAs (59%) had no bicycling infrastructure. In contrast, all areas in Kelowna (100%) and most in Victoria (82%) had access to bicycling infrastructure. In terms of equity, in Victoria and Kelowna there was greater access to bicycling infrastructure in lower income areas. In both cities, the lower income areas are typically within or nearby the city core and have higher densities. The higher income areas are more suburban, further from the core. In these cities, were bicycling infrastructure investment prioritized in more dense areas (based on population and employment), this would result in investment in lower income neighborhoods and have positive impacts and equity in spatial access. As a cross-sectional analysis, our study does not make inference on gentrification. However, two longitudinal studies in large U.S. cities provide evidence that areas with increases in bicycling infrastructure investment have also experienced demographic shifts to more privileged populations over time (*13, 14*). Differences in our study cities may result from contrasting land use patterns across cities, related political and historical contexts, but also to size. Compared with U.S. cities, Canadian cities tend to have higher densities, stronger central urban cores, and shorter trip distances due to compact, mixed land use (*37*). Mid-sized Canadian cities have less profound demographic segregation than larger cities, especially when contrasted with the United States (*15*).

Mapping spatial patterns is a helpful method to identify where to prioritize future investments. Our spatial pattern analysis identified areas that have low infrastructure and low income. If equity is a goal when planning bicycling infrastructure investments, these areas may be targets for investment to increase spatial access for disadvantaged populations (*2, 5*). These spatial methods can be applied to other cities, and also to different transportation infrastructure, such as sidewalks or public transit (*5*) to identify strategic investment opportunities. Spatial pattern approaches have also been found effective for monitoring change through time in other phenomena such as environmental toxins (*38*), wildfire ignitions (*39, 40*), and neighborhood change (*14*), and may be useful for assessing policy intervention impacts of equitable access to bicycling infrastructure.

Limitations and Future Work

This analysis focused on three mid-sized Canadian cities as part of the IBIMS project. To assess more generalized trends, future research may include a larger number of mid-sized North American cities, an effort that would require a standardized source for bicycling infrastructure data, such as OpenStreetMaps. Second, we did not capture quality of the bicycling infrastructure, although we included route types that are attractive and safe for cyclists (*19*) and excluded route types such as sharrows or mountain bike paths. In our study cities, the majority of bicycling infrastructure was on-street bicycle lanes. A complementary approach to our work would be to use level of traffic stress measures, which has not been done previously in these cities, perhaps due to data limitations. Third, our measure of socioeconomic status was median household income, an intuitive and interpretable indicator. Similar analyses could be undertaken with deprivation indices or indicators for race/ethnicity, which may be more applicable in certain settings. Fourth, this analysis focused on access to bicycling infrastructure in the nearby area, as measured by the density of infrastructure. This is a common metric in past literature (*41, 42*), although others have normalized by population (*43*), or by road (*44*). Network connectivity is also an important consideration (*45–47*), and as origin–destination surveys are undertaken in the study cities we may be better able to characterize transportation flows and ease of access between origins and destinations. Finally, this work is a complement to ongoing research on social equity, which may characterize the racial, gender, and age trends in who is bicycling (*2, 5, 13, 14, 48*).

Conclusion

In many cities bicycling infrastructure planning is not driven by equity considerations, but rather by population and employment density, traffic demand management, targets for ridership or safety, political conditions, and strategic opportunities. However, there are increasing calls to action in North American settings to prioritize equity considerations in transportation planning, as highlighted by a 2017 special issue of Transport Reviews dedicated to the topic (*49*), recent practice guidance (*5, 50, 51*), recognition by national professional bodies (the American Public Health Association [*52*], and the Canadian Institute of Planners [*53*]), as well as strong and continued efforts from the non-governmental sector, through organizations such as the Better Bike Share Initiative (*54*) and the League of American Bicyclists (*55*). Measuring and mapping trends in access to transportation resources from an equity perspective are requisite steps in the pathway toward healthy, sustainable cities for all.

Acknowledgments

We acknowledge partners at the City of Victoria, the Capital Regional District, the City of Kelowna and the Halifax Regional Municipality for informing the design of this study protocol and sharing spatial data on bicycling infrastructure, and Suzanne Therrien, Simon Fraser University for coordination. This work was funded by the Canadian Institutes for Health Research (Award nos. #365011 and #377333). MW holds a Scholar Award from the Michael Smith Foundation for Health Research.

Author Contributions

The authors confirm contribution to the paper as follows: study conception and design: MW; data collection: MW, JF; analysis and interpretation of results: MW, JF, TN, DF, DW; draft manuscript preparation: MW, JF, TN. All authors reviewed the results and approved the final version of the manuscript.

References

1. Winters, M., M. Brauer, E. M. Setton, and K. Teschke. Mapping Bikeability: A Spatial Tool to Support Sustainable Travel. *Environment and Planning B: Planning and Design*, Vol. 40, No. 5, 2013, pp. 865–883. https://doi.org/10.1068/b38185.
2. Cahen, C. "More than Paint on Concrete": The Winding Path toward Bike Equity. *Global Journal of Community Psychology Practice*, Vol. 7, No. 3, 2016, pp. 1–9.
3. Buehler, R., and J. Pucher. Cycling to Work in 90 Large American Cities: New Evidence on the Role of Bike Paths and Lanes. *Transportation*, Vol. 39, No. 2, 2012, pp. 409–432. https://doi.org/10.1007/s11116-011-9355-8.
4. Winters, M., K. Teschke, M. Brauer, and D. Fuller. Bike Score®: Associations between Urban Bikeability and Cycling Behavior in 24 Cities. *International Journal of Behavioral Nutrition and Physical Activity*, Vol. 13, No. 1, 2016, p. 18. https://doi.org/10.1186/s12966-016-0339-0.
5. Lee, R. J., I. N. Sener, and S. N. Jones. Understanding the Role of Equity in Active Transportation Planning in the United States. *Transport Reviews*, Vol. 37, No. 2, 2016, pp. 211–226. https://doi.org/10.1080/01441647.2016.1239660.
6. Lee, R. J., I. N. Sener, and S. N. Jones-Meyer. A Review of Equity in Active Transportation. Presented at 95th Annual Meeting of the Transportation Research Board, Washington, D.C., 2016.
7. Dill, J., and N. McNeil. Revisiting the Four Types of Cyclists: Findings from a National Survey. *Transportation Research Record: Journal of the Transportation Research Board*, 2016. 2587: 80–99.
8. Ursaki, J., and L. Aultman-Hall. Quantifying the Equity of Bikeshare Access in U.S. Cities. Presented at 95th Annual Meeting of the Transportation Research Board, Washington, D.C., 2016.
9. Manaugh, K. and A. M. El-Geneidy. Who Benefits from New Transportation Infrastructure? Using Accessibility Measures to Evaluate Social Equity in Transit Provision. InAccessibility and Transport Planning: Challenges for Europe and North America (K. T. Geurs, K. J. Krizek, and A. Reggiani, eds.). Cheltenham, UK: Edward Elgar, 2012, 211–227.
10. Fuller, D., L. Gauvin, and Y. Kestens. Individual- and Area-Level Disparities in Access to the Road Network, Subway System and a Public Bicycle Share Program on the Island of Montreal, Canada. *Annals of Behavioral Medicine*, Vol. 45, No. Suppl. 1, 2013, pp. 95–100. https://doi.org/10.1007/s12160-012-9433-y.
11. Pistoll, C., and A. Goodman. The Link between Socioeconomic Position, Access to Cycling Infrastructure and Cycling Participation Rates: An Ecological Study in Melbourne, Australia. *Journal of Transport and Health*, Vol. 1, No. 4, 2014, pp. 251–259. https://doi.org/10.1016/j.jth.2014.09.011.
12. King, K. E., and P. J. Clarke. A Disadvantaged Advantage in Walkability: Findings from Socioeconomic and Geographical Analysis of National Built Environment Data in the United States. *American Journal of Epidemiology*, Vol. 181, No. 1, 2015, pp. 17–25. https://doi.org/10.1093/aje/kwu310.
13. Flanagan, E., U. Lachapelle, and A. El-Geneidy. Riding Tandem: Does Cycling Infrastructure Investment Mirror Gentrification and Privilege in Portland, OR and Chicago, IL? *Research in Transportation Economics*, Vol. 60, 2016, pp. 14–24. https://doi.org/10.1016/j.retrec.2016.07.027.
14. Hirsch, J. A., G. F. Green, M. Peterson, D. A. Rodriguez, and P. Gordon-Larsen. Neighborhood Sociodemographics and Change in Built Infrastructure. *Journal of Urbanism*, Vol. 10, No. 2, 2017, pp. 181–197. https://doi.org/10.1080/17549175.2016.1212914.
15. Walks, R. A., and L. S. Bourne. Ghettos in Canada's Cities? Racial Segregation, Ethnic Enclaves and Poverty Concentration in Canadian Urban Areas. *Canadian Geographer*, Vol. 50, No. 3, 2006, pp. 273–297. https://doi.org/10.1111/j.1541-0064.2006.00142.x.
16. Winters, M., M. Branion-calles, S. Therrien, D. Fuller, L. Gauvin, M. Branion-calles, S. Therrien, and D. Fuller. Impacts of Bicycling Infrastructure in Mid-Sized Cities (IBIMS): Protocol for a Natural Experiment Study in

Three Canadian Cities. *BMJ Open*, Vol. 8, No. 2, 2018, p. e019130. http://doi.org/10.1136/bmjopen-2017-019130.
17. Statistics Canada. 2016 Census of Canada: Census Profiles Tables/Profile of Census Dissemination Areas. *Using CHASS [distributor]*. http://datacentre.chass.utoronto.ca.proxy.lib.sfu.ca/cgi-bin/census/2016/displayCensus.cgi?year=2016&geo=da. Accessed July 27, 2017.
18. Statistics Canada. *Census Dictionary*. 2011. http://www12.statcan.gc.ca/census-recensement/2011/ref/dict/98-301-X2011001-eng.pdf. Accessed July 27, 2017.
19. City of Kelowna. *Open Data Catalogue|Active Transportation*. http://opendata.kelowna.ca/datasets/active-transportation. Accessed July 4, 2017.
20. Winters, M., and K. Teschke. Route Preferences among Adults in the near Market for Bicycling: Findings of the Cycling in Cities Study. *American Journal of Health Promotion*, Vol. 25, No. 1, 2010, pp. 40–47. https://doi.org/10.4278/ajhp.081006-QUAN-236.
21. Teschke, K., M. A. Harris, C. C. O. Reynolds, M. Winters, S. Babul, M. Chipman, M. D. Cusimano, J. R. Brubacher, G. Hunte, S. M. Friedman, M. Monro, H. Shen, L. Vernich, and P. A. Cripton. Route Infrastructure and the Risk of Injuries to Bicyclists: A Case-Crossover Study. *American Journal of Public Health*, Vol. 102, No. 12, 2012, pp. 2336–2343. https://doi.org/10.2105/AJPH.2012.300762.
22. Statistics Canada. *National Household Survey Dictionary*. 2011. http://www12.statcan.gc.ca/nhs-enm/2011/ref/dict/99-000-x2011001-eng.pdf. Accessed July 27, 2017.
23. Pampalon, R., D. Hamel, P. Gamache, M. D. Philibert, G. Raymond, and A. Simpson. An Area-Based Material and Social Deprivation Index for Public Health in Quebec and Canada. *Canadian Journal of Public Health*, Vol. 103, No. 8, Supp. 2, 2012, pp. S17–S22. http://www.jstor.org.ezproxy.library.uvic.ca/stable/41995684.
24. Ewing, R., S. Hamidi, J. B. Grace, and Y. D. Wei. Does Urban Sprawl Hold Down Upward Mobility? *Landscape and Urban Planning*, Vol. 148, 2016, pp. 80–88. https://doi.org/10.1016/j.landurbplan.2015.11.012.
25. Matheson, F. I., J. R. Dunn, K. L. W. Smith, R. Moineddin, and R. H. Glazier. Development of the Canadian Marginalization Index: A New Tool for the Study of Inequality. *Canadian Journal of Public Health*, Vol. 103, No. 8, Supp. 2, 2012, pp. S12–S16.
26. Anselin, L. Local Indicators of Spatial Association—LISA. *Geographical Analysis*, Vol. 27, No. 2, 1995, pp. 93–115. https://doi.org/10.1111/j.1538-4632.1995.tb00338.x.
27. Nelson, T. A., and B. Boots. Detecting Spatial Hot Spots in Landscape Ecology. *Ecography*, Vol. 31, No. 5, 2008, pp. 556–566. https://doi.org/10.1111/j.0906-7590.2008.05548.x.
28. O'Sullivan, D., and D. Unwin. *Geographic Information Analysis*. Wiley, New York, 2010.
29. ESRI. *ArcGIS Desktop: Release 10.4*. Environmental Systems Research Institute, Redlands, Calif., 2016.
30. R Development Core Team. *R: A Language and Environment for Statistical Computing*. R Foundation for Statistical Computing, Vienna, Austria. https://www.r-project.org/.
31. Anselin, L., I. Syabri, and Y. Kho. An Introduction to Spatial Data Analysis. *Geographical Analysis*, Vol. 38, No. 1, 2006, pp. 5–22. https://doi.org/10.1111/j.0016-7363.2005.00671.x.
32. Martens, K., A. Golub, and G. Robinson. A Justice-Theoretic Approach to the Distribution of Transportation Benefits: Implications for Transportation Planning Practice in the United States. *Transportation Research Part A: Policy and Practice*, Vol. 46, No. 4, 2012, pp. 684–695. https://doi.org/10.1016/j.tra.2012.01.004.
33. Foth, N., K. Manaugh, and A. M. El-Geneidy. Towards Equitable Transit: Examining Transit Accessibility and Social Need in Toronto, Canada, 1996-2006. *Journal of Transport Geography*, Vol. 29, 2013, pp. 1–10. https://doi.org/10.1016/j.jtrangeo.2012.12.008.
34. Flatt, J., and L. Sotomayor. *Making Mid-Sized the Right Size: Re-Envisioning Success in Ontario's Mid-Sized Cities*. Discussion paper prepared for Evergreen, 2015. https://www.evergreen.ca/downloads/pdfs/2017/MSC_Discussion_Paper_-_April_10.pdf. Accessed November 6, 2017.
35. Bunting, T., P. Filion, H. Hoernig, M. Seasons, and J. Lederer. Density, Size, Dispersion: Towards Understanding the Structural Dynamics of Mid- Size cities. *Canadian Journal of Urban Research*, Vol. 16, No. 2, 2007, pp. 27–52.
36. Statistics Canada. *Population and Dwelling Counts, for Census Metropolitan Areas and Census Agglomerations, 2011 and 2006 Censuses*. http://www12.statcan.gc.ca/census-recensement/2011/dp-pd/hlt-fst/pd-pl/Table-Tableau.cfm?T=201&S=3&O=D&RPP=150. Accessed July 28, 2017.
37. Pucher, J., and R. Buehler. Why Canadians Cycle More than Americans: A Comparative Analysis of Bicycling Trends and Policies. *Transport Policy*, Vol. 13, No. 3, 2006, pp. 265–279. https://doi.org/10.1016/j.tranpol.2005.11.001.
38. Finnis, S., N. Krstic, L. McIntyre, T. A. Nelson, and S. B. Henderson. Spatiotemporal Patterns of Paralytic Shellfish Toxins and Their Relationships with Environmental Variables in British Columbia, Canada from 2002 to 2012. *Environmental Research*, Vol. 156, 2017, pp. 190–200. https://doi.org/10.1016/j.envres.2017.03.012.
39. Gralewicz, N. J., T. A. Nelson, and M. A. Wulder. Spatial and Temporal Patterns of Wildfire Ignitions in Canada from 1980 to 2006. *International Journal of Wildland Fire*, Vol. 21, No. 3, 2011, pp. 230–242. https://doi.org/10.1071/WF10095.
40. Bourbonnais, M. L., T. A. Nelson, and M. A. Wulder. Geographic Analysis of the Impacts of Mountain Pine Beetle Infestation on Forest Fi Re Ignition. *The Canadian Geographer*, Vol. 58, No. 2, 2014, pp. 188–202. https://doi.org/10.1111/j.1541-0064.2013.12057.x.
41. Leclerc, M. *Bicycle Planning in the City of Portland: Evaluation of the City's Bicycle Master Plan and Statistical Analysis of the Relationship between the City's Bicycle Network and Bicycle Commute*. Portland State University, 2002. http://web.pdx.edu/˜jdill/LeClerc.pdf. Accessed November 5, 2017.
42. Dill, J., and T. Carr. Bicycle Commuting and Facilities in Major U.S. Cities: If You Build Them, Commuters Will

Use Them. *Transportation Research Record: Journal of the Transportation Research Board*, 2003. 1828: 116–123.
43. Nelson, A., and D. Allen. If You Build Them, Commuters Will Use Them: Association Between Bicycle Facilities and Bicycle Commuting. *Transportation Research Record: Journal of the Transportation Research Board*, 1997. 1578: 79–83.
44. Parkin, J., M. Wardman, and M. Page. Estimation of the Determinants of Bicycle Mode Share for the Journey to Work Using Census Data. *Transportation*, Vol. 35, No. 1, 2008, pp. 93–109. https://doi.org/10.1007/s11116-007-9137-5.
45. Schoner, J. E., and D. M. Levinson. The Missing Link: Bicycle Infrastructure Networks and Ridership in 74 US Cities. *Transportation*, Vol. 41, No. 6, 2014, pp. 1187–1204. https://doi.org/10.1007/s11116-014-9538-1.
46. Lowry, M., and T. H. Loh. Quantifying Bicycle Network Connectivity. *Preventive Medicine*, Vol. 95, 2017, pp. S134–S140. https://doi.org/10.1016/j.ypmed.2016.12.007.
47. Furth, P. G., M. C. Mekuria, and H. Nixon. Network Connectivity for Low-Stress Bicycling. *Transportation Research Record: Journal of the Transportation Research Board*, 2016. 2587: 41–49.
48. Aldred, R., J. Woodcock, and A. Goodman. Does More Cycling Mean More Diversity in Cycling? *Transport Reviews*, Vol. 36, No. 1, 2016, pp. 28–44. http://www.tandfonline.com/doi/abs/10.1080/01441647.2015.1014451.
49. Di Ciommo, F., and Y. Shiftan. Transport Equity Analysis. *Transport Reviews*, Vol. 37, No. 2, 2017, pp. 139–151. https://doi.org/10.1080/01441647.2017.1278647.
50. Litman, T. *Evaluating Transportation Equity: Guidance for Incorporating Distributional Impacts in Transportation Planning*. Victoria Transport Policy Institute, Victoria, British Columbia. http://www.vtpi.org/equity.pdf. Accessed November 5, 2017.
51. Manaugh, K., M. G. Badami, and A. M. El-Geneidy. Integrating Social Equity into Urban Transportation Planning: A Critical Evaluation of Equity Objectives and Measures in Transportation Plans in North America. *Transport Policy*, Vol. 37, 2015, pp. 167–176. https://doi.org/10.1016/j.tranpol.2014.09.013.
52. American Public Health Association. *Health Equity*. https://www.apha.org/topics-and-issues/health-equity. Accessed July 27, 2017.
53. Canadian Institute of Planners. *Healthy Communities*. https://www.cip-icu.ca/Issues/Healthy-Communities. Accessed July 27, 2017.
54. Better Bike Share Partnership. Better Bike Share Partnership. Better Bike Share, 2017. http://betterbikeshare.org/. Accessed July 4, 2017.
55. The League of American Bicyclists. 2017 Bicycle Friendly State Ranking. *The League of American Bicyclists*. http://www.bikeleague.org/content/ranking. Accessed July 27, 2017.

The Pedestrians and Cycles Section (ANF00) peer-reviewed this paper (18-01466).

Article

TRR
JOURNAL OF THE TRANSPORTATION RESEARCH BOARD

Transportation Research Record
2018, Vol. 2672(36) 33–41

DOI: 10.1177/0361198118783109
journals.sagepub.com/home/trr
SAGE

Lowering Bicycle Stress One Link at a Time: Where Should We Invest in Infrastructure?

Sarah K. Moran[1], William Tsay[2], Sean Lawrence[3], and Gregory R. Krykewycz[4]

Abstract

This paper presents a new, regional-scale application of low-stress bicycle connectivity analysis. While prior network-based analyses have focused on the overall improvement in connectivity that could be achieved by implementing a package of projects from a comprehensive bike plan, the purpose of this project was to wholly evaluate potential improvements in connectivity through individual improvements at the street segment level. Using scripts and database tools, levels of traffic stress were assigned to the road network. Incorporating numerous computational optimization measures, shortest paths were calculated between millions of origin and destination pairs to identify the road segments that could most benefit low-stress connectivity. The resulting ranked list of links providing the greatest connectivity benefit allows planners to more efficiently prioritize locations for further investigation and analysis.

For the past several years, the Delaware Valley Regional Planning Commission (DVRPC) has worked with a variety of stakeholders representing the greater Philadelphia suburbs to look for opportunities to make bicycle facility improvements on state roads in Pennsylvania as part of the resurfacing process. State routes in Pennsylvania include most collector and arterial roadways with relatively high speeds, making it difficult to accommodate comfortable biking. Due to the current lack of bicycle infrastructure in the suburbs, there is a plethora of places throughout the Philadelphia metro region that would benefit from improvements, making prioritization difficult. The impetus for this project came from the working group's desire to identify and rank critical places which would provide the maximum benefit to the bicycle network if improvements were made.

Level of traffic stress (LTS) is a road classification scheme based on the comfort of bicyclists in the traffic stream (*1*). One commonly used LTS schema ranges from LTS 1 to LTS 4, classifying road segments that would be comfortable for any bicyclist to segments that only the fearless would brave on a bicycle, respectively. Over the past several years, a variety of papers have been written about different schemas for assigning LTS values to road segments and how these LTS values have been used to inform planning and infrastructure decisions through connectivity analysis. Since demand modeling for bicycling is absent from many regional travel demand models (including DVRPC's), a measure of the potential use of a bicycle facility is difficult to obtain. Instead of focusing on demand, bicycle network connectivity is often used as an accessibility metric. Much of the connectivity analysis in prior research compared existing conditions with a "bike-plan buildout" scenario, or similar, to compare project merits.

Instead, this paper provides an example of a different type of LTS and network connectivity application. While the finished tool will be able to compare existing conditions and potential future scenarios, it was developed for a broader perspective. The primary purpose of this analysis was to inform changes in road design that can occur alongside resurfacing work. For example, if 50 road segments are scheduled to be resurfaced in a given paving season, which would have meaningful impacts on low-stress bike connectivity and would be worth investing design effort in? Even projects that might seem inconsequential on the surface, such as adding bike lanes to a short stretch of roadway where no other bike facilities exist nearby, can have a meaningful local impact if they provide connections between low-stress residential streets, and contribute to building a low-stress regional network over time.

[1]Office of Transit, Bicycle, and Pedestrian Planning, Delaware Valley Regional Planning Commission, Philadelphia, PA
[2]Office of Modeling and Analysis, Delaware Valley Regional Planning Commission, Philadelphia, PA
[3]Office of Geospatial Information Systems, Delaware Valley Regional Planning Commission, Philadelphia, PA
[4]Delaware Valley Regional Planning Commission, Philadelphia, PA

Corresponding Author:
Address correspondence to Sarah K. Moran: smoran@dvrpc.org

Table 1. LTS Categories and Rider Types

LTS	Comfortable enough for (cyclist type)	Characteristics
1	Most people	• Lowest stress • Comfortable for most ages and abilities
2	Interested, but concerned	• Suitable for most adults • Presenting little traffic stress
3	Enthused and confident	• Moderate traffic stress • Comfortable for those already riding bikes in American cities
4	Strong and fearless	• High traffic stress • Multilane, fast moving traffic

Source: (*1*, *2*).

Rather than examine already proposed projects, this analysis identifies the potential impact on network connectivity for every individual road segment that is not already considered comfortable for bicycling. The result will be a ranking of road segments that, if improved to a lower LTS score, would provide the greatest benefit to low-stress bicycle connectivity. Based on the ranked results, planners can more efficiently prioritize areas for further investigation and analysis.

Literature Review

Assigning LTS

One of the most commonly used LTS schemas was developed by the Mineta Transportation Institute. Their research proposed a method "for classifying road segments by one of four levels of traffic stress" (*1*), based on the four types of cyclists identified by Roger Geller at the Portland Office of Transportation (*2*), as shown in Table 1. The Mineta study and related 2016 Transportation Research Record paper (*3*) relied on a variety of road attributes such as the presence of a bike lane, the presence and location of on street parking, the number of travel lanes per direction, and the speed limit to determine the level of stress on a road segment. It also examined the stress at intersections based on the presence of a signal, a right-turn lane, and the speed limit and number of lanes on the street being crossed. While this analysis provides a robust, data driven prediction of cyclist comfort on a road segment, it requires a large amount of data, including attributes that are difficult to obtain at a regional scale.

Montgomery County, Maryland, published bicycle planning guidance which included design guidelines for two types of cyclists: interested but concerned cyclists and confident cyclists (*4*). Their guidance provided graphs for determining the recommended *type* of bicycle facility based on roadway attributes, including vehicular volume and speed limit. While traffic counts are available for many roads throughout the DVRPC region, it would be time and cost prohibitive to collect data on vehicle volume on every single road in the region. To fill the gaps in the volume data in the region, this project would have had to rely on estimated volumes from DVRPC's regional travel demand model. While possible, using model-estimated volumes for this project would have introduced an additional level of complexity and potential for error.

In 2016, Lowry, Furth, and Hadden-Loh presented their marginal rate of substitution (MRS) based method for LTS assignment (*5*). Their method assigned LTS using a few relatively easy to acquire road attributes. The initial stress level was a function of the number of lanes and the speed limit. Then the stress level was reduced based on stress reduction factors developed for five levels of bicycle accommodation from the least protected bike route to the most protected, or comfortable, bike lane. "The stress factors are marginal rates of substitution with respect to distance cycling on a multi-use trail. For example, for a 6-lane road with a 35 mph speed limit and no bicycle accommodation, the stress factor of 140% implies that a cyclist would prefer traveling up to 140% farther on a multi-use trail" (*5*). Given the simplicity of the data requirements and the straightforward method of assigning LTS, this method was adapted for use in the DVRPC region.

Connectivity Analysis

Early research on LTS networks suggested that bicyclists not only seek the most direct route but also prefer not to deviate more than two blocks out of their way to use a known bicycle facility (*6*). Later research compared the "low-stress route," one comprised of only low-stress road segments, with the most direct route. If the low-stress route between an origin and destination was more than 25% longer than the most direct route, the origin and destination would not be considered "connected" by a low-stress route (*3*).

The Lowry et al. paper proposed using "the percent of residents that can reach a majority of important destinations via low-stress bikeways" as a metric to quantify accessibility (*5*). The important destinations included a variety of places that residents would need or want to access on a regular basis. Shortest paths were calculated from residential parcels in their study area to key destinations. The paper introduced a modified version of the classic centrality formula to determine each road segment's "relative importance to the network," specifically, how many paths utilized certain road segments (*5*). Origins and destinations were weighted by population and employment, respectively, to ensure that the links connecting the most people to the most destinations resulted in a high overall importance to the bicycle network. Due to the regional scale of this project, obtaining the level of detail required for each destination was a difficult prospect. Additionally, in places where destinations were dense, it would require the calculation of numerous shortest paths to essentially the same location, which would increase processing time. Therefore, this method was modified for application to the DVRPC region.

Data Preparation

DVRPC's regional travel demand model was selected as the base road and trail network for this analysis. The modeled network, based on Open Street Map, contains critical attributes for link and turn LTS assignment such as number of lanes, bicycle facilities, and speed. The network is updated on a regular basis and is consistent, in terms of attribute completion and accuracy, across the region. Since routine accuracy checks and updates are completed with traffic forecasting in mind, attributes contributing to LTS assignment were scrutinized further.

To verify the number of lanes attribute, the modeled network was visually compared with Pennsylvania Department of Transportation's Road Management System lane count data. Results from the QA/QC revealed the modeled network to be very accurate, with only a few minor changes needed. These changes were also verified using 2015 aerial imagery.

Off-road bicycle facilities in the modeled network were checked using DVRPC's Circuit Trail GIS layer. This layer is maintained by staff and is rectified using aerial imagery on a frequent basis. On-road facilities were verified using data and feedback from steering committee members.

The speed attribute relied on the model's effective speed. Effective speed is based on a road segment's functional classification and area type (urban, suburban, rural, etc.). It is meant to represent free flow speed. One reason effective speed was chosen over speed limit is because of the lack of consistent speed limit data for the entire study area. Additionally, many drivers travel at speeds above the speed limit when traffic is moving freely, which impacts how comfortable a cyclist would feel on that roadway. It was hoped that effective speed would capture this effect. In some areas, the effective speed is lower than the posted speed, while in others, it is higher.

The effective speed was compared with speed limit data provided by the Pennsylvania Department of Transportation (PennDOT). Approximately 21% of the road segments in the study area overlapped with the PennDOT data and could be compared. Based on steering committee feedback and the average differences between the two datasets, speed limit was used in place of effective speed where the modeled speed was lower than the speed limit. Since this analysis focused on cyclist comfort, it was important to err on the side of caution and use the higher speed and therefore a higher LTS. The segments where the effective speed was higher than the PennDOT speed limit were examined closely and were ultimately not modified.

Methodology

The methodology for this analysis relied on scripts that pushed and pulled information to and from a geospatial database. Scripts were written to prepare the road network by assigning LTS. Shortest paths were calculated between millions of origin and destination pairs to identify the road segments that enabled low-stress connections between the most places throughout the region. This section describes this process in detail.

Assigning LTS

Road segment LTS assignments relied on the MRS based stress values presented in Lowry et al. (*7*). Based on the number of lanes, effective modeled speed, and the presence and type of bicycle facility, LTS values were assigned to road segments using a script. The following upper limits were used to distinguish between the four levels of traffic stress:

- LTS 1: 10%
- LTS 2: 30%
- LTS 3: 60%
- LTS 4: no limit

Once LTS values were assigned to road segments, they were used to assign LTS values to turning movements. Lowry et al. considered traffic signal, functional priority, number of lanes, and speed limit of the cross street when assigning LTS to turns (*7*). Traffic signal and functional priority data were not consistently available at the regional scale. Therefore, this analysis relied on a modified turn LTS assignment method, somewhat based on the "weakest link" logic presented in the Mineta report (*1*). Each turn direction was treated differently:

- Right turns were assigned the LTS of the road segment to which the turn was being made.
- Left turns were assigned the LTS of the road segment to which the turn was being made, multiplied by two.
  - Left turns are typically more difficult, and therefore more stressful, than right turns.
  - Doubling the LTS value for left turns ensures this stress is reflected in the turn cost.
- Straight movements were assigned the maximum LTS value of all the links at the intersection.
  - This way, the impact of crossing a high-stress street was accounted for.
  - Crossing a comfortable street while on a stressful street did not significantly lower the cost of the route.
- U-turns were excluded.

The LTS assigned road network (Figure 1) was based on DVRPC's regional travel demand model network supplemented with the attributes necessary for connectivity analysis. Symbolizing the road segments by LTS values provided a picture of existing cyclist comfort throughout the study area. Since turns occur at the intersection of road segments, they were not visible in the network.

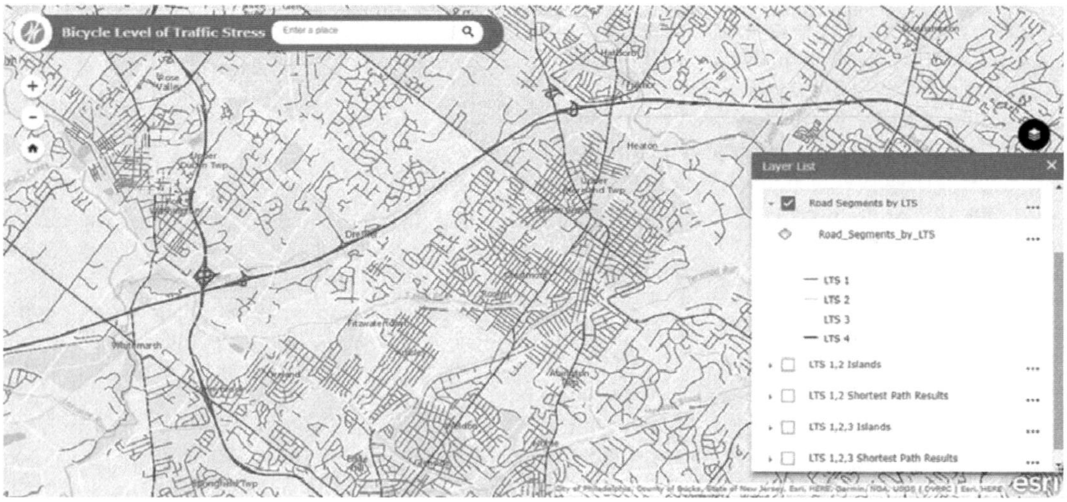

Figure 1. LTS assigned road network.

After LTS values were assigned to the road network, the following base input data were loaded into a database:

- LTS assigned road network
- Road network nodes
- Census block centroids

Census block centroids functioned as origins and destinations for connectivity analysis. A script identified the closest node to each block centroid to serve as the network start and end points of each shortest path calculation. A list of combinations of these closest nodes was compiled as inputs to the shortest path searches.

Next, the network data were prepared for shortest path searches. First, a script created a subset of road segments to be included in the "tolerable network." For the first analysis, "tolerable" links are those considered to have an acceptable LTS for most bicycle riders—in this case LTS 1 and 2 roads. Since the shortest paths in this analysis were intended to be able to be used by everyone, the more stressful LTS 3 and 4 roads were excluded from the network to ensure that shortest paths would not traverse them. In the second analysis, the tolerable network included LTS 1, 2, and 3 roads.

Overall cost fields were added to the road segments and turns for use as the path search impedance. For road segments, the overall cost was a function of the link length and assigned LTS value:

$$Cost = Link\ Length \times (1 + Link\ LTS)$$

The turn cost was a function of the assigned turn LTS, the direction of the intersection movement, and the base turn distance of 30 ft (0.005 miles), as used in the Lowry et al. paper.

$$Straight\ Cost = 0.005 \times (1 + Turn\ LTS)$$

$$Right\ Turn\ Cost = 0.005 \times (1 + 1 + Turn\ LLTS)$$

$$Left\ Turn\ Cost = 0.005 \times (1 + 2 + Turn\ LTS)$$

Therefore, turns were accounted for, but had a minimal impact on route selection since the overall cost of turning was small. In the future, additional research could be done to determine the sensitivity of the shortest path search results to turn cost.

Then, the script created a table to hold a subset of the nodes that correspond with the tolerable link subset. Turns were imported from the model and also subseted to correspond with the tolerable links and nodes.

Creating the Routable Network

The next critical step was to combine turns and links in the same table to create a complete routable network. The process was to explode every node in the network into repeated nodes at the same location; one for each approach of the intersection. These exploded nodes were termed "geoffs." In an effort to simplify discussions about how to create and use them, rather than referring to them as "turn nodes," an easily confused combination of other component names, geoff will be used to describe the exploded nodes going forward.

Looking at the "node based network" diagram in Figure 2*a*, nodes are numbered, starting with 100. Links are identified using a "from-node" and a "to-node." For example, the upward direction of the bottom link is identified by from-node 100 and to-node 101. A right turn starting at node 100 and ending at node 103 would be identified using the following node sequence: from-node 100, via-node 101, to-node 103.

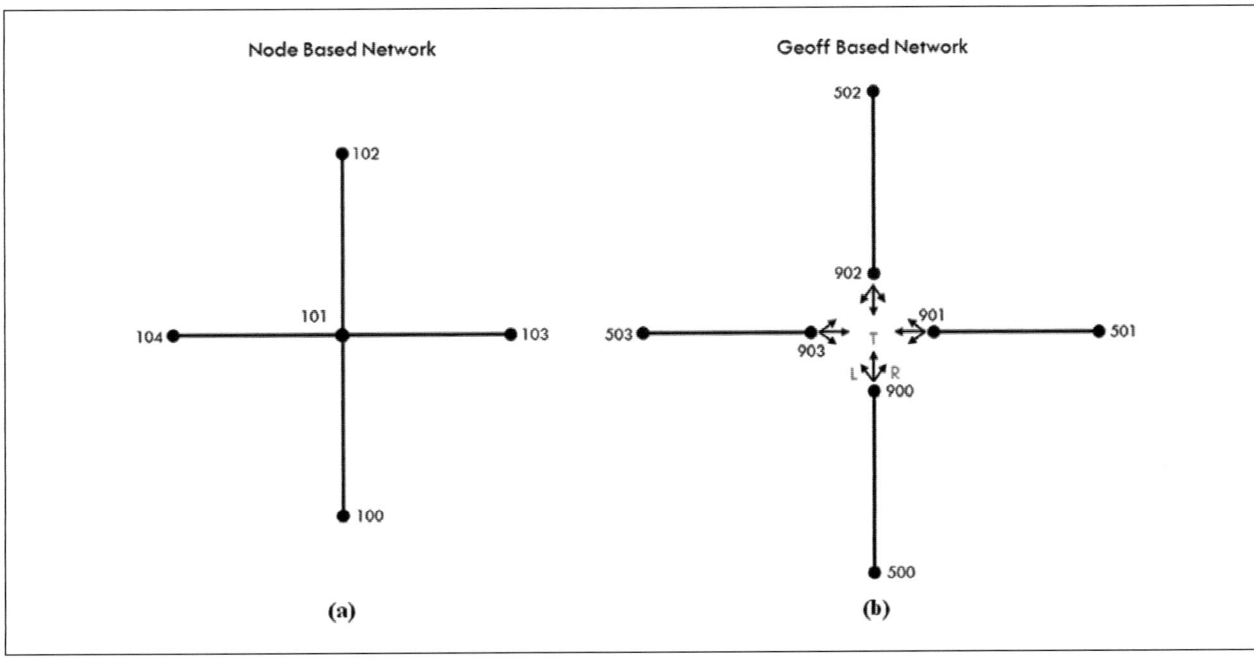

Figure 2. Node based network vs. geoff based network.

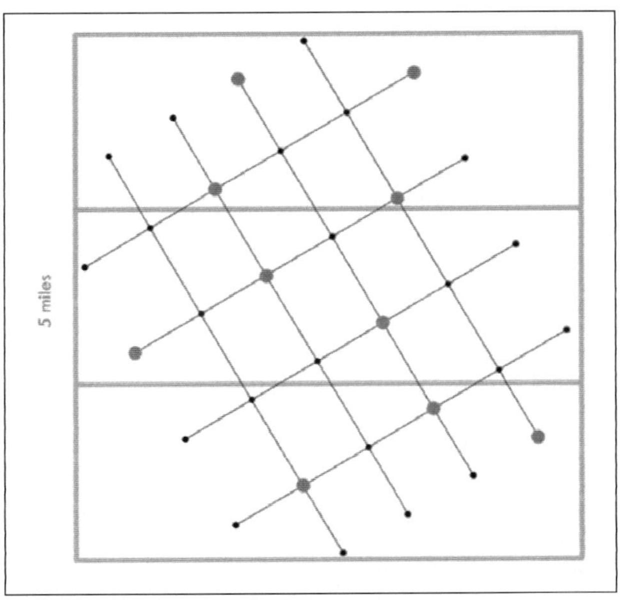

Figure 3. Moving frame network division.

The "geoff based network" diagram in Figure 2*b* represents the same intersection, this time based on geoffs. When combining the links and turns into a master network table, all nodes were converted to sets of geoffs. Instead of being represented by a different number and combination of node identifiers (from and to for links; from, via, and to for turns), the creation of geoffs allowed for each network component to be identified by a from-geoff and a to-geoff. This made it possible to combine what were previously links and turns into a single, routable, master network table with both types of components. For example, the bottom link is still identified using a from and to location; from-geoff = 500, to-geoff = 900. Once at 900, to make a right turn, one would travel along "R," from geoff 900 to geoff 901. Then one could continue from geoff 901 to geoff 501. On reaching 501, there would be another turning movement before connecting to the next link.

Island Analysis and Moving Frame

Testing revealed that the large routable network comprised of links and turns resulted in unacceptably long processing times. Island analysis and moving frames were introduced to divide the network dataset into more manageable subnetworks. The routable network was analyzed as a graph and divided into connected components, or groups of link "islands," to streamline analysis. A field was added to the master network table to identify the island to which each road and turn belonged. A subnetwork table was created for each individual island. Since an origin on one island would not be connected to a destination on another island, it was possible to calculate shortest paths on one island at a time, drastically reducing the number of records being searched through for each calculation.

The moving frame concept was introduced to further divide large routable islands into even smaller subnetworks for more efficient calculations. The moving frame started with a minimum bounding box around the island. The box was then divided into 5-mile vertical frames, starting at the top as shown in Figure 3. If the island's bounding box was not evenly divided by 5, the remainder was given its own frame.

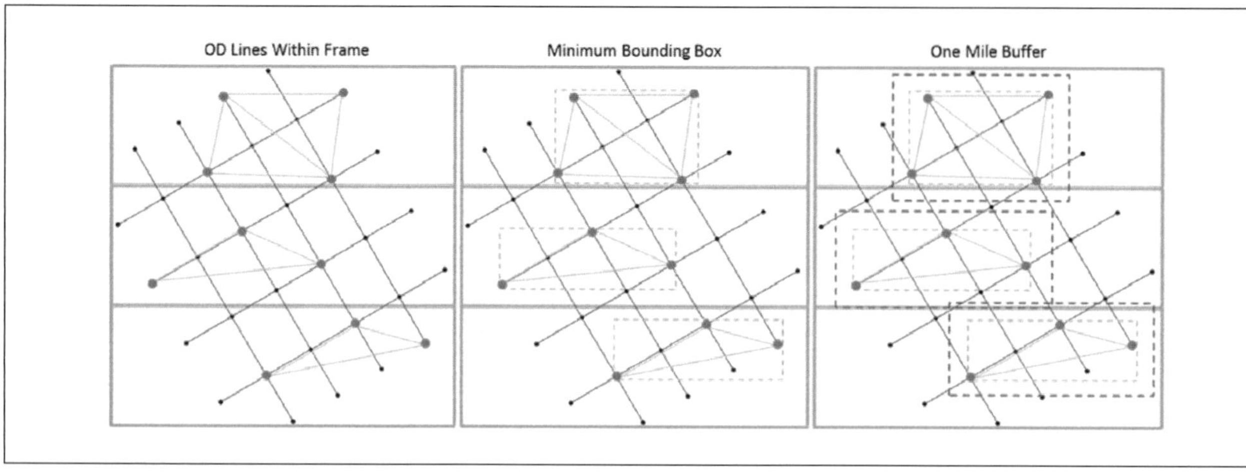

Figure 4. Origin–destination pairs connected within a frame.

In the example in Figure 3, the larger gray geoffs—which appear similar to nodes—represent those that were identified as closest to imaginary block centroids and will therefore serve as the origins and destinations for the example. Once the boundaries of the frames were determined, straight lines were drawn between each origin and destination pair. First, the lines that were completely contained within frame boundaries were selected, as shown by the blue lines in Figure 4. A minimum bounding box was drawn around the group of lines in each frame. Then, a buffer of 1 mile was added before clipping the network, to allow for shortest paths that are required to travel slightly out of the way to connect the origin and destination. Analysis was done on a full network to help determine an adequate buffer size while maintaining the benefit of smaller network chunks for shortest path calculations.

Then, the lines that intersected frame boundaries were selected, as shown by the purple and green lines in Figure 5. Similar to the lines within the frames, a minimum bounding box was drawn around the intersecting lines and an additional 1-mile buffer was added before clipping the network. The clipped networks were saved into temporary tables to be used in shortest path calculations for the island in question. The origin and destination pairs for the lines in the selection were also saved into a corresponding temporary table to ensure that only shortest paths for the pairs connected in each frame were calculated for the correct corresponding subnetwork.

Shortest Paths

Shortest paths between each origin–destination (O-D) pair were calculated using the directed Dijkstra algorithm. Inputs included the subnetwork, or island on which the O-D pair was located, the geoff ID of the origin and destination, and the weight, or calculated cost of each network component. For each O-D pair, if a path was found, the algorithm returned the length of the shortest path and a list of components (links and turns) used to connect the origin and destination. The components of each path were added to an output table for post processing. Finally, the components were tallied to determine the number of times each unique component served to connect a shortest path.

Discussion

According to Roger Geller's analysis of the four types of cyclists, supported by national surveys conducted by Portland State University (*7*), the slight majority of people fall into the "interested but concerned group (*2*)." This means that a lot of potential cyclists would ride their bicycles more if they had more comfortable places to ride. It was important to focus on this group when prioritizing infrastructure improvements.

The first step in identifying and ranking the critical places which would most benefit the bicycle network was to assign LTS values to road segments and turns. LTS was used along with distance to evaluate the overall cost of a route between two census blocks. High-stress roads (LTS 3 and 4), considered uncomfortable for most cyclists, were eliminated to create a picture of the existing tolerable road network. As mentioned earlier, the tolerable network (LTS 1 and 2) served as the basis for the first round of shortest path calculations between all census blocks in the study area.

Another network was created for comparison with the existing conditions and to enable the prioritization of places that are not currently comfortable for most cyclists. This network only eliminated LTS 4 roads, leaving LTS 3 segments to be used as network options for the shortest paths connecting census blocks. The results show which LTS 3 segments, if made more comfortable for cyclists, would allow for the most connections between census blocks and, in turn, provide the greatest benefit to the overall bicycle network.

The size of the study area and the computationally intensive analysis presented several challenges. Methods were iteratively modified throughout the process to overcome these issues. When preparing the origin and destination pairs for

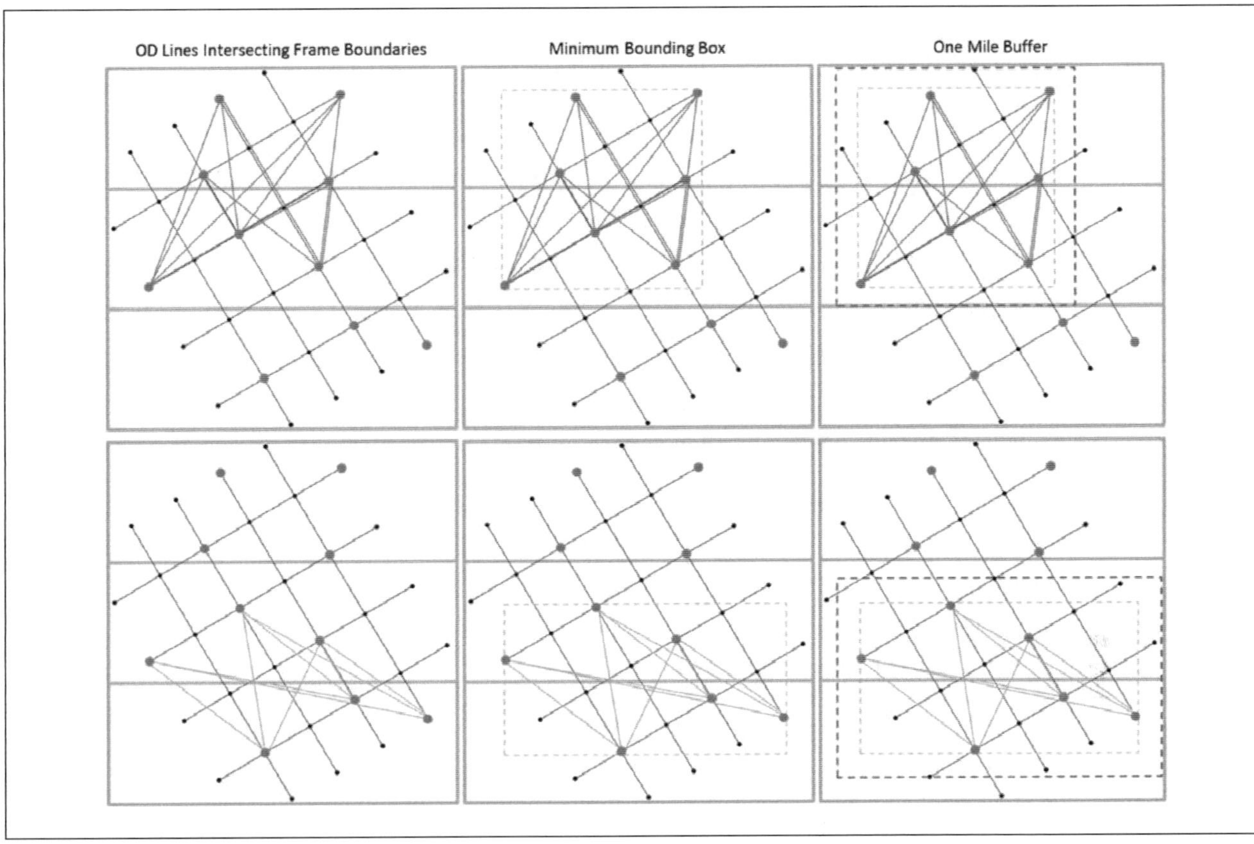

Figure 5. Origin–destination pairs connected across frames.

shortest path searches, steps were taken to ensure that only unique pairs of differing origin and destination nodes were calculated. In some cases, the same node was the closest to multiple block centroids. This resulted in the possibility of duplicate shortest paths being calculated between the same node pairs. Further analysis was required in post-processing to account for the number of duplicates that had been removed and to add them back in to ensure the components that connected the most block centroids had the highest counts.

Due to the size of the study area, the connectivity analysis yielded extremely large tables, with the largest containing upwards of 10 billion records. Analyzing datasets this large can tax even the most powerful workstations, resulting in unacceptably slow processing times. Routinely working with large datasets throughout this project underscored the importance of implementing an optimized relational database using keys and indices.

Combining links and turns into one large routable network table quadrupled the number of records being used to calculate shortest paths, significantly slowing processing time. Link islands were identified in an effort to increase efficiency. As shown in Figure 6, islands are comprised of connected parts of the low-stress network. If a bicyclist, only comfortable on low-stress (LTS <3) roads, begins a trip on a certain island, they can comfortably travel to any destination on their island. However, if this cyclist needs to reach a destination located on a different island, they will be faced with uncomfortable conditions and therefore be less likely to ride their bicycle. In terms of the implemented algorithm, no path will be found because one of the trip ends exists on a separate island.

Network islands were counted and initially visualized to compare their relative size. Although breaking the shortest path calculations into islands and running the algorithm on smaller tables drastically reduced processing time, some larger islands were considered candidates for further processing optimizations—moving frame analysis, as described in the methodology section. However, breaking up a connected island could result in some O-D pairs not being connected in the network when they would be in reality. Measures were put in place to preserve these connections. Tradeoffs between processing time and maintaining completeness of the connected shortest paths were carefully balanced.

After the network islands were separated and large islands were divided into smaller frames, the subnetworks were ready for shortest path calculations. To further increase the processing speed, the shortest path searches were conducted in parallel, maximizing the utilization of available computational resources. For each island or frame, O-D pairs were re-examined to ensure they were on the same island,

Figure 6. Bicycle network islands based on LTS 1 and 2 roads.

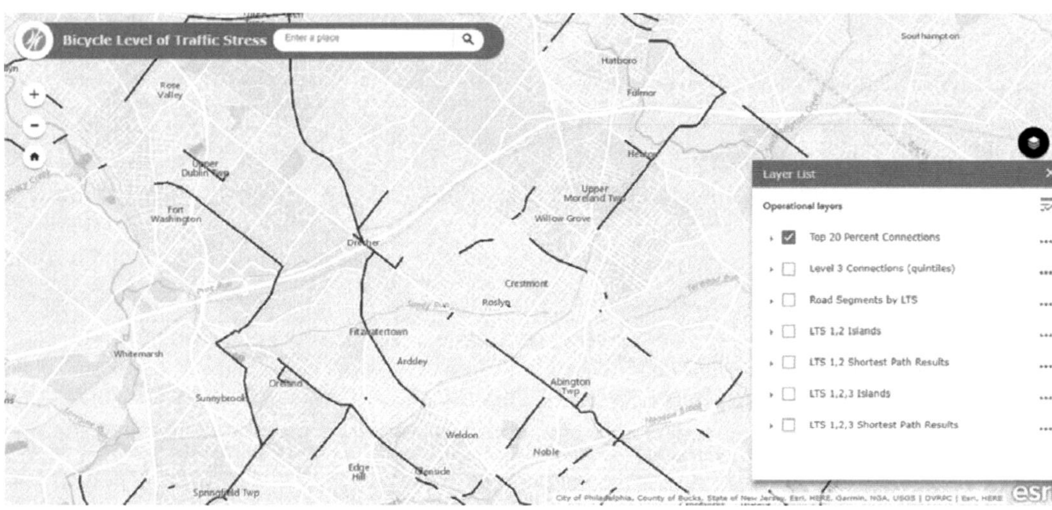

Figure 7. Prioritization results.

and therefore connected. The pairs that met this qualification were added to a list of pairs that would be run through the shortest path algorithm. Initially, the processing time was so great that the connection to the database was lost; this was solved by reopening a connection to the database periodically. Additionally, if a connection could not be made for any reason, the work performed was still dumped to disk.

In the first few iterations of the shortest path calculations, it was estimated to take over 400 days to process the entire study area, which was unacceptable and would not meet the needs of the project. Island analysis, moving frames for large islands, and multiprocessing, added through weeks of optimization, reduced the processing time to about two weeks—still significant, but acceptable.

Results

The LTS 3 roads with the highest counts, or traversed by the highest number of shortest paths, represent those that have the potential to allow the most connections between census blocks. However, being LTS 3 roads, they are not currently considered comfortable for most bicyclists. Therefore, these high-count LTS 3 roads represent those where improvements to bicycle facilities, reduction in speed limit, or reduction in number of lanes to reduce the LTS would have the biggest impact on improving overall low-stress network connectivity.

The idea of high-count LTS 3 segments is similar to the idea of high "network centrality" applied by Lowry et al. to prioritize links (*5*). However, when examining the results in this regional application, it is important to remember that the link's network importance is relative to the rest of the LTS 3 roads. Since the shortest paths were calculated on a network including all level 3 roads, the number of connections enabled by each segment is based on every LTS 3 road being part of the low-stress network. It would be impossible to make improvements to all LTS 3 roads at the same time. Therefore, these results serve to rank LTS 3 roads by their relative potential impact, as a resource to create a pipeline of realistic and beneficial projects. Figure 7 shows the top 20% of connections in Montgomery County, Pennsylvania, identifying the LTS 3 road segments that rank in the top 20% in terms of the number of shortest path connections enabled by that segment. This figure reveals a mix of long and short segments, each of which could have significant local benefit if made low stress. Alternatively, a local application of this analysis, including only one LTS 3 road segment at a time, would reveal the link's individual network importance.

Another metric used to measure network connectivity was the total number of connected block centroids in the network. This becomes most useful when comparing before and after changes in localized analysis. While the analysis was run on the full five-county study area as a baseline for setting priorities for infrastructure investments, the process will allow for comparison between specific network improvements, processed in small sub-areas, to determine the network impact of the change.

Next Steps

Currently, the results are available in a webmap and as shapefiles for download. The hope is that planners in the study area will use the webmap highlighting the LTS 3 roads that have the potential to connect the most census blocks to identify areas for further study. Site visits, community outreach, engineering, and cost estimates can help to narrow down the type of improvement possible to reduce the level of stress on the segment.

Once an improvement is proposed, pre- and post-improvement shortest path analyses on a selection of census blocks around the site can identify the number of connections the improvement would enable when implemented on its own. When visually compared, these analyses will also allow planners to see the expected growth in the low-stress island because of the proposed improvement, showing the extent of the potential network impact.

Analytically, future goals include the incorporation of origin and destination attributes to allow for more specific analysis, such as identifying the improvements that will help connect people to train stations or parks. Additionally, weights will be incorporated to highlight connections that are important for environmental justice or other purposes.

On the products side, webmap improvements are needed, specifically in terms of displaying directional results in an easily understandable way. Another future goal is to create an interactive web-based tool which will allow users to modify the LTS network and calculate the impact of their modifications on a sub-area.

Author Contributions

The authors confirm contributing to the paper as follows: study conception and design: Moran, Krykewycz; data collection: Lawrence, Tsay, Moran; analysis and interpretation of results: Moran, Tsay, Krykewycz; draft manuscript preparation: Moran. All authors reviewed the results and approved the final version of the manuscript.

References

1. Mekuria, M., P. Furth, and H. Nixon. *Low-Stress Bicycling and Network Connectivity*. Report no. 11-19. Mineta Transportation Institute, 2012.
2. Geller, R. *Four Types of Cyclists*. Portland Bureau of Transportation, Portland, OR, 2006. www.portlandoregon.gov/transportation/article/264746. Accessed August 11, 2016.
3. Furth, P., M. Mekuria, and H. Nixon. Network Connectivity and Low-Stress Bicycling. *Transportation Research Record: Journal of the Transportation Research Board*, 2016. 2587: 41–49.
4. Kittleson & Associates, Inc. and Toole Design Group. *Montgomery County Bicycle Planning Guidance*. Montgomery County Planning Department, The Metropolitan Washington Council of Governments, July 2014. http://www.montgomeryplanning.org/transportation/bikeways/documents/FINALBicyclePlanningGuidance.pdf. Accessed June 6, 2016.
5. Lowry, M., P. Furth, and T. Hadden-Loh. Low-Stress Neighborhood Bikeability Assessment to Prioritize Bicycle Infrastructure. Presented at 95th Annual Meeting of the Transportation Research Board, Washington, D.C., 2016.
6. Sorton, A., and T. Walsh. Bicycle Stress Level as a Tool to Evaluate Urban and Suburban Bicycle Compatibility. *Transportation Research Record: Journal of the Transportation Research Board*, 1994. 1438: 17–24.
7. Dill, J., and N. McNeil. Four Types of Cyclists? *Transportation Research Record: Journal of the Transportation Research Board*, 2013. 2387: 129–138.

The Pedestrians and Cycles Section (ANF00) peer-reviewed this paper (18-01979).

Article

TRR
JOURNAL OF THE TRANSPORTATION RESEARCH BOARD

Transportation Research Record
2018, Vol. 2672(36) 42–50

DOI: 10.1177/0361198118783107
journals.sagepub.com/home/trr
SAGE

Who Are Public Bicycle Share Programs Serving? An Evaluation of the Equity of Spatial Access to Bicycle Share Service Areas in Canadian Cities

Kate Hosford[1,2] and Meghan Winters[1,2]

Abstract
Public bicycle share users are predominantly Caucasian, employed, and have higher incomes and education levels, as compared to the general population. This has prompted bicycle share operators and researchers to increasingly consider equity in bicycle share program access and uptake. The location of bicycle share docking stations has been cited as a major barrier to uptake among lower socioeconomic groups. This study aimed to assess spatial access to bicycle share programs in Canadian cities by comparing the socioeconomic characteristics of dissemination areas inside and outside the bicycle share service areas. We obtained locations of bicycle share stations for the five existing programs in Canada: Vancouver, Hamilton, Toronto, Ottawa-Gatineau, and Montréal. We used the material component of the Pampalon Deprivation Index (2011) as a measure of socioeconomic status for each dissemination area, calculating city-specific quintiles. We compared the distribution of deprivation for dissemination areas inside the bicycle share service area, compared with outside the service area. We found that advantaged areas have better access to bicycle share infrastructure in Vancouver, Toronto, Ottawa-Gatineau, and Montréal, and conversely, that disadvantaged areas have better access in Hamilton. This analysis indicates that in most cities, substantial effort is needed to expand service areas to disadvantaged areas in order to increase spatial access for lower socioeconomic populations.

Public bicycle share programs are quickly gaining in popularity worldwide. These programs, now operating in over 1,300 cities (*1*), increase access to bicycles by providing a network of bicycles at designated docking stations throughout a city. Potential benefits for bicycle share users include improved health, expanded transportation networks, and reduced travel times and costs (*2*). However, these benefits are not experienced equally across the population. Trends across several studies indicate that bicycle share users are predominantly Caucasian, employed, and have higher incomes and education levels, as compared to the general population (*2–4*). This has prompted bicycle share operators and researchers to increasingly consider equity in bicycle share program access and uptake.

The geographical coverage of bicycle share service areas has been cited as a major barrier to uptake among lower socioeconomic populations. Two recent studies evaluating spatial access to bicycle share in U.S. cities found that docking stations were not equitably distributed across the population (*5*, *6*). Smith et al. evaluated 35 bicycle share programs in the United States and found that 53% of docking stations were located in census block groups in the lowest economic hardship quintile (i.e., most advantaged) and only 5% of stations were located in census block groups in the highest economic hardship quintile (i.e., most disadvantaged) (*5*). Ursaki and Aultman-Hall's evaluation of seven U.S. public bicycle share programs arrived at a similar conclusion. This evaluation compared the social and economic characteristics, including race, education, and income, of census block groups inside and outside the bicycle share service areas. In all cities, the census block groups located inside the service area were found to have higher household incomes than the census block groups outside the service area (*6*). Disparities in access based on race and education were also found in the majority of cities. In addition to the location of docking stations, lower socioeconomic populations face other barriers to accessing bicycle share, including the cost and pricing structure, credit card ownership, and limited access to a computer or smartphone, which is often required for registration (*7*).

Lower socioeconomic populations may have the most to gain from bicycle share programs. These populations tend to have lower levels of access to vehicles and are more

[1]Simon Fraser University, Faculty of Health Sciences, Canada
[2]Centre for Hip Health and Mobility, University of British Columbia, Canada

Corresponding Author:
Address correspondence to Kate Hosford: khosford@sfu.ca

Table 1. Implementation and Operation of Canadian Bicycle Share Programs

	Implemented	Operation	Operator	Operator type
Montréal	May 2009	Seasonal	BIXI Montréal	Non-profit organization
Toronto	May 2011	Year-round	Toronto Parking Authority	Public
Hamilton	January 2015	Year-round	Hamilton Bike Share Inc.	Non-profit organization
Ottawa	July 2015	Seasonal	Cycle Hop	Private
Vancouver	June 2016	Year-round	Cycle Hop	Private

dependent on alternative transportation modes such as walking and public transit (*8*). Bicycle share can fill transportation gaps and provide a fast and convenient way of commuting, especially through "last mile" linkages. Bicycle share also increases access to bicycles for those who do not own a personal bicycle because of financial barriers or lack of storage space. Additionally, the cost of using bicycle share can be very affordable relative to the cost of public transit. For example, in Vancouver, Canada, the cost of an annual bicycle share membership is 129 Canadian dollars (Can$) (*9*), whereas the cost of an adult public transit pass for a year would be Can$ 1,092 (*10*).

There is evidence to suggest that individuals living in deprived areas would use public bicycle share if it was more accessible (*11*, *12*). Bicycle share users in London, England, living in the most deprived areas were found to make more bicycle share trips per month than users in the least deprived areas, after adjusting for the fact that users in deprived areas had lower levels of access to bicycle share (*11*). Additionally, the proportion of trips made by users from highly deprived areas increased from 6% to 12% in 2014 after London's public bicycle share system expanded their service area to more deprived areas (*12*). Moreover, there is significant interest in using bicycle share among low income populations. In a survey of residents living in low income neighborhoods in four U.S. cities, 56% of respondents who had lower incomes and were of color reported that they would like to use bicycle share more often and 44% indicated that they would like more information about using bicycle share (*7*).

The operating model of a bicycle share program may influence the extent to which equity is considered (*13*). Operating models for bicycle share programs are typically structured in one of the following ways: publicly owned and operated, publicly owned/privately operated, or privately owned and operated. Private operators can be either for-profit companies or non-profit organizations. The first priority for privately owned and operated bicycle share programs is likely to generate revenue to cover operating costs, especially in programs that are starting out. This may result in prioritizing expansion to areas that will increase usage of the program, rather than expanding to low use areas for equity reasons (*13*).

In the North American context, equity has been a primary consideration for public bicycle share programs in many U.S. cities. Specifically, a survey of owners and operators from 55 U.S. bicycle share programs, found that 72% of programs considered equity with respect to fee structure and payment systems, 68% with respect to station citing, and 57% with respect to marketing and outreach (*13*). An example of a program that is considering equity in their system's operations is Chicago's public bicycle share program, which in 2015 expanded their service area to better serve low income neighborhoods (station citing), and launched their campaign, "Divvy for Everyone" (marketing and outreach), which offered low income residents an annual membership for US$5 and waived the usual requirement for a credit card (fee structure and payment systems) (*14*).

Equity considerations have not taken hold in Canadian public bicycle share programs to the same extent that they have in the United States. As of June 2017, SoBi in Hamilton, Ontario, is the only Canadian program with an explicit equity initiative. SoBi's "Everyone Rides" initiative includes subsidized memberships for low income residents, education and outreach initiatives, and the addition of 14 new stations and 75 bicycles in low income neighborhoods (*15*). There is no documented evidence of similar initiatives in other Canadian cities, moreover, there has been no evaluation of the equity of bicycle share programs in Canadian cities. Assessing spatial access to bicycle share services is necessary to understand the extent to which programs are or are not providing equitable access, and can inform strategic expansion efforts.

Bicycle share programs have been implemented in five Canadian cities since 2009. These programs are: Mobi by Shaw Go (Vancouver, British Columbia); SoBi Hamilton (Hamilton, Ontario); Bike Share Toronto (Toronto, Ontario); VeloGo (Ottawa, Ontario and Gatineau, Quebec); and BIXI Montréal (Montréal, Quebec) (see Figure 1). Programs in Canada have a variety of operating models (see Table 1). Ottawa-Gatineau has the smallest program with a fleet of ~80 bicycles across 18 stations (*16*), and Montréal the largest, with ~6,200 bicycles across 532 stations (*17*). Bicycle share programs in Canada are young programs, with plans of expanding their service areas (*18–20*), thus an evaluation of the programs is timely.

This paper examines the equity of spatial access to bicycle share programs in Canadian cities by comparing the socioeconomic characteristics of the populations inside and outside the bicycle share service areas. This analysis is focused on equity in spatial access to bicycle share, and not

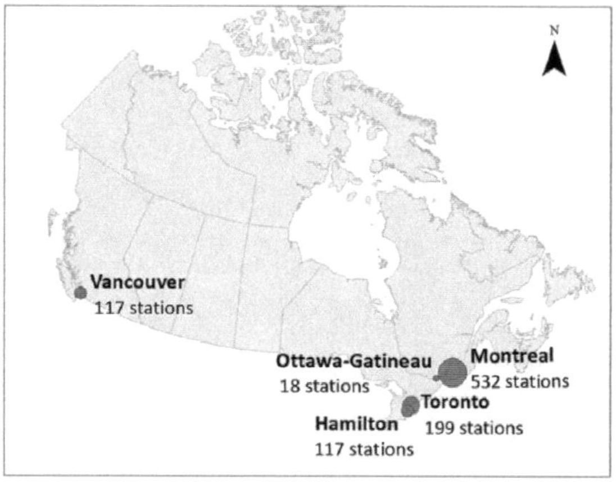

Figure 1. Location and size of Canadian bicycle share programs as of June 2017.

on marketing and outreach initiatives, or pricing and cost structures, which would require different methodologies.

Methods

Data

Data for the bicycle share docking station locations in Vancouver was obtained by contacting the Mobi bicycle share operators. Hamilton and Ottawa-Gatineau's station locations were obtained from the open data platform Bike Share Research (*21*), and Toronto and Montréal's station locations were obtained from the City of Toronto Open Data Catalogue (*22*) and BIXI's website (*23*), respectively. Each dataset included the latitude and longitude of bicycle share docking station locations, which were geocoded in ArcGIS 10.3 using the NAD 1983 coordinate system.

Defining the Study Area and Bicycle Share Service Area Boundaries

Study areas in each city were based on administrative boundaries that reflect local decision making. In Vancouver, Hamilton, and Toronto, this was the census subdivision boundary (equivalent to city boundaries). Ottawa-Gatineau's bicycle share program is implemented across two cities, therefore, the study area boundary in this case includes both the Ottawa and Gatineau census subdivision boundaries. In Montréal the census division boundary was used, as the bicycle share service area already extends beyond the Montréal (Ville) census subdivision boundary.

Within each city, dissemination areas (DAs) were chosen as the geographical unit of analysis. DAs are composed of one or more neighboring dissemination blocks, and include a population of 400–700 people (*24*). In Canada, this is the smallest geographical unit for which socioeconomic data are disseminated. The geographical size of DAs varies with population density; more densely populated areas will have smaller DAs compared to areas with low population density.

The bicycle share service area was defined as the area within 500 meters of a bicycle share docking station. Buffers were created in ArcGIS and DAs that were fully or partially located in the bicycle share service area were categorized as being inside the bicycle share service area. We only considered bicycle share docking stations that were located inside the defined study area boundaries. This resulted in the exclusion of seven docking stations from the Montréal bicycle share service area.

Measure of Socioeconomic Status

The material component of the Pampalon Deprivation Index was used as the measure of socioeconomic status. This index provides an area-based measure of material deprivation derived from Canadian census data (*25*), and has been widely used in health research (*26*). The index includes three measures: the proportion of people without a high school diploma, the employment–population ratio, and the average personal income for people aged 15 and older. This index has been calculated for all DAs using principal component analysis, which calculates a factor score for each DA, for which a lower factor score represents a lower level of deprivation (i.e., more advantaged) and a higher factor scores represents a higher level of deprivation (i.e., less advantaged) (*26*). More details on how this index is calculated are found elsewhere (*26*). This data is available for download from the Institut National de Santé Publique du Québec's website (*25*). In this analysis, the factor scores were used to assign DAs in each city to a quintile, ranging from least deprived (quintile 1) to most deprived (quintile 5), where each quintile represents 20% of the total number of DAs in that city. When DAs are missing census data for at least one of the three variables included in the index, a factor score cannot be generated, and thus they were not included in this analysis. The percentage of DAs in cities that were missing deprivation data ranged from 4.3% to 5.3%.

Analysis

The location of bicycle share docking stations were mapped for each city, and integrated with the deprivation data for DAs. The population, area (km^2), population density, and the number of DAs inside and outside the bicycle share service area were summarized. To assess the equity of bicycle share service areas, the proportion of DAs inside the service area for each deprivation quintile was calculated. As DAs were categorized into city-specific deprivation quintiles, bicycle share service areas that include DAs that are more evenly distributed across deprivation quintiles were considered more equitable.

Table 2. Selected Characteristics of Vancouver, Hamilton, Toronto, Ottawa-Gatineau, and Montréal and the Areas Inside and Outside the Bicycle Share Service Area

	Service area	# of docking stations[a]	Population (% of total)	Area (km^2) (% of total)	Population density (persons/km^2)	# of DAs[b]
Vancouver	Inside	117	193,180 (32.0)	18.5 (16.1)	10,453.5	273
	Outside	–	410,322 (68.0)	96.5 (83.9)	4,252.5	719
	Total	–	603,502	115.0	5,249.2	992
Hamilton	Inside	117	133,212 (25.6)	28.8 (2.6)	4,620.6	245
	Outside	–	386,737 (74.4)	1,088.4 (97.4)	355.3	629
	Total	–	519,949	1,117.2	465.4	874
Toronto	Inside	199	474,184 (18.1)	43.0 (6.8)	11,017.3	686
	Outside	–	2,140,876 (81.9)	587.2 (93.2)	3,646.1	2,999
	Total	–	2,615,060	630.2	4,150.1	3,685
Ottawa-Gatineau	Inside	18	71,572 (6.2)	10.3 (0.3)	6,982.6	127
	Outside	–	1,077,168 (93.8)	3123.0 (99.7)	344.9	1,642
	Total	–	1,148,740	3,133.3	366.6	1,769
Montréal	Inside	525	878,525 (46.6)	96.2 (19.3)	9,135.1	1,510
	Outside	–	1,007,956 (53.4)	403.4 (80.7)	2,498.8	1,691
	Total	–	1,886,481	499.6	3,776.4	3,201

[a]Only docking stations inside the defined study area boundaries are counted.
[b]Dissemination areas with missing 2011 Pampalon deprivation data: Vancouver (inside service area: 25, outside service area: 22); Hamilton (21, 93); Toronto (53, 115); Ottawa-Gatineau (15, 64); Montréal (77, 93).

Results

Table 2 shows selected characteristics of the cities and the areas inside and outside the bicycle share service areas for each city. Of the five Canadian bicycle share programs, Ottawa-Gatineau has the smallest program both in terms of the number of docking stations and the bicycle share service area, and Montréal the largest program. In each city, bicycle share stations are located in the more densely populated areas. The percentage of the total population that live within the bicycle share service area is 32.0% in Vancouver, 25.6% in Hamilton, 18.1% in Toronto, 6.2% in Ottawa-Gatineau, and 46.6% in Montréal.

Figure 2 shows the distribution of bicycle share docking stations across deprivation quintiles, and the data is summarized in Figure 3. More advantaged areas have greater access to bicycle share service areas in Vancouver, Toronto, Ottawa-Gatineau, and Montréal. Hamilton was the exception, with the bicycle share service area disproportionately serving the most deprived areas. In Hamilton, DAs in the most deprived quintiles (quintiles 4 and 5) comprise 48.6% of the DAs inside the bicycle share service area, compared with 3.6% in Vancouver, 26.4% in Toronto, 23.3% in Ottawa, and 34.1% in Montréal. Vancouver's bicycle share program is the least equitably distributed, with 86.7% of DAs inside the service area located in the least deprived quintiles (quintiles 1 and 2) and only 3.6% of DAs located in the most deprived quintiles (quintiles 4 and 5).

Discussion

This study evaluated the equity of spatial access to public bicycle share programs, based on an area-level index for deprivation, for all of the Canadian public bicycle share programs. The findings suggest that advantaged areas have greater access to bicycle share infrastructure in Vancouver, Toronto, Ottawa-Gatineau and Montréal, and conversely, disadvantaged areas have greater access in Hamilton. We did see that even within the four cities where advantaged areas had greater access, there was considerable variation in the equity of access. Vancouver and Ottawa-Gatineau's bicycle share service areas provide far better access to advantaged areas, with 86.7% of DAs inside the service area located in the least deprived quintiles (quintiles 1 and 2) in Vancouver, and 69.7% in Ottawa. In contrast, the bicycle share service areas in Toronto and Montréal were closer to a more balanced distribution across deprivation quintiles.

Notably, Vancouver's and Ottawa-Gatineau's programs are the two in Canada that are the least equitably distributed across deprivation quintiles and are also both privately operated. The programs in Hamilton and Montréal, which are more equitably distributed across deprivation quintiles, are operated by a non-profit organization, and Toronto's program is publicly operated. In the United States, a survey on equity policies and initiatives was conducted with owners and operators from 55 bicycle share programs (*13*). The programs surveyed included a mix of publicly and privately owned and operated programs. For the majority of programs, the main barrier to implementing equity initiatives was limited funding and staffing (*13*). Although these are concerns for both publicly and privately operated bicycle share programs, programs that are privately operated may be under greater pressure to install stations in areas that will maximize use.

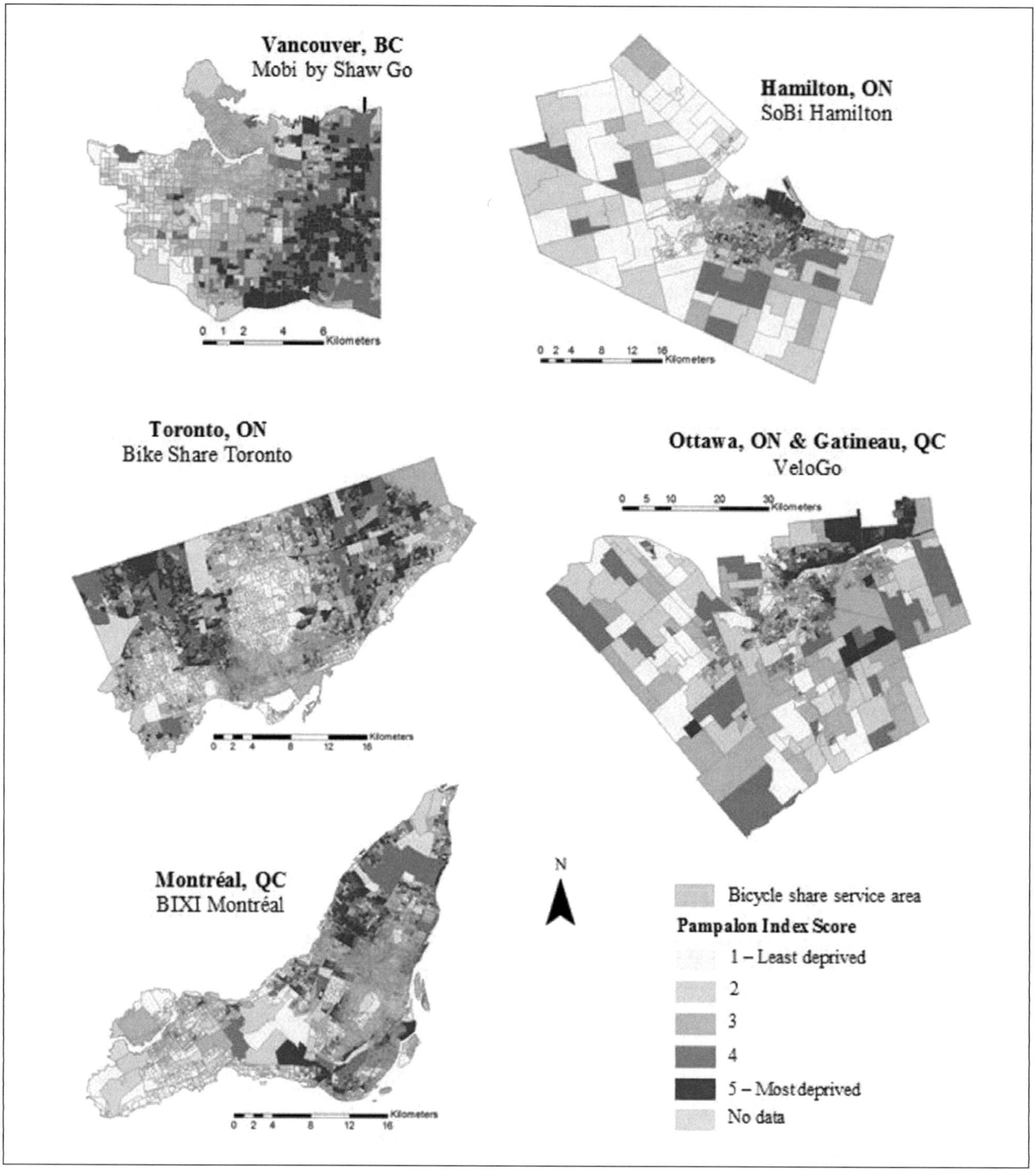

Figure 2. Bicycle share docking stations by deprivation quintile for Canadian cities with public bicycle share programs.

The different trends across cities may also in part be explained by socioeconomic patterning of populations in each of the cities. Public bicycle share programs are typically rolled out within and nearby city centers, where there is high population density and economic activity (*27*). If wealthier neighborhoods are located in the city center, as is the case in Vancouver, Toronto, and to some extent Montréal, then it is less likely that a bicycle share program will be implemented

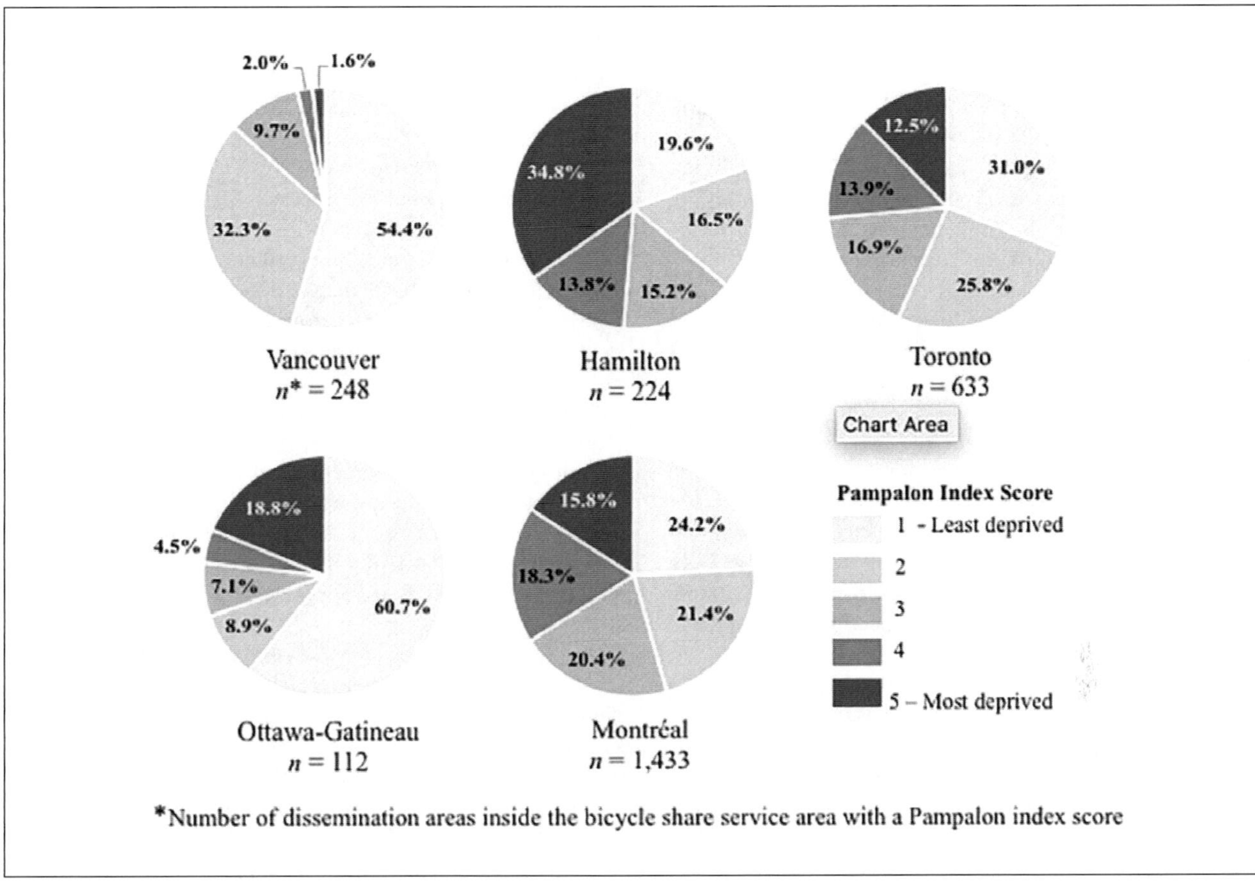

Figure 3. The proportion of DAs inside the bicycle share service area by deprivation quintile in Vancouver, Hamilton, Toronto, Ottawa-Gatineau, and Montréal.

in neighborhoods of diverse socioeconomic status. Hamilton stands out in that the lower income neighborhoods are located near the city center and wealthier neighborhoods are in the surrounding suburban areas (*28*). Thus, relative to other Canadian programs, Hamilton's bicycle share program could more easily be established in lower income areas while still being near high population density areas.

There are many other considerations that influence station siting, including existing bicycle infrastructure, topography, public transportation networks, and the bicycle share program size and target market (*27*). Higher bicycle mode shares correspond to areas with a higher density of bicycle facilities (i.e., bicycle lanes, bicycle paths) (*29*), and often public bicycle share programs install stations near a city's pre-existing bicycling network. Two previous studies have found positive associations between the presence of bicycle facilities near bicycle share docking stations and bicycle share usage (*30, 31*). Another factor that may influence station siting is the topography of the area. Bicycle share bicycles are typically heavier bicycles, which makes them more challenging to ride up hills. In addition, bicycle share stations are often placed near major transit hubs to facilitate the first and last miles of transit trips. The size and target market of a program are also important considerations in station siting. Programs that primarily target tourists and recreational users are typically located near tourist destinations (i.e., parks, historical sites), whereas, programs that target commuters are usually placed near major transit hubs, and are located in both residential and commercial areas (*27*). Equity may be less of a consideration for programs that are small or that primarily target tourists and recreational users. For example, it may not be feasible for Ottawa-Gatineau's program to increase equitable access until their program substantially increases the number of stations and expands their service area beyond the city center.

Bicycle share programs in Canada are still in the early stages of implementation. Montréal was the first Canadian city to adopt a bicycle share program in 2009, while Vancouver's launched in 2016. These programs are continuing to evolve and expand over time, and a primary consideration is long-term financial viability. As bicycle share programs in Canada become more established, it may be that more attention will be given to making access (both spatial and social) more equitable. For example, since it has first been implemented, Montréal's program expanded such that more disadvantaged areas now have access to the program. In 2009, only 9.5% of the DAs in the bicycle share service area were in the most deprived quintile, and in 2017, 15.8% of DAs are in the most deprived quintile. As other programs in Canada expand, particularly in Vancouver and Ottawa-Gatineau, they may want to prioritize expanding into disadvantaged neighborhoods. Incorporating equity considerations into bicycle share planning could help ensure that transportation inequities are not further exacerbated.

The results from this study are similar to the findings from Ursaki and Aultman-Hall's evaluation of bicycle share service areas in U.S. cities (*6*). The U.S. evaluation compared socioeconomic variables (e.g., % White, % with college degree, % with household income >$100,000, etc.) of census block groups inside and outside the bicycle share service areas. We used a similar study design, however, instead of comparing individual socioeconomic variables we used a composite measure of socioeconomic status. In addition, we provide contextual information about the operating models of programs and how this may influence equity outcomes. Together, these studies provide evidence that many public bicycle share programs in Canada and the United States are underserving lower socioeconomic status neighborhoods.

In addition to considering spatial access, other barriers to accessing bicycle share must be addressed in order to increase equitable access to bicycle share programs. These barriers include the cost and pricing structure, credit card ownership, limited access to a computer or smartphone, and a lack of awareness about how bicycle share works (*7*). A number of public bicycle share programs have adopted initiatives to address these barriers, such as the subsidized memberships and community outreach initiatives offered by Chicago's and Hamilton's bicycle share programs (*14*, *15*). Since 2016, the Better Bike Share Partnership has administered US$830,000 in grants to 16 public bicycle share programs across the United States to help support their equity initiatives (*32*). There has been limited research on the effectiveness of these interventions in increasing bicycle share usage among lower socioeconomic populations. Future work examining the impacts of these interventions could provide bicycle share operators with practical information on effective strategies for improving access to bicycle share for underserved populations.

Study Strengths and Limitations

This research on disparities in access to bicycle share programs is part of a larger area of inquiry on inequalities in transportation opportunities, perhaps especially active transportation opportunities that may carry positive health benefits (*8*, *33*). For example, a recent study on accessibility to bicycle infrastructure in eight Canadian cities found that low income areas have less access to bicycle infrastructure than high income areas (*34*). Yet other research has shown that disadvantaged areas have greater access to active transportation infrastructure (*35–37*). The mixed evidence in this area suggests that local context plays an important role in access to transportation opportunities.

This is the first study to evaluate spatial access to bicycle share programs in all Canadian cities with bicycle share programs. The results provide bicycle share operators with valuable information about the accessibility of their programs across areas of different socioeconomic status, and how programs within Canada compare in this regard. Bicycle share operators can use these maps to inform their expansion efforts. For example, operators can identify areas for expansion that are practical, such as low socioeconomic areas that are on the edge of the bicycle share service area boundary. In addition, we use the material component of the Pampalon Deprivation Index as a measure of deprivation, which includes the conventional measure of income, but also incorporates metrics for education and employment to more holistically consider deprivation. This measure of deprivation is well suited to this analysis because income, education, and employment variables are all associated with lower bicycle share uptake (*2–4*).

We acknowledge several study limitations. First, this ecological analysis uses area-level deprivation, but not all individuals living in deprived neighborhoods have low socioeconomic status. Second, there were higher than normal non-response rates for the 2011 Canadian census due to a change from a mandatory census to a voluntary survey in 2011, which could have affected measures of area-level deprivation (*38*). Census data was suppressed for DAs with a non-response rate exceeding 50%, and thus, deprivation scores could not be calculated for these DAs (4.5% to 5.3% of DAs in each of our cities). Despite missing census data, the 2011 index has been validated and is considered a good measure of area-level socioeconomic status (*39*). In our own dataset, we repeated the analysis with the 2006 deprivation index, which had less missing deprivation data, and found similar results; thus, the missing data was not a concern. Third, this evaluation did not consider other barriers to accessing bicycle share, such as cost and language barriers, which are also important considerations to ensure equitable access. And lastly, this study did not assess the usage or financial success of bicycle share programs. The long-term sustainability and viability of bicycle share programs is critical to ensuring that these programs can continue to operate.

Conclusion

Bicycle share programs have the potential to bridge gaps in a city's transportation network, while offering an active and sustainable mode of transportation for commuters. However, the benefits of bicycle share programs are not experienced equally across the population. Among Canadian bicycle share programs, we found that advantaged areas have greater access to bicycle share in Vancouver, Ottawa-Gatineau, Toronto, and Montréal, and conversely, disadvantaged areas have greater access in Hamilton. As bicycle share programs expand, more consideration should be put into expanding service areas to disadvantaged areas.

Acknowledgments

We thank Mobi by Shaw Go, Bike Share Research, Toronto Open Data, and BIXI Montréal for providing bicycle share station location data. We also acknowledge Pampalon et al. for the development of the Pampalon Deprivation Index and the Institut National de Santé Publique du Québec for providing access to the index. Meghan Winters is supported by a Scholar Award from the Michael Smith Foundation for Health Research.

Author Contributions

The authors confirm contribution to the paper as follows: study conception and design: MW, KH; data collection: KH; analysis and interpretation of results: KH; draft manuscript preparation: KH, MW. All authors reviewed the results and approved the final version of the manuscript.

References

1. Meddin, R. *The Bike-Sharing World Map*, June 2017. www.bikesharingmap.com. Accessed June 1, 2017.
2. Ricci, M. Bike Sharing: A Review of Evidence on Impacts and Processes of Implementation and Operation. *Research in Transportation Business & Management*, Vol. 15, 2015, pp. 28–38.
3. Shaheen, S. A. *Public Bikesharing in North America: Early Operator and User Understanding*. MTI Report 11-19. Mineta Transportation Institute Publications, 2012.
4. Fishman, E. Bikeshare: A Review of Recent Literature. *Transport Reviews: A Transnational Trans-disciplinary Journal*, Vol. 36, No. 1, 2016, pp. 92–113.
5. Smith, C. S., J.-S. Oh, and C. Lei. *Exploring the Equity Dimensions of US Bicycle Sharing Systems*. TRCLC 14-01. Transportation Research Center for Livable Communities (TRCLC), 2015.
6. Ursaki, J., and L. Aultman-Hall. *Quantifying the Equity of Bikeshare Access in U.S. Cities*. Transportation Research Board, Washington, D.C., 2016.
7. McNeil, N., J. Dill, J. MacArthur, J. Broach, and S. Howland. *Breaking Barriers to Bike Share: Insights from Residents of Traditionally Underserved Neighborhoods*. NITC-RR-884b. National Institute for Transportation and Communities (NITC), 2017.
8. Lee, R. J., I. N. Sener, and S. N. Jones-Meyer. Understanding the Role of Equity in Active Transportation Planning in the United States. *Transport Reviews*, Vol. 37, No. 2, 2016, pp. 211–226. http://dx.doi.org/10.1080/01441647.2016.1239660
9. Vancouver Bike Share Inc. *Offers*. https://www.mobibikes.ca/en/offers-subscription. Accessed June 15, 2017.
10. TransLink. *Monthly Pass*, 2017. http://www.translink.ca/en/Fares-and-Passes/Monthly-Pass.aspx. Accessed June 15, 2017.
11. Ogilvie, F., and A. Goodman. Inequalities in Usage of a Public Bicycle Sharing Scheme: Socio-Demographic Predictors of Uptake and Usage of the London (UK) Cycle Hire Scheme. *Preventative Medicine*, Vol. 55, No. 1, 2012, pp. 40–45.
12. Goodman, A., and J. Cheshire. Inequalities in the London Bicycle Sharing System Revisited: Impacts of Extending the Scheme to Poorer Areas but Then Doubling Prices. *Journal of Transport Geography*, Vol. 41, 2014, pp. 272–279.
13. Howland, S., N. McNeil, J. Broach, K. Rankins, J. MacArthur, and J. Dill. Current Efforts to Make Bikeshare More Equitable: Survey of System Owners and Operators. *Transportation Research Record: Journal of the Transportation Research Board*, 2017. 2662: 160–167.
14. Greenfield, J. *City Launches "Divvy for Everyone" Bike Share Equity Program*, July 2015. http://chi.streetsblog.org/2015/07/07/city-launches-divvy-for-everyone-bike-share-equity-program/. Accessed June 15, 2017.
15. Hamilton Spectator. *SoBi Bikes to Offer Subsidized Memberships*, April 2017. https://www.thespec.com/news-story/7259556-sobi-bikes-to-offer-subsidized-memberships/. Accessed June 15, 2017.
16. VeloGo. *Map*. https://velogo.ca/#map. Accessed June 15, 2017.
17. BIXI Montréal. *About BIXI Montréal*, 2017. https://montreal.bixi.com/en/who-we-are. Accessed June 15, 2017.
18. Global News. *Bike Share Toronto Doubles in Size with Over 1,000 New Bicycles and Stations*, April 2016. http://globalnews.ca/news/2631671/bike-share-toronto-doubles-in-size-with-over-1000-new-bicycles-and-stations/. Accessed June 15, 2017.
19. Mobi by Shaw Go. *Our Company*. https://www.mobibikes.ca/en/our-company. Accessed June 15, 2017.
20. CBC News. *Montreal to Ramp Up Bixi Service with 1,000 New Docking Stations*, October 2016. http://www.cbc.ca/news/canada/montreal/bixi-montreal-stations-1.3811816. Accessed June 15, 2017.
21. Bike Share Research. *Open Bike Share Data*. https://bikeshare-research.org. Accessed June 1, 2017.
22. City of Toronto. *Open Data - Bicycle Stations (Bike Share Toronto)*. http://www1.toronto.ca/wps/portal/contentonly?vgnextoid=ad3cb6b6ae92b310VgnVCM10000071d60f89RCRD. Accessed June 1, 2017.
23. BIXI Montréal. *Open Data*. https://www.bixi.com/en/open-data. Accessed June 13, 2017.
24. Statistics Canada. *Dissemination Area (DA)*, November 2015. http://www12.statcan.ca/census-recensement/2011/ref/dict/geo021-eng.cfm. Accessed June 1, 2017.
25. Institut National de Santé Publique du Québec. *Deprivation Index, Canada*, 2011. https://www.inspq.qc.ca/expertises/gestion-et-analyse-de-l-information/diffusion-d-une-information-adaptee/santescope/indice-de-defavorisation. Accessed June 1, 2017.
26. Pampalon, R., D. Hamel, P. Gamache, M. D. Philibert, G. Raymond, and A. Simpson. An Area-Based Material and Social Deprivation Index for Public Health in Québec and Canada. *Canadian Journal of Public Health*, Vol. 103, 2012, pp. S17–S22.
27. García-Palomares, J. C., J. Gutiérrez, and M. Latorre. Optimizing the Location of Stations in Bike-Sharing Programs: A GIS Approach. *Applied Geography*, Vol. 35, No. 1, 2012, pp. 235–246. https://doi.org/10.1016/j.apgeog.2012.07.002
28. Harris, R., J. Dunn, and S. Wakefield. *A City on the Cusp: Neighbourhood Change in Hamilton Since 1970*. Neighbourhood Change Research Partnership, University of Toronto, 2015.
29. Buehler, R., and J. Pucher. Cycling to Work in 90 Large American cities: New Evidence on the Role of Bike Paths and Lanes. *Transportation*, Vol. 39, No. 2, 2012, pp. 409–432. http://doi.org/10.1007/s11116-011-9355-8
30. Buck, D., and R. Buehler. Bike Lanes and Other Determinants of Capital Bikeshare Trips. Presented at 91st Annual Meeting of the Transportation Research Board, Washington, D.C., 2012.

31. Faghih-Imani, A., N. Eluru, A. M. El-Geneidy, M. Rabbat, and U. Haq. How Land-Use and Urban Form Impact Bicycle Flows: Evidence from the Bicycle-Sharing System (BIXI) in Montreal. *Journal of Transport Geography*, Vol. 42, 2014, pp. 306–314.
32. Better Bike Share Partnership. *Grants*, 2017. http://betterbike-share.org/grants/. Accessed June 28, 2017.
33. Lee, R. J., I. N. Sener, and S. N. Jones-Meyer. A Review of Equity in Active Transportation. Presented at 95th Annual Meeting of the Transportation Research Board, Washington, D.C., 2016.
34. Fuller, D., and M. Winters. Income Inequalities in Bike Score and Bicycling to Work in Canada. *Journal of Transport & Health*, Vol. 7, 2017, pp. 264–268. https://doi.org/10.1016/j.jth.2017.09.005
35. King, K. E., and P. J. Clarke. A Disadvantaged Advantage in Walkability: Findings from Socioeconomic and Geographical Analysis of National Built Environment Data in the United States. *American Journal of Epidemiology*, Vol. 181, No. 1, 2015, pp. 17–25. https://doi.org/10.1093/aje/kwu310
36. Turrell, G., M. Haynes, L. A. Wilson, and B. Giles-Corti. Can the Built Environment Reduce Health Inequalities? A Study of Neighbourhood Socioeconomic Disadvantage and Walking for Transport. *Health & Place*, Vol. 19, 2013, pp. 89–98. https://doi.org/10.1016/j.healthplace.2012.10.008
37. Fuller, D., L. Gauvin, and Y. Kestens. Individual- and Area-Level Disparities in Access to the Road Network, Subway System and a Public Bicycle Share Program on the Island of Montreal, Canada. *Annals of Behavioral Medicine*, Vol. 45, No. 1, 2013, pp. 95–100. https://doi.org/10.1007/s12160-012-9433-y.
38. Statistics Canada. *NHS: Data Quality*, December 2015. http://www12.statcan.gc.ca/nhs-enm/2011/ref/about-apropos/nhs-enm_r005-eng.cfm. Accessed June 20, 2017.
39. Gamache, P., and D. Hamel. *The Challenges of Updating the Deprivation Index with Data from the 2011 Census and the National Household Survey (NHS)*, January 2017. https://www.inspq.qc.ca/en/publications/2207. Accessed June 15, 2017.

The Pedestrians and Cycles Section (ANF00) peer-reviewed this paper (18-00589).

Article

TRR
JOURNAL OF THE TRANSPORTATION RESEARCH BOARD

Transportation Research Record
2018, Vol. 2672(36) 51–62

DOI: 10.1177/0361198118790637
journals.sagepub.com/home/trr
SAGE

Empirical Analysis of the Macroscopic Characteristics of Bicycle Flow during the Queue Discharge Process at a Signalized Intersection

Bernat Goñi-Ros[1], Yufei Yuan[1], Winnie Daamen[1], and Serge P. Hoogendoorn[1]

Abstract

Signalized intersections are one of the most common types of bottleneck in urban cycling networks. Gaining knowledge on the macroscopic characteristics of bicycle flow during the queue discharge process is crucial for developing ways to reduce the delay experienced by cyclists at intersections. This paper aims to determine these characteristics (including jam density, shockwave speed, and discharge flow), and to unveil possible relationships between them, particularly whether and to what extent discharge flow is correlated with jam density and shockwave speed (which is of high relevance from a traffic management viewpoint). To this end, the study analyzes high-resolution bicycle trajectories derived from video footage on a one-direction cycle path leading to an intersection in Amsterdam (the Netherlands). Linear regression analysis is used to investigate the relationships between macroscopic variables. The results indicate that jam density, shockwave speed, and discharge flow vary considerably across traffic-signal cycles, which highlights the stochastic nature of bicycle flow. Furthermore, the results show that discharge flow is strongly positively correlated with jam density and shockwave speed. It is hypothesized that there is a causal relationship between these variables, which would imply that traffic engineers can increase discharge flows (thus reducing delay) at signalized intersections if they find effective ways to increase jam densities and shockwave speeds.

Bicycle use is increasing in cities all around the world owing to the advantages that cycling offers to travelers and to the implementation of policies promoting this transportation mode (*1*). Although this trend has important benefits for society from the environmental, accessibility, and public health viewpoints (*2*), it also comes with challenges, such as safety issues and congestion at bottlenecks. One of the most common types of bottleneck in urban cycling networks is the signalized intersection, where bicycle flow is interrupted (so cyclists need to stand in a queue) during red-signal phases. Long queues may not dissolve completely during one green-signal phase, thus some cyclists may need to wait for more than one red-signal phase to access an intersection. In order to mitigate this problem, it is necessary to maximize the queue outflow during green-signal phases. Some cities are currently testing innovative street-layout concepts and/or bicycle traffic management measures to achieve this goal (*3*). For that, knowledge concerning the characteristics of bicycle flows and, more specifically, the characteristics of the bicycle queue-formation and discharge processes, is crucial. Key characteristics (from a macroscopic perspective) are jam density, shockwave speed, and queue discharge flow. The *jam density* of a standing-still queue can be defined as the average number of cyclists per unit of space, and is an outcome of the queue-formation process. Knowing typical jam density values one can estimate queue lengths. When the traffic light turns green, a shockwave between two traffic states (i.e., cyclists who are standing still, and cyclists who are starting to move) emerges. The speed at which this shockwave propagates upstream, which is determined by the gaps cyclists keep when they start to move, is called *shockwave speed* (or wave speed). Knowing typical wave

[1]Department of Transport and Planning, Faculty of Civil Engineering and Geosciences, Delft University of Technology, Delft, The Netherlands

Corresponding Author:
Address correspondence to Yufei Yuan: y.yuan@tudelft.nl

speed values one can estimate the time that any cyclist will start moving given its location within the queue. The *queue discharge flow* (or simply discharge flow) is generally defined as the average number of cyclists that pass the stop line per unit of time. Importantly, the discharge flow determines the delay experienced by cyclists at the intersection (given the red- and green-phase lengths, and a certain demand profile).

Although knowledge concerning the characteristics of the queue discharge process is very important for the development of effective ways to reduce the delay and discomfort experienced by cyclists at intersections, currently there is only a partial understanding of these characteristics. To the authors' knowledge, empirical measurements of wave speeds in bicycle queues have not been reported in the literature. And although several studies present empirical measurements of jam density and discharge flow, the reported values vary considerably: jam densities range between 0.27 and 0.65 bicycles/m^2 (*4–8*), whereas discharge flows range between 1900 and 4300 bicycles/h/m (*7–12*). Multiple causes may explain this high variation. Regarding jam density, the variation may be as a result of differences in bicycle dimensions (*13*, *14*), presence of other types of vehicles (e.g., scooters, e-bikes), and cultural differences in perception of personal space, as well as differences in the methods used to calculate jam density. Also, it has been shown that jam density is influenced by queue size (*15*) and cycle path width (*8*). Moreover, variation in jam density may be partially intrinsic to the queue-formation process (*15*). On the other hand, it has been shown that variation in discharge flow measurements can partially be the result of differences in time of the day (which is related to trip purpose) (*12*), cycle path width (*11*, *12*, *16*) and grade (*17*), as well as methods used to calculate discharge flow (*18*). Also, it seems that the capacity of uninterrupted cycle paths is influenced by the presence of vehicles other than bicycles (*18*, *19*), so it is reasonable to believe that the discharge flow at intersections may also depend on that. However, another reason explaining the high variation in discharge flows reported in the literature may be a possible dependency on jam density (which itself also shows a considerable amount of variation, as mentioned above) and/or wave speed (which may also be highly variable). First-order traffic flow theory (*20*) states that $q_d = \omega \cdot (k_j - k_c)$, where q_d, ω, k_j and k_c denote discharge flow, wave speed, jam density, and critical density, respectively. Critical density is defined as the maximum density achievable under uncongested traffic conditions, and is usually estimated as the average number of cyclists per unit of space observed right downstream of the stop line during the queue discharge process (*7*). Note that the equation presented above already assumes that q_d, k_j and ω are interrelated.

The main objective of this study is to determine the macroscopic characteristics of bicycle flow during the queue discharge process at signalized intersections (including jam density, wave speed, and discharge flow) based on empirical data, and to unveil possible relationships between these characteristics, particularly whether and to what extent discharge flow is correlated with jam density and wave speed. As mentioned above, discharge flow is a very relevant macroscopic variable from a traffic management viewpoint, as it determines the delay experienced by cyclists at an intersection. For this purpose, the study analyzes bicycle trajectories on a one-way cycle path leading to a signalized intersection in Amsterdam (the Netherlands). The trajectories were derived from top-view video images. Jam density, wave speed, discharge flow, and critical density were measured for 57 queue discharge periods. Linear regression analysis was used to investigate the relationships between these macroscopic flow characteristics. Thus this paper takes into account and investigates the theoretical relationship $q_d = \omega \cdot (k_j - k_c)$, but it also establishes simpler relationships in order to confirm and highlight correlations between macroscopic characteristics based on the empirical observations. Also, note that the paper focuses on the analysis of the queue discharge process only when: (a) the queue is formed mainly by bicycles; and (b) cyclists stay within the cycle path and are not disturbed by any traffic agents external to the queue while moving. An analysis of the queue discharge process with mixed traffic and/or complex traffic interactions is beyond the scope of this paper.

The rest of this paper is structured as follows. The next section describes the site and explains the procedure used to derive the trajectory data, followed by a description of the methods used to calculate the macroscopic variables and explore the relationships between them. The results of the analysis are then presented and their possible implications for traffic management discussed. The final section contains the conclusions of this study.

Data Collection

Bicycle trajectories along a cycle path leading to a signalized intersection in Amsterdam were derived from top-view video images. The cycle path is 2 m wide and unidirectional; it is segregated from car traffic and is used by both bicycles and scooters. Traffic demand in this path is relatively high in peak hours. Access to the intersection is regulated by a traffic light. In order to obtain the video footage, a 10-m pole with two video cameras on top (*21*) was placed next to it. The camera views can be seen in Figure 1, *a* and *b*. The front camera recorded the last 9 m of the cycle path upstream of the stop line, whereas the back camera recorded the previous 11 m

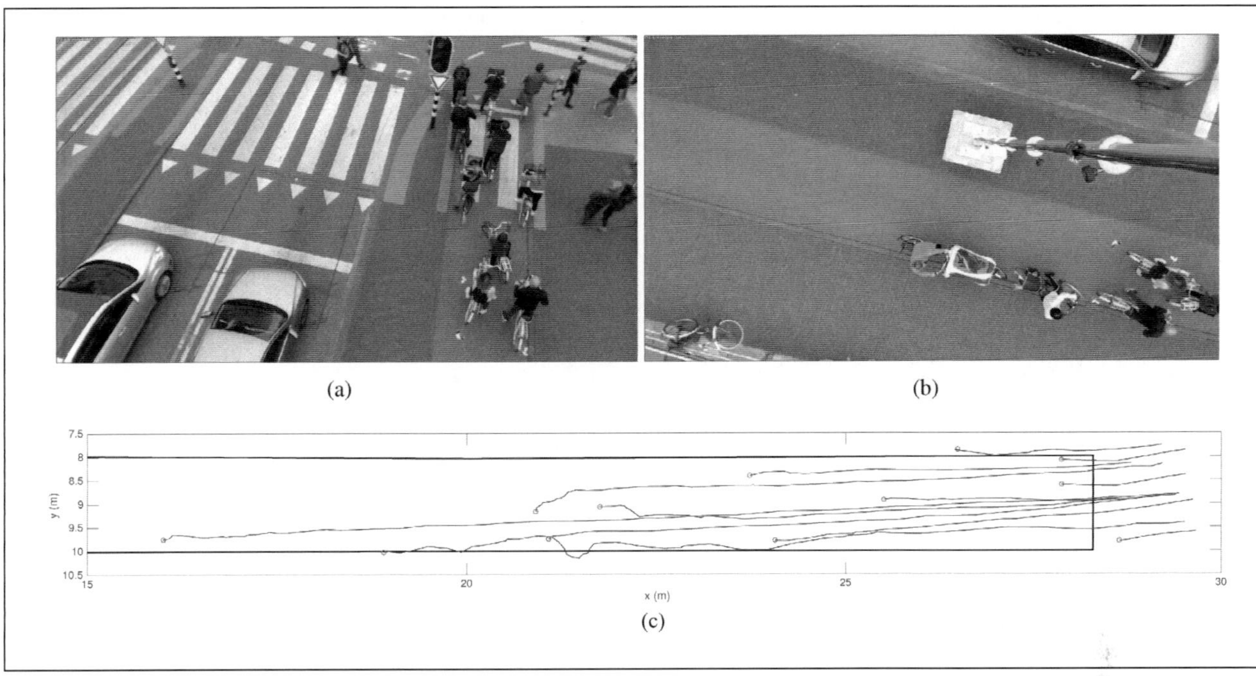

Figure 1. View of the two cameras installed on the 10-m pole at 13:34:23h, when the traffic light was red and 12 bicycles were standing still forming a queue, and derived trajectories: (*a*) front camera; (*b*) back camera; (*c*) derived trajectories.

(there is some overlap between the two camera views). Therefore, the two cameras combined covered 20 m of cycle path upstream of the traffic light. The frame rate varied between 5 and 10 fps during the recordings.

Bicycle traffic was recorded between 12:45 and 19:00 h on June 6, 2016. By looking at the video footage, queue discharge periods were selected that met the following criteria: (a) at least seven bicycles/scooters are standing still inside the path before the traffic light turns green; (b) there are no more than two scooters within the queue; (c) all the bicycles/scooters move toward the stop line and pass it without getting out of the cycle path; (d) no pedestrian crosses the cycle path during the discharge process; (e) the discharge process is not affected by downstream traffic conditions; and (f) all the bicycles/scooters pass the stop line before the end of the green phase. The threshold established by criterion (a) aims to ensure that the selected queues are long enough to allow for a proper analysis of the macroscopic flow characteristics during the discharge process. Criterion (b) defines a threshold to guarantee that the selected queues are formed mainly by bicycles (instead of scooters). Criterion (c) ensures that, in the selected periods, no cyclist leaves the cycle path during the queue discharge process. This is important because sometimes cyclists do not respect the traffic rules and move to the road or to the sidewalk before crossing the stop line (see the site in Figure 1*a*). This irregular behavior may have a significant influence on the macroscopic characteristics of bicycle flow. Criteria (d), (e), and (f) ensure that the queue discharge process is not disturbed by the traffic signal nor by agents external to the bicycle queue.

The total number of selected discharge periods was 57. The trajectories of all bicycles and scooters forming part of the queue were derived for every period (see example in Figure 1*c*). The complete procedure to derive the trajectories for a given period is as follows:

1. *Video clip decomposition*. The video clip corresponding to the selected period is decomposed into its individual frames.
2. *Manual cyclist tracking*. TrajectoryViewer, a software package developed at TU Delft, is used to manually track over time (i.e., in each video frame) the cyclists that form part of the queue. The exact point being tracked is the center of the head. In this step, space coordinates (x and y) are expressed in number of pixels, and time (t) is given by the frame number. The vehicle type is also recorded. The main input required by TrajectoryViewer are the individual frames of a given video clip.
3. *Height transformation*. Space coordinates are transformed in order to project the head positions to the ground. This is done using linear functions that give the x and y coordinates (in pixels) of the

ground position directly below the head of any cyclist in the video frames. These functions are derived by manually selecting several head points in the frames and identifying their projected points on the ground, and then applying linear regression. The linear functions are different for the front and back cameras.

4. *Orthorectification.* The x and y coordinates of the ground points are transformed in order to obtain cyclist trajectories as seen from a top view (see Figure 1c) instead of a side view. This is achieved by performing a wedge transformation of the coordinates, which corrects for the distortion caused by the cameras not pointing vertically downwards to the cycle path. The parameters of the wedge transformation function are determined using ImageTracker (*22*). Also, in this step, x and y coordinates are scaled into a unified Cartesian coordinate system (in meters). The scaling parameters are determined using ImageTracker. Orthorectification is necessary to ensure that distances between all points of a given trajectory are equivalent to real distances between cyclist heads (projected to the ground).
5. *Time coding.* The time units of each trajectory point are changed from frame number to seconds. To do this, it is necessary to first assign a time instant in seconds to every frame in accordance with the video file metadata.
6. *Trajectory merging.* Trajectories from the two cameras that correspond to the same cyclist are merged into consistent full trajectories. This is done by coupling pairs of trajectories that are very close in space and time within the cycle path area where the two camera views overlap.

Data Analysis Methods

This section presents the way macroscopic flow characteristics were calculated for each selected discharge period, and describes the methodology used to establish relationships between these characteristics.

Measured Quantities

An example of the trajectories during one of the selected discharge periods is shown in Figure 2*a*. The three following quantities were measured for every selected period: jam density (k_j), wave speed (ω), discharge flow (q_d), critical density (k_c) and critical speed (v_c). This was done as follows. First, a count area was defined between the stop line (see stop line in Figure 1*a*) and a parallel line located 0.4 m downstream of the stop line (which we called line a, see Figure 2*b*). The count area stretches across the full width of the cycle path (2 m). Then, in every period, two indices (i and j) were assigned to every cyclist within the queue (Figure 2*b*). Index i gives the rank of each cyclist regarding their distance to line a at standstill (i = 1 corresponds to the closest cyclist to that line). Note that the x-coordinate of line a is constant (x_a). Index j gives the rank of each cyclist regarding the time they pass line a (j = 1 corresponds to the cyclist that crosses it first). Note that a certain cyclist can have different i and j indices. For instance, in Figure 2*b*, the third closest cyclist to line a at standstill (i = 3) is the second cyclist to pass it (j = 2). Once cyclists were assigned these indices, the macroscopic quantities were calculated as follows (see also Figure 2*b*).

The jam density (k_j) was calculated by dividing the number of bicycles in the queue (N) minus one by the distance at standstill between the bicycles that are closest and farthest away from line a at standstill (i = 1 and $i = N$, respectively), which is denoted by L (see Figure 2*b*). The jam density is standardized by dividing it by the cycle path width (W).

$$k_j = \frac{N-1}{L \cdot W} \tag{1}$$

where:

$$L = d_N(t_0) - d_1(t_0) \tag{2}$$

In Equation 2, d_i denotes the distance from line a (i.e., $d_i = x_a - x_i$); and t_0 denotes the first time instant for which the positions of all cyclists are known. Note that at t_0 all cyclists are standing still. In this definition of jam density, cyclist i = 1 is not counted as being inside area $L \cdot W$ (this is why the numerator of the right-hand term of Equation 1 is $N - 1$). This is done for consistency with the definitions of discharge flow, critical density, and critical speed, which exclude the first vehicle to pass line a (see Equations 3–5).

The wave speed (ω) was estimated by fitting a line (using linear regression) to the last points in the td plane before bicycles $j = 1, \ldots, N$ start moving, which are denoted by (t_j^m, $d_j(t_j^m)$), where t_j^m is the last time instant before cyclist j starts moving. The intercept of this line is fixed to $d_1(t_1^m)$, and the t coordinates are transformed to substract t_1^m. The wave speed (ω) is defined as the slope of the fitted line, which can be expressed as $d = d_1(t_1^m) + \omega \cdot (t - t_1^m)$ (see Figure 2*b*).

Discharge flow (q_d), critical density (k_c) and critical speed (v_c) are defined as the average flow, density and speed within the count area (right downstream of the stop line), and they were calculated using Edie's definitions (*23*) (see also Figure 2*b*). Discharge flow was defined as a flow through an area instead of a line to ensure consistency with the definitions of critical density and speed. Note that q_d and k_c were standardized by

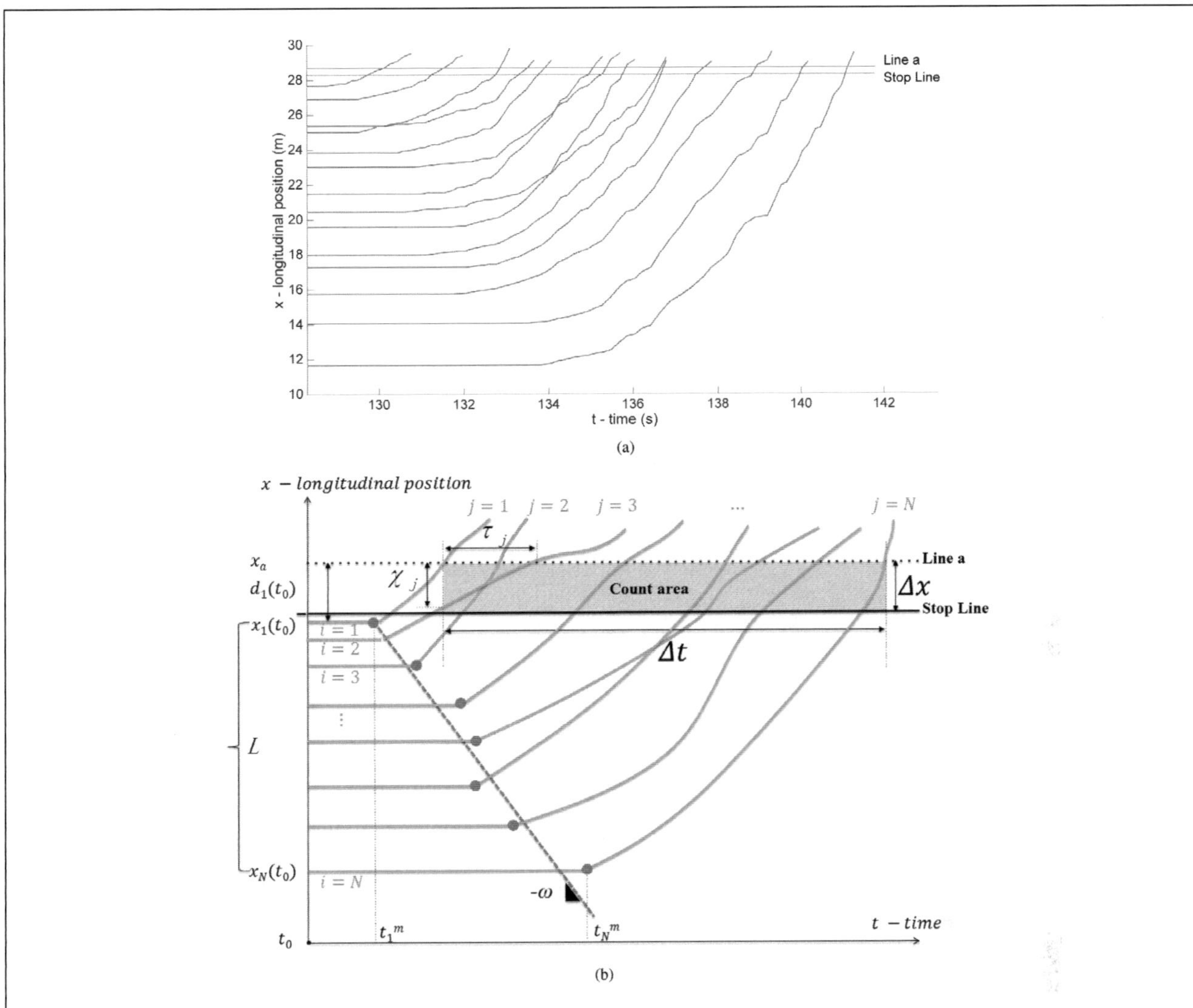

Figure 2. Real trajectories and conceptual plot showing the calculation methods: (*a*) real bicycle trajectories; (*b*) calculation of the measured quantities.

dividing them by the cycle path width (W). Discharge flow (q_{d}) was calculated as follows:

$$q_{\mathrm{d}} = \frac{\sum_{j=2}^{N} \chi_j}{\Delta x \cdot \Delta t \cdot W} \tag{3}$$

where: Δx is the distance in the x -dimension between the stop line and line a (0.4 m); Δt is the period between the time instants when cyclists $j = 1$ and $j = N$ cross line a (i.e., when d_1 and d_N become zero, respectively); and χ_j is the distance that bicycle j travels through the count area during period Δt.

Note that, since the time instant when cyclist $j = 1$ crosses line a is used to define the period Δt (see Figure 2*b*), cyclist $j = 1$ is not counted when calculating q_{d}. This is done to exclude the period with zero flow between the beginning of the green-signal phase and the passing time of the first cyclist. The length of this period may vary substantially because of variation in the reaction times of cyclists $j = 1$ to the green signal, so including it may add error to the discharge flow measurements. For consistency, the first cyclist is not counted when calculating the critical density (k_{c}) and the critical speed (v_{c}) either. These quantities were calculated as follows (see also Figure 2*b*):

$$k_{\mathrm{c}} = \frac{\sum_{j=2}^{N} \tau_j}{\Delta x \cdot \Delta t \cdot W} \tag{4}$$

and

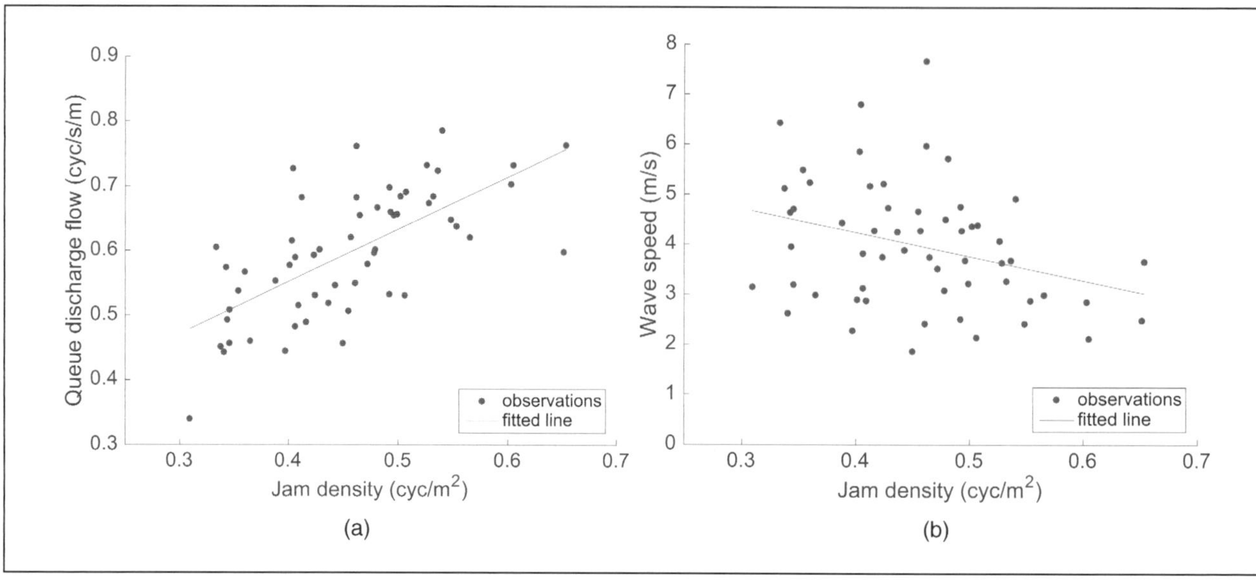

Figure 3. Observations and fitted lines for some of the relationships investigated in this research: (*a*) Relationship between q_d and k_j (Model 1); (*b*) Relationship between ω and k_j (Model 4).

$$v_c = \frac{\sum_{j=2}^{N} X_j}{\sum_{j=2}^{N} \tau_j} = \frac{q_d}{k_c} \quad (5)$$

where τ_j denotes the travel time of cyclist j within the count area during period Δt.

Data Analysis

The empirical measurements of macroscopic flow characteristics (jam density, wave speed, discharge flow, critical density, and critical speed) described above were analyzed in two different ways. First, the variability of the measurements across the 57 selected periods was analyzed. Histograms were created and descriptive statistics calculated (such as the average and the percentiles) in order to gain insight into the shapes of their distributions. Note that the analyzed periods are only a selection of all the observed periods (the selection criteria were explained earlier). Therefore, the histograms and descriptive statistics do not represent all observed periods but only those periods for which it is appropriate to measure the macroscopic flow characteristics given the scope of this research project.

Second, the relationships between different macroscopic flow characteristics were analyzed. The main goal was to determine whether and to what extent discharge flow (which can be seen as an outcome of the queue discharge process) is correlated with jam density and wave speed (which are features of that process). After visual inspection of the data points (see, for example, Figure 3), linear regression analysis was selected to explore these relationships. Nevertheless, this approach is considered as a first step in the analysis of the relationships between macroscopic flow characteristics. The use of more complex regression techniques may provide additional insights, but that will be addressed in further research. In this paper, five linear regression models were estimated based on the empirical measurements:

- Model 1: $q_d = \alpha_1 + \beta_1 \cdot k_j$;
- Model 2: $q_d = \alpha_2 + \beta_2 \cdot k_j + \gamma_2 \cdot \omega$;
- Model 3: $k_c = \alpha_3 + \beta_3 \cdot k_j + \gamma_3 \cdot \omega$;
- Model 4: $\omega = \alpha_4 + \beta_4 \cdot k_j$;
- Model 5: $q_d = \beta_5 \cdot (k_j - k_c)$.

F-tests and t -tests were used to determine the significance of the estimated models and their parameters (a 5% significance level was used for all models and parameters). Models 1 and 2 were estimated to investigate the relationship between discharge flow (q_d), jam density (k_j), and wave speed (ω). An incremental approach was used to gain insight into the individual correlation of q_d with k_j and ω. For this reason, two separate linear regression models were estimated: one with only k_j as regressor (Model 1), and one with both k_j and ω as regressors (Model 2). Model 3 was estimated to investigate the relationship between critical density (k_c) on one hand, and jam density (k_j) and wave speed (ω) on the other. Model 4 was estimated to determine whether and to what extent jam density and wave speed are linearly correlated.

Table 1. Descriptive Statistics (57 Observations)

Variable	Avg	StD	p10	p25	p50	p75	p90	Min	Max
Queue size (cyclists)	12.12	2.99	8.20	10	12	13.25	16.80	7	19
Percentage of scooters (%)	2.08	4.05	0.00	0.00	0.00	0.00	9.09	0.00	14.29
Jam density (cyc/m^2)	0.45	0.08	0.34	0.40	0.46	0.50	0.55	0.31	0.65
Wave speed (m/s)	3.98	1.25	2.43	2.99	3.81	4.71	5.68	1.86	7.67
Discharge flow (cyc/s/m)	0.60	0.10	0.46	0.53	0.60	0.68	0.73	0.34	0.79
Critical density (cyc/m^2)	0.26	0.07	0.19	0.20	0.25	0.30	0.35	0.15	0.48
Critical speed (m/s)	2.43	0.46	1.86	2.09	2.46	2.80	2.94	1.34	3.45
Start-up time (s)	0.70	0.75	0.04	0.34	0.59	1.02	1.33	-1.55	3.55

Finally, Model 5 was estimated to determine the wave speed according to first-order traffic flow theory—which states that parameter β_5 is analogous to the wave speed *(20)*—and to see whether the empirical data match this theory.

Models 1–5 were first estimated using raw scores (i.e., the k_j, ω, q_d and k_c measurements expressed in their own units). Then, the models were also estimated using standard scores (*Z*-scores) in order to obtain their standardized coefficients. These coefficients indicate how many standard deviations the regressand is expected to change per standard deviation increase in the regressors, according to a given regression model. In the case of these models, standardization of the coefficients is useful because the variables involved are measured in different units.

Results

Descriptive Statistics

The total number of queue discharge periods analyzed in this study is 57. The average queue size in these periods is 12.12 cyclists, and in 75% of them the queue consists of 10 or more cyclists (see Table 1). In 44 out of 57 periods (77.2%), there are no scooters in the queue; in 11 periods (19.3%), there is one scooter, and in the remaining 2 periods (3.5%), there are two. The maximum percentage of scooters is 14.29% (see Table 1), corresponding to a period in which 2 out of 14 cyclists are riding a scooter.

The measurements of the macroscopic quantities defined earlier vary substantially across different queue discharge periods. Figure 4, *a–e*, shows the histograms of these measurements, and Table 1 shows various descriptive statistics. Jam density measurements (k_j) range between 0.31 and 0.65 cyclists/m^2, with an average equal to 0.45 cyclists/m^2 and a standard deviation equal to 0.08 cyclists/m^2. Visual inspection of the video images indicates that some of the factors contributing to this high variation in jam density may be the presence of social groups, and the preference of some cyclists to stand on the right side of the cycle path, where they can put their right foot on the sidewalk (see Figure 1*b*). Also, some cyclists seem to avoid standing on the pedestrian crossing area. Wave speeds range between 1.86 and 7.67 m/s (6.69 and 27.63 km/h); the average is 3.98 m/s (14.32 km/h), and the standard deviation is 1.25 m/s (4.51 km/h). Discharge flow (q_d) measurements range between 0.34 and 0.79 cyclists/s/m (1227 and 2829 cyclists/h/m) (see Table 1). The average discharge flow is 0.60 cyclists/s/m (2148 cyclists/h/m) and the standard deviation is 0.10 cyclists/s/m (351 cyclists/h/m). Regarding the critical density (k_c), the average is 0.26 cyc/m^2 and the standard deviation is 0.07 cyc/m^2. The average critical speed is 2.43 m/s (8.75 km/h) and the standard deviation is 0.46 m/s (1.66 km/h).

Table 1 and Figure 4*f* also show the distribution of start-up times, which correspond to the length of the time period between the beginning of the green phase and the last time instant before the first vehicle to pass line *a* starts moving (t_1^m, see earlier). The average start-up time is 0.70 s, and in 25% of the analyzed periods it is equal to or greater than 1.02 s. In four cases (7%), it is negative, meaning that cyclist $j = 1$ starts moving before the traffic light turns green.

Relationships between Macroscopic Characteristics

The regression analysis results show that there is a positive linear relationship between discharge flow (q_d) and jam density (k_j) (see Table 2, Model 1, and Figure 3*a*). According to this model, k_j accounts for 45% of the variance in q_d (R^2 is 0.453). If wave speed (ω) is added to jam density as regressor, a multiple linear model with a better fit is achieved (see Table 2, Model 2b). According to this model, k_j and ω combined account for 71% of the variance in q_d (R^2 equals 0.709), 26 percentage points more than k_j alone. Note that this model was estimated with the intercept (α_2) fixed to zero, since the *t*-tests suggest that the intercept is not significantly different than this value (see Table 2, Model 2a). Furthermore, there seems to be a positive linear relationship between critical density (k_c), jam density, and wave speed. The latter two variables account for 31% of the variance in k_c (see Table 2, Model 3b). Note that this model was also estimated

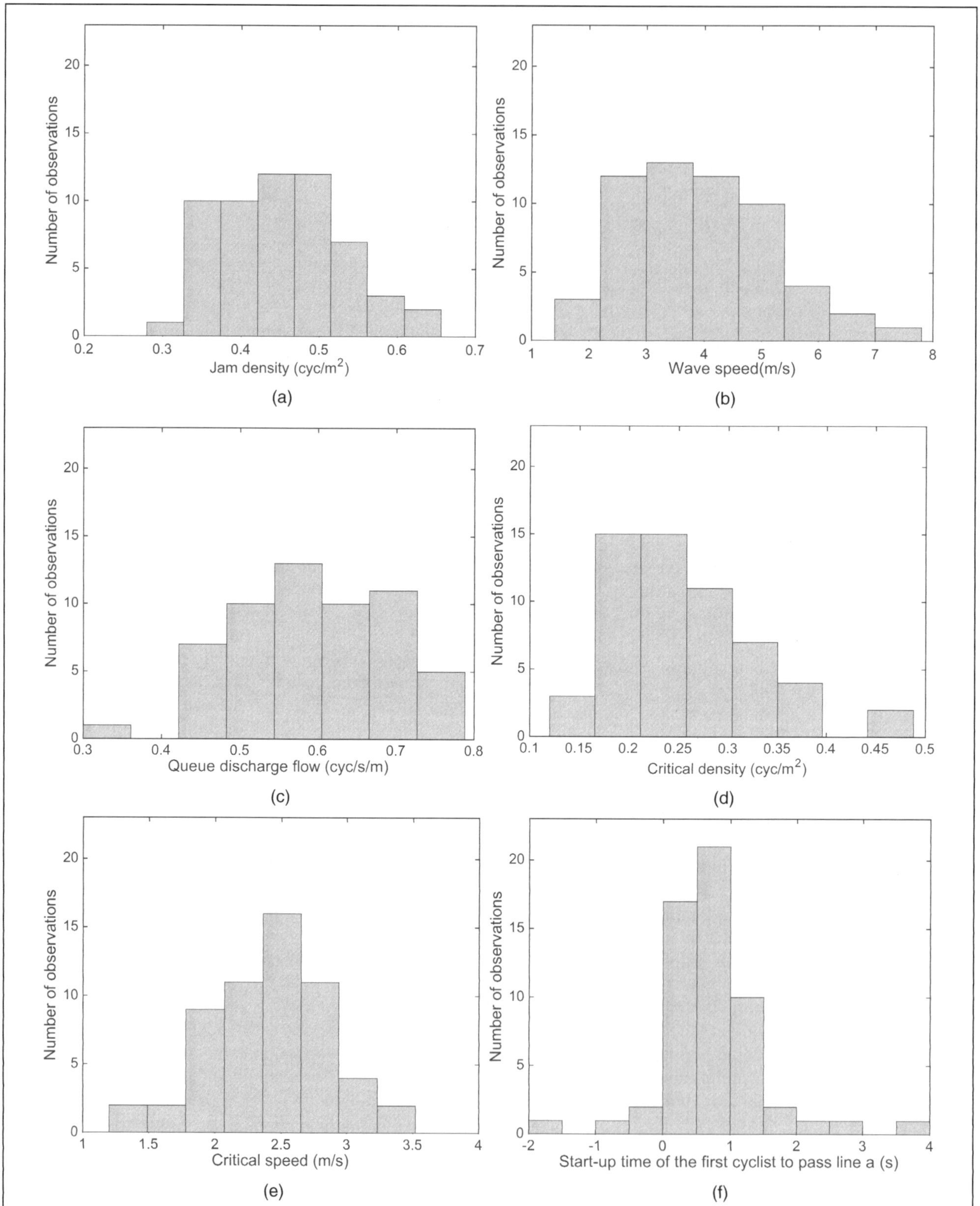

Figure 4. Frequency distributions of the measured quantities: (*a*) jam density; (*b*) wave speed; (*c*) queue discharge flow; (*d*) critical density; (e) critical speed; (*f*) start-up time.

Table 2. Model Estimation Results

	F-stat	*p*-value	R^2	Adj. R^2	RMSE	Estimate	SE	*t*-stat	*p*-value	Standardized estimate
Model 1: $q_d(\text{cyc/s/m}) = \alpha_1 + \beta_1 \cdot k_j\ (\text{cyc/m}^2)$										
Model	45.6	9.57e-09	0.453	0.443	0.073					
α_1						0.300	0.055	4.169	1.09e-04	0
β_1						0.807	0.119	6.753	9.57e-09	0.673
Model 2a: $q_d(\text{cyc/s/m}) = \alpha_2 + \beta_2 \cdot k_j\ (\text{cyc/m}^2) + \gamma_2 \cdot \omega\ (\text{m/s})$										
Model	74.8	2.77e-16	0.735	0.725	0.051					
α_2						−0.038	0.052	−0.716	0.477	
β_2						1.016	0.088	11.487	4.04e-16	
γ_2						0.043	0.006	7.567	4.95e-10	
Model 2b: $q_d(\text{cyc/s/m}) = \beta_2 \cdot k_j\ (\text{cyc/m}^2) + \gamma_2 \cdot \omega\ (\text{m/s})$										
Model	70.0	1.83e-15	0.709	0.704	0.051					
β_2						0.959	0.038	25.174	7.25e-32	0.848
γ_2						0.041	0.004	9.630	2.12e-13	0.558
Model 3a: $k_c\ (\text{cyc/m}^2) = \alpha_3 + \beta_3 \cdot k_j\ (\text{cyc/m}^2) + \gamma_3 \cdot \omega\ (\text{m/s})$										
Model	16.7	2.23e-06	0.382	0.360	0.056					
α_3						−0.054	0.057	−0.946	0.348	
β_3						0.558	0.097	5.763	4.08e-07	
γ_3						0.014	0.006	2.252	0.028	
Model 3b: $k_c\ (\text{cyc/m}^2) = \beta_3 \cdot k_j\ (\text{cyc/m}^2) + \gamma_3 \cdot \omega\ (\text{m/s})$										
Model	12.1	4.46e-05	0.305	0.293	0.056					
β_3						0.476	0.042	11.360	4.76e-16	0.649
γ_3						0.010	0.005	2.187	0.033	0.254
Model 4: $\omega(\text{m/s}) = \alpha_4 + \beta_4 \cdot k_j\ (\text{cyc/m}^2)$										
Model	5.93	0.018	0.097	0.081	1.200					
α_4						6.164	0.912	6.760	9.32e-09	0
β_4						−4.812	1.975	−2.436	0.018	−0.312
Model 5: $q_d(\text{cyc/s/m}) = \beta_5 \cdot (k_j\ (\text{cyc/m}^2) - k_c\ (\text{cyc/m}^2))$										
Model	51.78	1.64e-09	0.480	0.480	0.209					
β_5						2.693	0.131	20.561	9.08e-28	

with the intercept (α_3) fixed to zero. Interestingly, wave speed is found to be negatively linearly correlated to jam density. However, this relationship does not seem to be very strong, as k_j only accounts for 10% of the variance in ω (see Table 2, Model 4, and Figure 3*b*).

Finally, there is a positive linear relationship between discharge flow (q_d) and the difference between jam density and critical density ($k_j - k_c$). This difference accounts for 48% of the variance in q_d (see Table 2, Model 5). According to first-order traffic flow theory, the slope of this linear relation (β_5) should correspond to the wave speed (*20*). However, the value of β_5 (2.69 m/s) is lower than the average observed wave speed (3.98 m/s), thus this theoretical model may tend to underestimate the discharge flow. One of the reasons for this may be related to the definitions of the macroscopic quantities. First-order traffic flow theory assumes infinite acceleration, which implies that after the traffic light turns green, bicycle traffic at the count area is immediately homogeneous. In reality, however, acceleration is not infinite. The first group of cyclists to cross the stop line are not moving at constant speed (see, for instance, Figure 2*a*), thus traffic at the count area is not immediately homogeneous. As this first group of cyclists is included in the calculation of flow and density at the count area, higher critical densities and lower discharge flows may be obtained than those assumed by first-order traffic flow theory. This implies that β_5 cannot be interpreted directly as the wave speed.

Possible Implications for Traffic Management

Although regression analysis can only establish correlation (not causation) between variables, it is hypothesized that discharge flow is actually influenced by jam density and wave speed (and possibly also by other factors not accounted for in this study, such as the way cyclists accelerate from standstill, and passing maneuvers during the queue discharge process). This hypothesis is assumed true in first-order traffic flow theory, as explained previously. It is argued that the hypothesis is plausible from a traffic-engineering viewpoint, at least within reasonable ranges of k_j and ω: if cyclists stand closer to each other in the queue (i.e., if k_j is higher) and/or they start moving earlier (i.e., if ω increases), they will probably cross the stop line earlier, which implies that the discharge flow (q_d) will increase. This hypothesis needs to be experimentally tested. However, a relevant implication follows from assuming it is true: traffic engineers can increase the discharge flow at signalized intersections (and thus reduce delay) if they find ways to influence the behavior of cyclists in such a way that queues become denser and wave speeds higher.

In this respect, Model 2b suggests that, within the range of observed values, if both k_j and ω increase by one standard deviation, the discharge flow increases by $(0.848 + 0.558) \cdot 100 = 140.6\%$ of its standard deviation, that is, 0.141 bicycles/s/m (508 bicycles/h/m). This can be concluded from the values of the standardized coefficients of k_j and ω (see Table 2, Model 2b). Let us analyze an imaginary case to illustrate the hypothetical influence of these two quantities on the discharge flow according to Model 2b. Assuming a 2-m wide cycle path and a traffic-light cycle with a 20-s long effective green time, if a queue has the average k_j and ω observed in our measurements, the maximum number of cyclists that can access the intersection during the green phase is $2\text{m} \cdot 20\text{s} \cdot (0.959\text{m/s} \cdot 0.45\text{cyc/m}^2 + 0.041\text{cyc/m}^2 \cdot 3.98\text{m/s}) = 23.8$ cyclists. If there are more cyclists in the queue, these need to wait one more red-signal phase. Instead, if the jam density and wave speed increase both by one standard deviation, this number increases to $2\text{m} \cdot 20\text{s} \cdot (0.959\text{m/s} \cdot (0.45\text{cyc/m}^2 + 0.08\text{cyc/m}^2) + 0.041\text{cyc/m}^2 \cdot (3.98\text{m/s} + 1.25\text{m/s})) = 28.9$ cyclists (a 22% increase). It is important to remark that Model 2b is not a validated predictive model. Nevertheless, these simple calculations give an indication of the potential benefits that could be obtained by increasing the jam densities and wave speeds of queues at signalized intersections.

Conclusions and Outlook

A proper understanding of the bicycle queue discharge process is crucial for developing ways to reduce delay at signalized intersections. The main goal of this study was to determine the macroscopic characteristics of bicycle flow during the queue discharge process, and to unveil possible relationships between them. To this end, the study analyzed bicycle trajectories from a one-way cycle path leading to a signalized intersection in Amsterdam. Jam density, wave speed, and discharge flow were measured for 57 queue discharge periods. Linear relationships between these macroscopic flow variables were established using regression analysis.

The results show that jam density, wave speed, and discharge flow vary considerably across traffic-signal cycles, although they are within the range of values reported in the literature. Further research is needed to establish the causes of this variability; however, its existence is a key finding that highlights the stochastic nature of bicycle flow. An important implication is that traffic engineers should take stochasticity into account when predicting bicycle flow dynamics at intersections. Moreover, the results confirm that discharge flow is positively linearly correlated with both jam density and wave speed, at least within the range of observed values. Also, discharge flow is positively linearly correlated to the difference between jam density and critical density, as expected from first-order traffic flow theory. However, it was found that this theory is not able to describe bicycle flow dynamics at intersections very accurately.

The data set analyzed in this paper is of limited size (57 discharge periods at only one intersection). An analysis of additional trajectory data, preferably from other sites (including intersections with two-directional cycle paths), is necessary to obtain more conclusive results. More specifically, data from other sites are crucial to identify the factors contributing to the variability of jam density, wave speed, and discharge flow at signalized intersections, as well as to establish more accurately the relationships between these variables. In this regard, it is necessary to investigate whether these relationships may be non-linear and also whether additional variables may need to be considered.

Although regression analysis cannot establish causation between variables (only correlation), it is hypothesized that discharge flow is actually influenced by jam density and wave speed, as generally assumed by traffic flow theory. Further experimental research may validate this hypothesis; nevertheless, two relevant implications follow from assuming it is true. The first one is that variability in jam density and wave speed in real bicycle queues may partially explain the differences in discharge flow reported in the literature. The second implication is that there is often room to fit additional cyclists in queues, and cyclists are able to start moving earlier, while increasing the discharge flow. This points to two potentially effective strategies to increase outflow at signalized intersections, and thus to reduce the delay experienced by cyclists (especially if queues are too long to dissolve completely during one green-signal phase): (a) increasing jam densities; and (b) increasing wave speeds.

However, more research is needed to determine whether it is possible to implement these strategies through specific traffic-engineering measures. To increase jam densities, one possibility could be to guide cyclists to certain locations during the queue-formation process by indicating ideal paths and/or stopping locations within the waiting area. A further possibility could involve installing static or dynamic information panels asking cyclists to stand close to each other. These panels could also be used to encourage cyclists to start moving early when the traffic light turns green, thus increasing wave speeds. In addition, it may be possible to increase wave speeds by installing traffic lights displaying the time left before the beginning of the green-signal phase; this may induce cyclists to get ready and start moving earlier. The potential effectiveness of these measures is not clear at this point and should be evaluated through field tests.

Acknowledgments

This research was supported by the ALLEGRO project (Unravelling slow mode travelling and traffic: with innovative data to create a new transportation and traffic theory for pedestrians and bicycles), which is funded by the European Research Council (Grant Agreement No. 669792), and the Amsterdam Institute for Advanced Metropolitan Solutions.

Author Contributions

The authors confirm contribution to the paper as follows: study conception and design: all authors; data collection: B. Goñi-Ros, W. Daamen; analysis and interpretation of results: B. Goñi-Ros, Y. Yuan, W. Daamen; draft manuscript preparation: B. Goñi-Ros, Y. Yuan; study supervision: S. P. Hoogendoorn. All authors reviewed the results and approved the final version of the manuscript.

References

1. International Transport Forum. *Cycling, Health and Safety*. International Transport Forum, Organisation for Economic Co-operation and Development, 2013.
2. Heinen, E., B. van Wee, and K. Maat, Commuting by Bicycle: An Overview of the Literature. *Transportation Reviews*, Vol. 30, No. 1, pp. 59–96, 2010.
3. Pucher, J., and R. Buehler. Making Cycling Irresistible: Lessons from The Netherlands, Denmark and Germany. *Transportation Reviews*, Vol. 28, No. 4, pp. 495–528, 2008.
4. Homburger, W. S. *Capacity of Bus Routes, and of Pedestrian and Bicycle Facilities*. Institute of Transportation Studies, University of California, Berkeley, 1976.
5. Navin, F. P. D. Bicycle Traffic Flow Characteristics: Experimental Results and Comparisons. *ITE Journal*, Vol. 64, No. 3, 1994, pp. 31–36.
6. Yang, J. M. Bicycle Traffic in China. *Transportation Quarterly*, Vol. 39, No. 1, 1985, 93–107.
7. Zamanov, M. *Estimation of Fundamental Diagram of Bicycle Traffic Flow Using Observations on Bicycle Lanes in Delft*, Master thesis. Delft University of Technology, 2012.
8. Deng, X., and J. Xu. Study on Bicycle Flow Characteristics at Intersections. In *Civil Engineering and Urban Planning III* (Mohammadian, K., K. G. Goulias, E. Cicek, J.-J. Wang, and C. Maraveas, eds.), Taylor & Francis Group, London, U.K., 2014.
9. *Highway Capacity Manual*. TRB, National Research Council, Washington, D.C., 2000.
10. Li, Z., M. Ye, Z. Li, and M. Du. Some Operational Features in Bicycle Traffic Flow: Observational Study. *Transportation Research Record: Journal of the Transportation Research Board*, 2015. 2520: 18–24.
11. Raksuntorn, W., and S. I. Khan. Saturation Flow Rate, Start-Up Lost Time, and Capacity for Bicycles at Signalized Intersections. *Transportation Research Record: Journal of the Transportation Research Board*, 2003. 1852: 105–113.
12. Seriani, S., R. Fernandez, and E. Hermosilla. Experimental Study for Estimating Capacity of Cycle Lanes. *Transport Research Procedia*, No. 8, 2015, pp. 192–203.
13. CROW. *Sign-Up for the Bike: Design Manual for a Cycle-Friendly Infrastructure*, CROW Record, Vol. 10. Centre for Research and Contract Standardization in Civil Engineering, The Hague, The Netherlands, 1993.
14. Allen, D. P., N. Rouphail, J. E. Hummer, and J. S. Milazzo, II. Operational Analysis of Uninterrupted Bicycle Facilities. *Transportation Research Record: Journal of the Transportation Research Board*, 1998. 1636: 29–36.
15. Cao, J. X., R. Liu, and B. Li. *Research on the Bicycle Flow in Signalizeds Intersections with Video-based Detection Technologies*. International Conference on Transportation Information and Safety (ICTIS), 2011.
16. Figliozzi, M., N. Wheeler, and C. M. Monsere. Methodology for Estimating Bicyclist Acceleration and Speed Distributions at Intersections. *Transportation Research Record: Journal of the Transportation Research Board*, 2013. 2387: 66–75.
17. Zhou, D., C. Xu, D. Wang, and S. Jin. Estimating Capacity of Bicycle Path on Urban Roads in Hangzhou, China. Presented at 94th Annual Meeting of the Transportation Research Board, Washington, D.C., 2015.
18. Chen, X., K. An, and L. Li. Study on the Influencing Factors of Bike Lane Capacity. Presented at 93rd Annual Meeting of the Transportation Research Board, Washington, D.C., 2014.
19. Jin, S., X. Qu, D. Zhou, C. Xu, D. Ma, and D. Wang, Estimating Cycleway Capacity and Bicycle Equivalent Unit for Electric Bicycles. *Transportation Research Part A: Policy and Practice*, Vol. 77, 2015, pp. 225–248.
20. Daganzo, C. F. The Cell Transmission Model: A Dynamic Representation of Highway Traffic Consistent with the Hydrodynamic Theory. *Transportation Research Part B: Methodological*, Vol. 28, No. 4, 1994, pp. 269–287.
21. Duives, D. C. *Analysis and Modelling of Pedestrian Movement Dynamic at Large-scale Events*. T2016/15, TRAIL thesis series. 2016.

22. Knoppers, P., J. W. C. van Lint, and S. P. Hoogendoorn. Automatic Stabilization of Aerial Traffic Images. Presented at 91st Annual Meeting of the Transportation Research Board, Washington, D.C., 2012.
23. Edie, L. Discussion on Traffic Stream Measurements and Definitions. *Proc., 2nd International Symposium on the Theory of Traffic Flow*, OECD, Paris, France, 1963, pp. 139–154.

The Standing Committee on Bicycle Transportation (ANF20) peer-reviewed this paper (18-04103).

TRR
JOURNAL OF THE TRANSPORTATION RESEARCH BOARD

Article

Transportation Research Record
2018, Vol. 2672(36) 63–72

DOI: 10.1177/0361198118798445
journals.sagepub.com/home/trr
SAGE

The Keys to Connectivity: The District of Columbia's Innovative Approach to Unlocking Low-Stress Bicycle Networks

Conor Semler[1], Meredyth Sanders[2], Darren Buck[3], Stephanie Dock[3], Burak Cesme[2], and Shi Wang[4]

Abstract

Washington, D.C. (the District) has led the way in the adoption of state-of-the-art bicycle facilities and the innovative application of bicycle level of traffic stress (LTS) mapping for the District. Although the District's LTS network map provides insight into bicycle network accessibility at the street level, the District needs a data-driven mechanism for identifying and prioritizing future investments to improve cycling at the system level. As part of its ongoing District Mobility Project, District Department of Transportation (DDOT) applied a geographic information system (GIS)-based network analysis to its bicycle LTS map to quantify network connectivity on low-stress District streets. This analysis was used to develop a bicycle network accessibility rating for District neighborhoods. The results revealed that though DDOT has invested in a range of innovative projects to improve accessibility in the District, only 27% of the overall network has been connected. By taking the analysis a step further and identifying the bottom 10% of accessible census blocks, DDOT both confirmed existing perceptions about the presence of disconnected low-stress "islands" in the District, and identified key gaps in the overall bicycle network. Using this information, DDOT is able to prioritize future bicycle infrastructure investments, while possessing a data-driven mechanism for assessing and communicating the accessibility benefits offered by individual investments. In addition, the methodology for developing the network accessibility rating provides a proof-of-concept for other jurisdictions looking to maximize the utility of their own LTS network maps.

The District Department of Transportation (DDOT), serving Washington, D.C. (the District), is nationally known as an innovator in the planning, development, and implementation of bicycle infrastructure. Across the District, DDOT has installed 75 mi of marked bicycle lanes, 7 mi of cycle tracks, deployed more than 3,000 bicycle racks, launched one of the first public bike sharing programs in the United States, and achieved peak-hour cycling volumes nearly quadruple those measured in 2004 (*1*). To prioritize future needs and identify strategic investments, DDOT has also completed a level of traffic stress (LTS) analysis on all District roads using an innovative geographic information system (GIS)-based approach (*1*).

Although DDOT's LTS network map provides a vital jumping-off point for identifying strategic missing gaps in the bicycle network, DDOT needs a data-driven tool for prioritizing system-wide needs and identifying future investments that will link disconnected bicycle networks in the District. To that end, DDOT leveraged work being done on the continuation of the District Mobility Project to develop a bicycle network accessibility rating for all District census blocks. The bicycle network accessibility rating uses the District's LTS map and a GIS-based network analysis to identify strategic opportunities to fill missing gaps in the District's bicycle network.

This paper provides an overview of DDOT's approach to developing a bicycle LTS map for the District, and explains why DDOT made the decision to take its LTS map a step further and conduct a network accessibility analysis. It offers precedent network accessibility studies conducted by researchers across the United States, and details DDOT's approach to conducting a network accessibility analysis for the District; presents two approaches to network accessibility: census blocks and jobs, and the

[1]Kittelson & Associates, Inc., Boston, MA
[2]Kittelson & Associates, Inc., Washington, DC
[3]District Department of Transportation, Washington, DC
[4]KPMG US, Seattle, WA

Corresponding Author:
Address correspondence to Conor Semler: csemler@kittelson.com

pros and cons of both; it explains how the chosen approach was used to identify the worst performing networks for project prioritization; and outlines lessons learned from initial analysis and opportunities for future refinement as DDOT continues to refine its bicycle project prioritization processes.

Background and Literature Review

In 2016, DDOT initiated the District Mobility Project in response to a request from the Council of the District of Columbia to assess the current state of multimodal congestion in the District. In addition to measuring multimodal mobility, the Council Action called for recommendations for remedying existing mobility issues (*2*). The District Mobility Project expanded the traditional definition of "congestion" to assess three areas related to system performance: intensity of use (congestion), reliability, and accessibility (*3*). As bicycle congestion remains uncommon even on the District's busiest bicycle corridors, and bicycle transportation is inherently reliable, DDOT honed in on defining and measuring bicycle accessibility in the District. The resulting bicycle LTS map helped DDOT to better understand existing opportunities and barriers to bicycle access.

Since the completion of the District Mobility Project report, DDOT has launched an interactive website (http://districtmobility.org) that provides annual data regarding the state of multimodal mobility in the District. In addition to collecting and visualizing new data for individual performance measures, including the District's LTS map, the District Mobility Project includes a biennial assessment of all performance measures to identify "focus areas" with a confluence of multimodal issues. The focus areas serve as foundations for DDOT to prioritize planning studies and projects to improve multimodal mobility in the District. Though the bicycle LTS map offers insights into accessibility along specific roadways, DDOT needs a bicycle accessibility performance measure that offers insights on network accessibility and connectivity and informs the focus area identification process.

The following section provides an overview of the District's 2016 approach to LTS mapping, outlines DDOT's justification for developing a holistic performance measure for bicycle accessibility, and summarizes network connectivity measures pioneered by DDOT's transportation peers.

DDOT's Unique Approach to Level of Traffic Stress Mapping

DDOT's decision to apply the LTS method (*4*) as an indicator of system accessibility built on the established theory that the appeal of bicycling as an uncrowded, predictable mode of transportation declines rapidly if District residents and visitors cannot access a comfortable route for their trips. As the District Mobility Project entailed a District-wide analysis of all residents and jobs, on all District streets, the project team also needed an accessibility measure that could be applied reliably and efficiently. Unlike the bicycle level of service methodologies applied in prior DDOT planning efforts, the LTS method would provide a network-level analysis with relatively little data (*5*, *6*).

The LTS method evaluates and classifies streets into one of four "stress levels" for bicycling based on characteristics of the roadway environment. These stress levels are built on the "four types of cyclists" pioneered by the City of Portland (*7*) and aligned with the theorized comfort level of different types of bicyclists:

- LTS 1: a level of traffic stress that most children can tolerate
- LTS 2: the level that would be tolerated by the mainstream adult population
- LTS 3: the level tolerated by American cyclists who are "enthused and confident" but prefer dedicated space for riding
- LTS 4: a level tolerated only by those characterized as "strong and fearless."

The LTS method acknowledges that cyclists are uniquely sensitive to their perceptions of traffic conditions, and are likely to limit their travel to streets that fall at or below their personal stress tolerance level. Given this sensitivity to roadway conditions, communities that provide more low-stress connections have the potential to improve bicycle accessibility and attract more riders (*8*).

The District LTS mapping process built on lessons learned through several precedent studies conducted in Montgomery County, Maryland, Oakland, California, and Baltimore, Maryland (*9–11*). Using similar data collection and analysis methods tested in the precedent studies, DDOT capitalized on the "weakest link" logic of the LTS method, wherein the stress level is assigned based on the lowest-performing attribute. For example, even if a segment has mostly low-stress characteristics, the occurrence of one high-stress attribute (e.g., a narrow bicycle lane) dictates the stress level for the link (*1*). From an analysis perspective, this means it is not always necessary to collect all of the data on each street. Based on a subset of information, streets can be efficiently scored according to LTS analysis categories.

After building on the groundwork provided by earlier studies, DDOT took the opportunity to refine the method even further. In addition to automatically

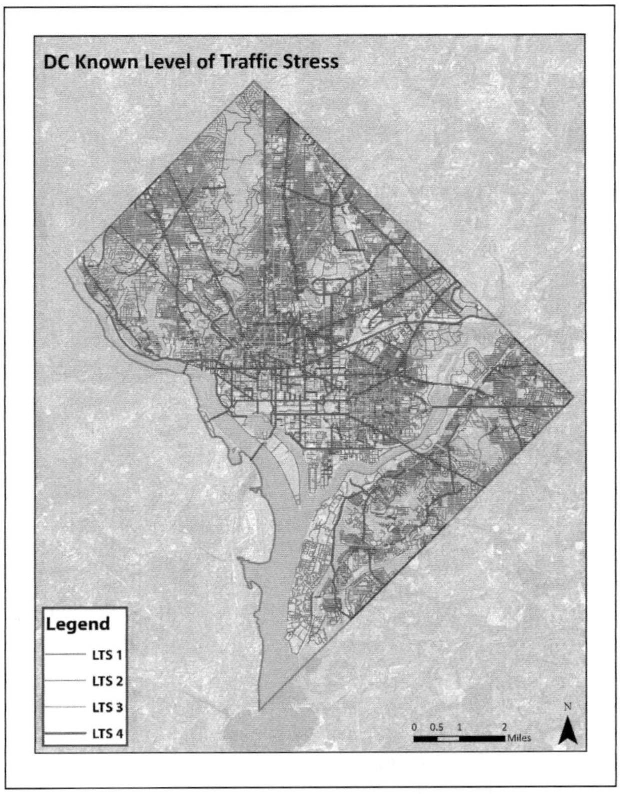

Figure 1. Preliminary results of District LTS analysis.

classifying the "easy to identify" LTS 1 and LTS 4 streets, the remaining streets could be organized in relation to what data are needed. DDOT applied data filtration, data collection, and analyses processes to develop a complete bicycle LTS map for the District in just under a month (*1*). By strategically filtering out streets that could be classified with existing information, the project team drastically condensed the most time-consuming process in the LTS mapping effort: data collection. The data collection and analysis processes were completed in three short weeks due to the advanced digitizing capabilities of DDOT's linear referencing system. The complete LTS network map, shown in Figure 1, provided the District with a data-driven means of identifying strategic missing gaps in the bicycle network. The linkage to DDOT's linear referencing system also provides a straightforward means to maintain the LTS classifications as new facilities are added, helping to keep the data current over time.

Defining and Measuring Bicycle Network Accessibility

The LTS map created for the 2016 Phase of District Mobility Project represented a significant step for DDOT in terms of identifying opportunities and barriers in the District's bicycle network. However, a uniform bicycle performance measure was still needed to inform District Mobility's focus areas identification process. The first focus area identification process for the District Mobility study used high-stress roadways (LTS 3 and LTS 4 streets) as a proxy for accessibility deficiencies. Though the District's LTS network map provided insight into network accessibility at the street level, the District needed a data-driven mechanism for identifying and prioritizing future investments at the system level. Therefore, the District Mobility team worked with members of DDOT's bicycle program to review precedent studies and identify an effective bicycle performance measure that would be applicable to the District. The team ultimately determined that a network connectivity analysis would form the foundation for a uniform, data-driven depiction of bicycle accessibility.

Precedent Studies

The concept of using a bicycle LTS network to conduct a network connectivity analysis was first introduced in the Mineta Transportation Institute's seminal report, "Low Stress Bicycling and Network Connectivity" (*4*). The report introduces the bicycle level of traffic stress concept and describes the San Jose case study in which the city's streets were classified by LTS in an attempt to understand network connectivity. Mekuria, Furth, and Nixon drove home the point that a bicycle network is only as accessible as its low-stress network of LTS 1 and LTS 2 streets (*4*). Visualizing an LTS map without its high-stress streets exposes a network divided into "islands of low-stress connectivity" where low-stress cycling is possible within each island, but travel between islands can only be achieved using high-stress streets.

Using a network connectivity analysis to identify and improve strategic links between islands of connectivity has the potential to unlock previously inaccessible areas. The Mineta report details two potential measures of network connectivity: "percent trips connected" and "percent nodes connected" (*4*). Both network connectivity measures use GIS network analysis tools to identify points in the network that are connected without undue detour. Undue detour is defined as a low-stress trip whose path length is 25% longer than the most direct path on the full street network.

The "percent trips connected" measure uses a regional trip table to determine the fraction of bicycle trips that can be made on an existing low-stress network without undue detour. This measure has substantive analysis requirements, in which a regional trip table must be disaggregated from the traffic analysis zone level to census blocks, origin trips are allocated in proportion to population, and destination trips are allocated based on land use data. The "percent nodes connected" measure offers a simpler approach to measuring network connectivity, and measures the fraction of nodes in the street network that are connected to each other. Applied to San Jose's LTS network, these network connectivity measures determined the fraction of connected census blocks on the existing low-stress network and tested the effect of a series of hypothetical roadway improvements on citywide connectivity.

The network connectivity analyses pioneered by Mekuria, Furth, and Nixon have since been tested and modified by researchers and practitioners alike. In 2016, Lowry, Furth, and Hadden-Loh used an LTS network map and GIS tool to measure network access to important destinations and rank proposed improvement scenarios based on the ability of projects to improve low-stress connectivity between homes and destinations (*12*). In 2017, the nonprofit organization PeopleForBikes released bicycle network accessibility scores for cities across the United States. PeopleForBike's bicycle network accessibility approach uses U.S. census and OpenStreetMap data to calculate network accessibility scores for census blocks based on the number of nearby census blocks that are accessible on the low-stress cycling network and the types of destinations that are located within each accessible census block (*13*).

These precedent network connectivity analyses all build on key concepts established in the original Mineta report, including the comparison of possible low-stress connections to direct connections using the full network, the use of census blocks or more fine-grained geographies as the core analysis unit, and the inclusion of logical trip length limits and trip detour criterion to capture trips that are reasonable and feasible to most cyclists. However, a key difference in each precedent study lies in the chosen accessibility measure, which ranges from access to census blocks to access to a wide variety of destination types that are ranked based on their assumed desirability.

Measuring Network Accessibility in the District

The researchers defined a meaningful and robust accessibility performance measure to quantify accessibility by bicycle in the District, which could be compared with the need for other urban mobility measures that were assessed as part of the District Mobility Project. The selected performance measure needs to be updateable annually by DDOT staff to inform the ongoing District Mobility Project system performance analysis and focus area identification processes. It also must provide DDOT's bicycle program with a data-driven tool to communicate existing conditions and trends in bicycle accessibility to DDOT staff and the public. To incorporate these requirements, the project team determined that a performance measure based on the number of connected census blocks would offer the most time- and cost-effective approach to measuring network connectivity in the District while offering insights on the bike network accessibility and connectivity in the District.

Analysis Assumptions and Data Requirements

After identifying "number of census blocks connected" as the baseline performance measure for measuring bicycle network accessibility in the District, the District Mobility team and members of DDOT's bicycle program established a set of key assumptions about the analysis. The group defined "number of census blocks connected" as the sum of connected census blocks on LTS 1 and LTS 2 streets. The team selected LTS 2 streets as the upper limit for its accessible network because LTS 2 streets are acceptable to most of the adult population. However, the team also agreed to measure accessibility on LTS 1 streets to assess existing islands of connectivity for the District's most vulnerable cyclists (e.g., most children). The group also measured connectivity on the District's full LTS network (LTS 1 through LTS 4 streets) to compare accessibility on the low-stress network with unhindered cyclist travel and provide a baseline for the results.

In addition to defining the low-stress network, it is also assumed that a 30% detour penalty would be incorporated into the analysis. The detour penalty acknowledges that although a bicycle trip may be feasible on the low-stress network, the trip may entail an unacceptable number of deviations compared with the most direct path. Research indicates that although bicyclists will deviate from a direct route for a perceived better route, few bicyclists will take a detour that is 30% longer than the most direct route (*14*). Consequently, a trip that would take 30% longer on LTS 1 and 2 streets compared with a direct trip on the full network would not be considered accessible to most cyclists.

A key benefit of the selected performance measure is that it requires less data compared with other network accessibility measures. Though past network accessibility measures have relied on origin–destination data from regional trip tables and census data, the "number of census blocks connected" measure simply requires LTS data for the District's street and trail networks, and publically available census block shapefiles from the U.S. Census Bureau.

Accessibility Analysis

The selected network connectivity measure, "number of census blocks connected," used census blocks as origin–destination pairs and a merged shapefile of the District's streets and off-road trails as the routable network. The analysis used ArcGIS Network Analyst to calculate the shortest path from each District census block to all other census blocks located within a set network path distance. Although DDOT plans to extract a routable network from its built out linear referencing system (LRS), the roadway distribution data for the LRS was not complete when the researchers conducted the first iteration of the bicycle network accessibility analysis. The researchers built a simple routable network for the preliminary analysis, with the understanding that future iterations of the District's bicycle network accessibility analysis would be able to capitalize on the dynamic editing capabilities of the District's LRS.

The routable network used in the bicycle network accessibility analysis was comprised of static shapefiles of the District's census blocks, street network and off-road trails. The street network included the LTS classifications assigned to the District's roads during the 2016 LTS mapping process, with key updates made to streets where DDOT had since implemented new bike lanes. Since the street and off-road trails shapefiles were developed independently, their vertices did not perfectly align. To create a connected network, the street and trails shapefiles were merged and cleaned using ET GeoWizards (*15*). A 30-ft tolerance distance was applied to merge features from the two shapefiles.

Using the routable network, the shortest path was calculated from each census block centroid to all other census block centroids located within a 3.3-mi network path distance using the full bicycle network (LTS 1 to LTS 4 streets and off-road trails). When routing trips between census block centroids, the network analyst tool snaps to the vertices on the origin census block that connect to the most efficient routes between other accessible census blocks. The 3.3-mi distance limit was selected by the project team because it encompasses a 20-min bike ride at a comfortable pace of 10 mph, which cyclists of most ages and abilities can complete with little difficulty.

After calculating the shortest path distance between District census blocks using the full bicycle network, high-stress streets were eliminated from the network so that the same analysis could be conducted using the LTS 1 and LTS 2 network. The trip lengths between each census block pair that could be reached on both networks (LTS 1 to 2 and the full network) were compared to determine whether the LTS 1 to 2 block pair was connected without any undue detour. If the trip length between a census block pair on the LTS 1 to 2 network was 30% longer than the trip length between the same census block pair on the full bicycle network, then those block pairs were considered inaccessible using the LTS 1 and LTS 2 network. Figure 2 shows an example route between points on the District's low-stress bicycle network

After identifying accessible census block pairs, the sum of accessible census blocks on the LTS 1 and LTS 2 network was calculated for each census block in the District. The sum of connected census block pairs on the full network was also calculated to provide an additional indicator of network connectivity. Using the existing low-stress network, approximately 3,257,937 unique trips can be made between census blocks in the District. Using the full network, approximately 11,921,566 unique trips can be made between District census blocks. Although DDOT has invested in a range of innovative projects to improve bicycle accessibility in the District, only 27% of the network has been connected. The network accessibility analysis began to identify opportunities to improve accessibility by highlighting key gaps.

The same GIS analysis was conducted to assess accessibility on the District's LTS 1 network. This analysis allowed DDOT to hone in on the role that LTS 1 streets play in connecting the District's bicycle network. Figure 3 illustrates the results of the network accessibility analyses, comparing the number of census blocks connected on the LTS 1 network, the number of blocks connected on the LTS 1 to 2 network, and the number of blocks connected on the full network.

Assessing Bicycle Accessibility to Jobs

Following the preliminary analysis of the number of connected census blocks, the project team conducted an additional assessment of bicycle accessibility to jobs in the District. Though the assessment of census block connectivity provides a clear picture of network connectivity, it does not necessarily reflect cyclists' ability to reach opportunities using the low-stress network. To address this issue and gain additional insights, accessibility to jobs is considered as another metric to assess bicycle connectivity in the District.

The assessment of District jobs connectivity used the same GIS network and processes to identify connected census blocks, and incorporated jobs data from the District's Longitudinal Employer-Household Dynamics (LEHD) Regional Employment Realignment dataset, a blended version of the Census LEHD employment data at the census block level and more accurate regional employment data from the Metropolitan Washington Council of Governments (MWCOG) at the traffic analysis zone level that incorporates some federal employment missing in the LEHD data. The LEHD dataset was used to identify the total number of jobs located within each District census block. The network analyst tool was then used to identify the total number of jobs that were accessible from each census block, excluding the jobs

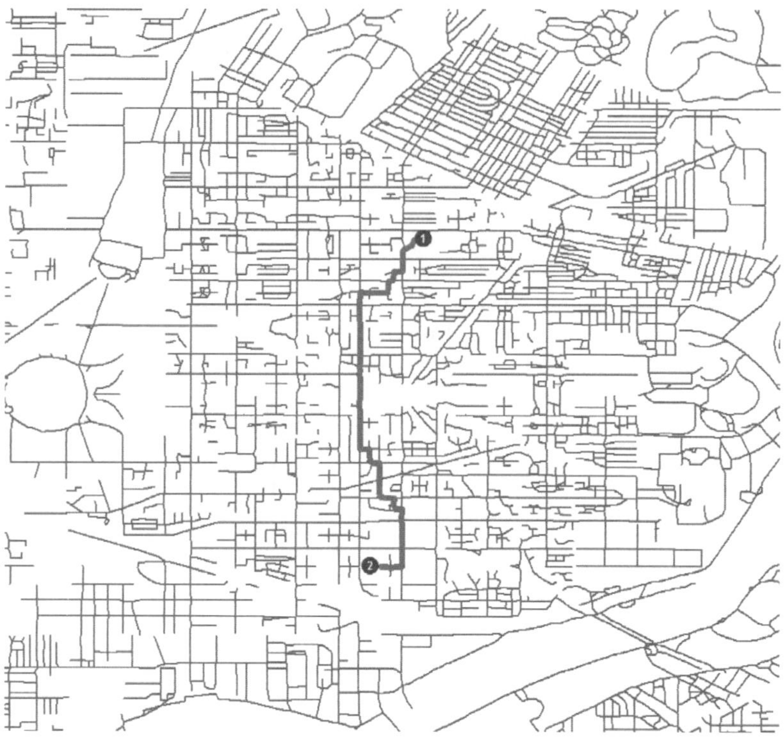

Figure 2. Illustrative trip route on the District's low-stress bicycle network.

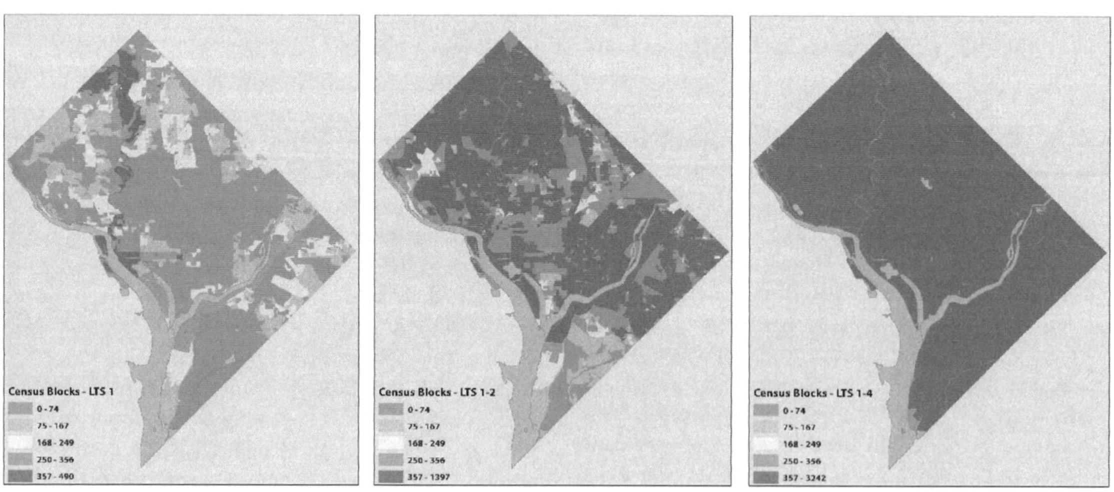

Figure 3. Comparison of bicycle accessibility on the low-stress networks (LTS 1 and LTS 1 to 2) and full network (LTS 1 to 4).

associated within each origin block (assuming those trips would most likely be taken on foot). The analysis was conducted on the LTS 1 network, LTS 1 to 2 network and the full network. Although census block connectivity can help inform internal DDOT planning processes, the District jobs connectivity measure provides a more tangible description of low-stress accessibility for members of the general public. As the District Mobility Project serves

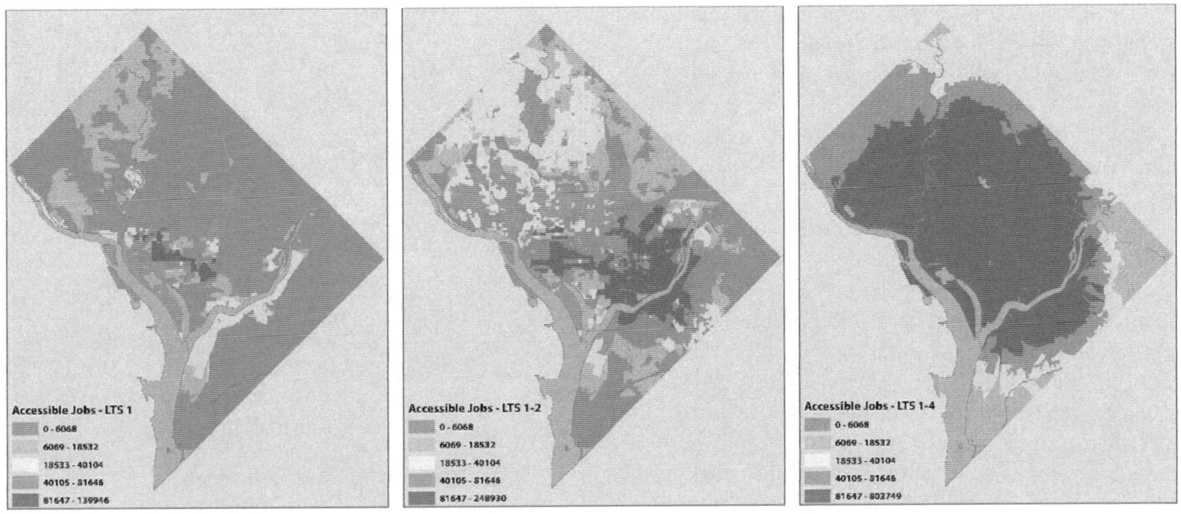

Figure 4. Comparison of bicycle access to District jobs on the low-stress networks (LTS 1 and LTS 1 to 2) and full network (LTS 1 to 4).

the dual purpose of informing the public on the state of multimodal mobility and informing DDOT efforts to implement performance-based improvements to the network, both accessibility measures have merit. Figure 4 shows the results of the analysis, comparing the number of accessible jobs on all three networks.

Although the accessibility to jobs analysis provided key connections between bicycle infrastructure and employment opportunities in the District, the analysis was inherently limited by the exclusion of job centers located just outside of the District border, that is, in Maryland and Virginia. Therefore, the results provided above underestimate the number of jobs accessible from some of the census blocks that are within close proximity to job centers located just outside of the District (e.g., Bethesda, Maryland). Several of the District's surrounding jurisdictions, including Arlington County and Montgomery County, have developed their own LTS networks. However, the work required to develop LTS networks for unmapped jurisdictions and merge the separate networks into a regional system was beyond the scope of the District Mobility analysis, and thus not included in this analysis. Figure 5 highlights jobs located within 3.3 mi of the District, which were not incorporated into District accessibility analyses. Future efforts will focus on a combined effort of neighboring counties and cities to develop a consistent, regional LTS network.

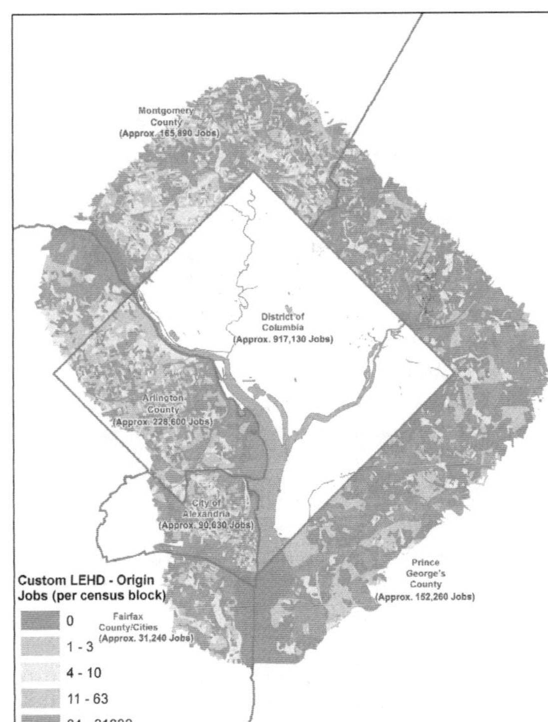

Figure 5. Illustration of jobs within and beyond the District's boundaries.

Establishing an Accessibility Threshold

Once the number of census blocks and job accessibility measures were analyzed, the authors assessed the benefits and drawbacks of each measure in order to develop an accessibility threshold for use in identifying focus areas for prioritizing investments and decision making. The thresholds highlight the worst performing corridors or networks in the District, so the bicycle accessibility threshold needed to provide a meaningful depiction of key gaps in the low-stress network.

Both measures provided a cost- and time-effective measure of network connectivity, with the processes required to develop the routable network and run the analysis taking approximately 8 h. The accessibility to jobs measure painted a more tangible picture for people seeking to understand the relationship between the District's bicycle network and employment opportunities. However, the lack of connected LTS data for the District's surrounding jurisdictions limited the efficacy of the accessibility to jobs measure. The census block connectivity measure provided a simpler, network-descriptive analysis that presented a straightforward depiction of accessibility in the District, and could provide a more meaningful measure of accessibility for non-commute trips. Given the caveats of the accessibility to jobs measure, the authors chose to develop the accessibility threshold from the census block connectivity analysis. The bicycle accessibility threshold identified the bottom 10% of accessible census blocks using the census block connectivity analysis of the low-stress network (number of connected census blocks on LTS 1 and LTS 2 streets). Census blocks that fall within the bottom 10% are the least connected on the low-stress bicycle network. Cyclists seeking to make trips to and from these blocks will have few comfortable options and may opt for another mode of transportation. Once the researchers had identified and mapped all census blocks within the bottom 10% of connected census blocks, they used the GIS dissolve function to merge groups of connected, inaccessible census blocks. Some of the inaccessible census blocks were isolated within a network of accessible census blocks, and did not pose a substantial connectivity challenge. However, many of the inaccessible census blocks were located in large groups that delineated clear barriers to bicycle accessibility in Downtown and Southeast D.C., among other locations. In order to eliminate outliers and focus on the key gaps in the District's bicycle network, the researchers excluded any census block or merged group of census blocks that had a total area below 0.05 mi^2. The resulting map of accessibility "hot spots" is shown in Figure 6. These areas with acute bicycle accessibility concerns will be included on the District Mobility website and incorporated into the revised focus area map to assist DDOT staff with the decision-making process.

The first phase of the District Mobility Project produced the District's LTS network map with the understanding that the District's bicycle accessibility measure would evolve over time, as would the accessibility data that is displayed on the District Mobility website. Though the census block connectivity measure and the associated 10% mobility threshold moves DDOT closer to developing a robust, sustainable performance measure for bicycle accessibility, the measure will undoubtedly experience ongoing refinements based on public feedback, the potential regional LTS network map, institutional goals, and the full buildout of DDOT's LRS capabilities.

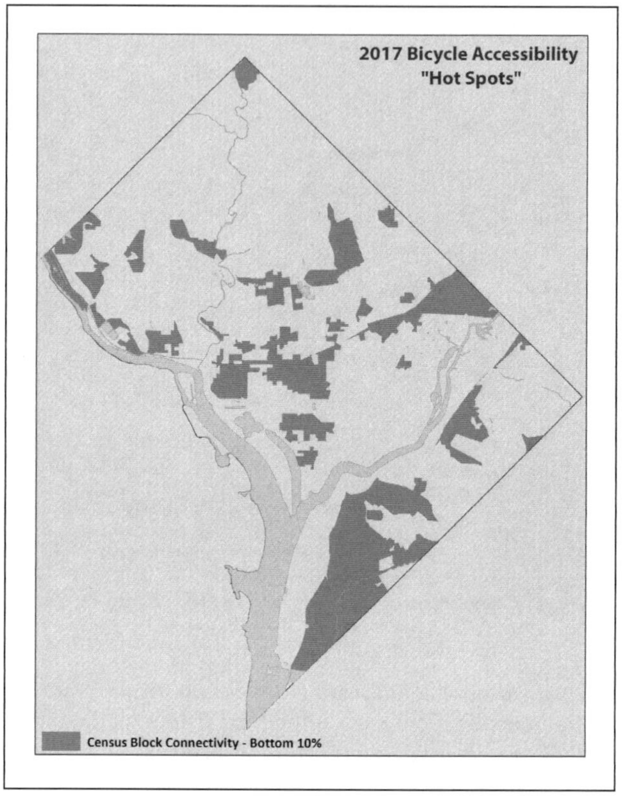

Figure 6. Key accessibility gaps in the District's low-stress network.

Conclusion

By capitalizing on the District's LTS network map and applying a network connectivity analysis to the low-stress network, the project team met a key goal of DDOT's District Mobility study: to develop a sustainable performance measure for bicycle accessibility in the District. The census block connectivity measure provided DDOT with a reliable, data-driven means of using the LTS network to identify islands of connectivity and network gaps in the District. The accessibility measure will support DDOT's efforts to prioritize future needs and invest strategically in projects and plans that will fill key gaps in the network. Just as DDOT will be able to easily update its LTS network map, DDOT will also be able to re-run the census block connectivity analysis efficiently and at low cost as the LTS network evolves, ensuring the data and performance measure will remain current in the long term.

Lessons Learned

The efforts of the project team confirmed that with a robust LTS network map, a network connectivity analysis can be applied to large urban areas in a condensed period of time. Leveraging the District's LTS network map, the process to build the routable network and run the connectivity analysis took about a day to complete. By selecting a connectivity measure that provided a network-descriptive analysis without incorporating the information and biases associated with origin–destination data, the project team reduced the time and cost associated with the analysis while achieving an informative result. The census block connectivity measure can continue to be refined as DDOT builds out its LRS.

The researchers' assessment of two potential connectivity measures, census block connectivity and District jobs connectivity, highlighted the value of institutional feedback and regional cooperation to bicycle accessibility analyses. Although both connectivity measures are limited by the lack of a cohesive LTS network map that expands beyond the District's boundaries, the District jobs connectivity measure is particularly affected by the absence of connected jobs data from major employment centers located just outside of the District. DDOT provided the project team with critical input on the benefits and limitations of both potential measures, and helped to select the measure that would provide the greatest value to DDOT planners and project managers in the near-term. Both connectivity measures may ultimately be improved in the long term if a regional LTS network map is produced.

Opportunities for Future Research

As with prior efforts to develop an LTS network map, the project team did not fully incorporate the traffic stress criteria for intersection approaches into the network connectivity analysis. By querying out high-stress streets during the analysis process, the team was still able to acquire an accurate approximation of the shortest distance paths on the LTS 1 and LTS 2 network. However, this does not take into account intersection treatments that have been incorporated into certain District intersections, and represents an opportunity for future refinement of the District's LTS accessibility analyses. Opportunities to incorporate crossing penalties into the routable network at unsignalized crossings of high-stress streets should be investigated and implemented as part of future research efforts.

DDOT's linear reference system offers additional untapped opportunities to further streamline the LTS accessibility analysis. The full buildout of the LRS will enable DDOT to extract a routable network that can be dynamically updated as changes are made to the District's roadways and bicycle facilities. Changes to the District's roadways should be reflected in the District's LTS map and accessibility analyses. Opportunities to automatically integrate dynamic changes to the District Roads and Highways feature class with DDOT's LTS network map and routable network should be evaluated during future analysis efforts.

Although the census block connectivity measure provides a valuable, network-descriptive depiction of bicycle accessibility in the District, the District jobs connectivity measure has the potential to provide a robust, tangible representation of bicycle accessibility to the public. MWCOG, DDOT and other regional Departments of Transportation should seek out opportunities to build a regional LTS network map. This map would allow DDOT and other jurisdictions to measure bicycle accessibility to jobs and communicate the benefits offered by proposed bicycle projects in relation to increased jobs accessibility both within and across jurisdictional borders.

With a robust, routable LTS network for the District, DDOT can also begin to test the connectivity benefits offered by proposed projects. Given the low cost and quick turnaround inherent in the census block connectivity measure, DDOT can conduct sensitivity analyses in which groups of planned or proposed bicycle projects are incorporated into the LTS network to calculate their effects on overall network connectivity. Such analyses can help DDOT to identify the most effective projects and to communicate the quantitative benefits offered by each project to stakeholders and the public at large.

Acknowledgments

The authors would like to acknowledge the contributions of Elizabeth Gordon and Alek Pochowski of Kittelson & Associates, Inc.

Author Contributions

The authors confirm contribution to the paper as follows: study conception and design: CS, MS, DB, SD, BC; data collection: MS, SW; analysis and interpretation of results: CS, MS, DB,

SD, BC; draft manuscript preparation: CS, MS, BC. All authors reviewed the results and approved the final version of the manuscript.

References

1. Semler, C., M. Sanders, D. Buck, J. Graham, A. Pochowski, and S. Dock. Low Stress LTS: The District of Columbia's Innovative Approach to Applying Level of Traffic Stress. Presented at 96th Annual Meeting of the Transportation Research Board, Washington, D.C., 2017.
2. *Congestion Management Study*. D.C. Code § 50-921.21. D.C. Council, 2016.
3. *District Mobility Project*. District Department of Transportation. Washington, D.C., 2017.
4. Mekuria, M. C., P. G. Furth, and H. Nixon. *Low-Stress Bicycling and Network Connectivity*. Mineta Transportation Institute, San Jose, 2012.
5. *District of Columbia Bicycle Master Plan*. District Department of Transportation, Washington, D.C., 2005.
6. *moveDC: The District of Columbia's Multimodal Long-Range Transportation Plan*. District Department of Transportation, Washington, D.C., 2014.
7. Geller, R. *Four Types of Cyclists*. City of Portland Bureau of Transportation. https://www.portlandoregon.gov/transportation/article/158497. Accessed July 8, 2017.
8. Dill, J., and N. McNeil. *Four Types of Cyclists? Testing a Typology to Better Understand Bicycling Behavior and Potential*. Portland State University, 2012.
9. Kittelson & Associates, Inc., and Toole Design Group. *Montgomery County Bicycle Planning Guidance*. Montgomery County Planning Department, Silver Spring, Md., 2014.
10. Patton, J. *A Bicyclist "Level of Traffic Stress" Analysis for Oakland's Streets*. City of Oakland Public Works Department, 2015.
11. Gordon, E., and C. Semler. *Building a Bike Network that Works: Level of Traffic Stress Methodology and Separated Facility Network Recommendations*. Baltimore City Department of Transportation. Baltimore, 2016.
12. Lowry, M. B., P. G. Furth, and T. Hadden-Loh. Prioritizing New Bicycle Facilities to Improve Low-Stress Network Connectivity. *Transportation Research Part A: Policy and Practice*. Vol. 86, 2016, pp. 124–140. http://dx.doi.org/10.1016/j.tra.2016.02.003.
13. Bicycle Network Analysis Methodology. *People for Bikes*. https://bna.peopleforbikes.org/#/methodology. Accessed July 8, 2017.
14. Winters, M., K. Tescheke, M. Grant, E. Setton, and M. Brauer. How Far Out of the Way Will We Travel? Built Environment Influences on Route Selection for Bicycle and Car Travel. *Transportation Research Record: Journal of the Transportation Research Record*, 2010. 2190: 1–10.
15. ET GeoWizards User Guide. ET GeoWizards. http://www.ian-ko.com/ET_GeoWizards/ETGW12UG_WEB/etgw/ETGeoWizards.htm. Accessed June 8, 2017.

The Standing Committee on Bicycle Transportation (ANF20) peer-reviewed this paper (18-05793).

Article

TRR
JOURNAL OF THE TRANSPORTATION RESEARCH BOARD

Transportation Research Record
2018, Vol. 2672(36) 73–82

DOI: 10.1177/0361198118776522
journals.sagepub.com/home/trr
SAGE

Park 'n' Roll: Identifying and Prioritizing Locations for New Bicycle Parking in Québec City, Canada

Marie-Pier Veillette[1], Emily Grisé[1], and Ahmed El-Geneidy[1]

Abstract

Promoting active modes of transportation, such as cycling, is an ongoing challenge faced by many cities around the world. Fostering a bicycle culture in an auto-dominant region is riddled with challenges, but success has been achieved with investments in bicycle infrastructure, including bicycle parking. This study presents a new methodology to identify the optimal locations to install short-term (bicycle racks) and long-term (bicycle lockers or indoor locking facilities) bicycle parking using a GIS-based approach that considers multiple criteria. Using Québec City, Canada, as a case study, our methodology considers multiple criteria related to the demand for bicycle parking, including the destinations of existing and potential cyclists and proximity to a frequent bus service. A prioritization index is developed to identify the optimal locations for long-term and short-term bicycle parking. This is followed by a recommendation of the number of bicycle parking spaces required to meet existing and potential demand. This paper aims to provide practitioners with an easy-to-use method to aid in the planning of new bicycle parking infrastructure, which is designed to be flexible and adaptable to other contexts.

A broad body of literature has explained the positive effects of the presence of bicycle infrastructure, such as bicycle lanes, parking, showers, and availability of bicycle-sharing systems on cycling levels in a region (*1–3*). In the Netherlands, Denmark, and Germany, measures related to bicycle parking supply and security, as well as interventions aimed at better integrating cycling with other modes, have promoted cycling as a safe and convenient mode of transport, and thereby increased bicycle usage (*2, 4*). Locating bicycle parking, however, has received little attention in the cycling literature. The aim of this study was to develop an easy-to-reproduce GIS-based method to identify and prioritize locations to add new short-term (bicycle racks) and long-term (bicycle lockers or indoor facilities) bicycle parking in Québec City, Canada.

This paper seeks to answer three main research questions: 1) how can a region prioritize the installation of new short-term bicycle parking in order to meet the needs of existing and potential cyclists; 2) how can the same region locate long-term bicycle parking; 3) what are the quantities of long-term and short-term bicycle parking spaces needed to accommodate existing and potential demand? We define short-term bicycle parking as any type of free-standing rack, similar to those presently provided by Québec City and many North American cities, and long-term bicycle parking as any bike storage, locker, or shed with restricted access and typically provided for a fee, similar to what is present at major public transport stations in European cities. Our study builds on a previous method developed by Larsen, Patterson, and El-Geneidy (*5*), who proposed the creation of a GIS-based prioritization index to identify high-priority grid cells to guide bicycle network improvements at a city-wide scale. The paper is divided in four sections. We will first provide an overview of the existing literature on bicycle parking, which will be followed by a brief description of our study area. The next section, the core of this paper, will be dedicated to describing our methodology and presenting our findings. Finally, we will discuss the implications and recommendations of our results.

Literature Review

The presence of end-of-trip facilities, such as bicycle parking and showers, has been identified as an important

[1]School of Urban Planning, McGill University, Montréal, Québec, Canada

Corresponding Author:
Address correspondence to Ahmed El-Geneidy: ahmed.elgeneidy@mcgill.ca

factor in increasing bicycle modal share for utilitarian purposes (*1*, *2*, *6–8*). However, the perceived risk of bicycle theft and vandalism can deter individuals from cycling (*9*, *10*). Van Lierop, Grimsrud, and El-Geneidy (*11*) studied the locations of stolen bicycles and found that 50% of reported stolen bicycles were locked in a fly-parking fashion. Fly-parking is a concept that refers to locking a bicycle to any type of street furniture that is not designated bicycle parking (*11*, *12*). Accordingly, providing cyclists with a sufficient supply of bicycle parking at destinations, especially in areas where the bicycle mode share is increasing, has the potential to reduce bicycle theft and subsequently encourage bicycle usage (*9*).

For optimal levels of security against bicycle theft, particularly for individuals parking their bike for long periods of time, demand and preference for bicycle parking that provides greater levels of security is rising. Long-term bicycle parking is typically referred to in the literature as a bicycle facility guarded by an individual or a facility that limits access through electronic keys (*12*). Bicycle lockers have also been added to the list of long-term bicycle parking as they are usually present near destinations, such as train stations (*1*). In Montréal, Canada, the Bicycle Theft Survey conducted in 2012 revealed that cyclists perceived bicycle lockers as more secure against theft compared with free-standing bicycle racks (*11*). The availability of long-term bicycle parking and showers at destinations was found to have a significant influence on bicycle usage (*13*). More specifically, Wardman, Tight, and Page (*7*) presented evidence that the availability of outdoor bicycle parking at work increased the share of trips to work by bicycle from 5.8% to 6.3%. However, the combination of both indoor bicycle parking, described as more secure than outdoor, and showers at work would increase bicycle mode share from 5.8% to 7.1% (*7*).

The presence of bicycle parking near the public transit network can lead to a better-integrated transportation system (*1*). In this respect, previous studies in the Netherlands have noted the key role of cycling as a means to access train stations, and thus highlighted the importance of providing bicycle spaces near stations (*10*, *14*). Van der Spek and Scheltema (*15*) reported, among other examples, the case of Zutphen Station, in the Netherlands, where a "guarded" bicycle parking facility was built in 2006 in proximity to the train station, and increased the number of train users who cycle to the station from 41% in 2004 to 58% in 2009. Similarly, the availability of short-term and long-term bicycle parking near train stations increased cyclists' satisfaction and bicycle usage for bike-and-ride purposes (*14*).

Krizek and Stonebraker (*16*) identified key challenges related to the integration of bicycle and transit use, and evaluated five potential ways of integrating bicycle with transit. For example, the Final Mile Initiative, implemented by Boulder County, is a program that supplies bicycle lockers and bicycle loans along their regional bus routes. The authors mentioned that the program aims to limit the number of bicycles aboard buses in order to improve the bus service by reducing the dwell time (*16*).

Bachand-Marleau, Larsen, and El-Geneidy (*17*) highlighted that bicycles on buses are the most common way to integrate these two modes together in North American cities. The authors found that 60% of respondents in their study preferred taking their bicycle aboard transit. This preference can potentially be explained by either the absence of secure bicycle parking at stations or by an actual need to bring a bicycle aboard a transit vehicle for access to their bicycle upon egress from the transit service. Despite the potential benefit of bringing bicycles aboard transit vehicles, longer dwell time associated with passengers bringing bicycles aboard buses has led to programs aimed at limiting the number of bicycles aboard buses (*16*). Although these two studies provide conflicting views on the integration of bicycles and transit use, better long- and short-term bicycle parking at transit stations would be beneficial for the integration of these two modes.

Although recent studies used a GIS-based approach to plan for new bicycle lanes and bike-sharing stations, no study has developed a methodology to specifically identify the optimal locations to install long-term and short-term bicycle parking (*5*, *18*, *19*). This study tries to fill this gap by using a multi-criteria GIS-based approach, which modifies and expands on a similar method developed by a previous study (*5*).

Study Area

Québec City is the second largest populated urban center in the province of Québec, Canada, with 531,902 inhabitants in 2016 (*20*). The study area is divided into five boroughs, for a total area of approximately 454 km^2 (*20*). Similar to many North American cities, the construction of highways between 1960 and 1980 encouraged the localization of employment nodes and housing developments at the periphery of the inner city, which in turn contributed to the creation of a polycentric-structured city (*21*). Nowadays, the presence of highways, in addition to railways, rivers, and territory steepness, are major barriers to the expansion and consolidation of the existing bicycle network (*22*).

In 2016, the City announced its Bicycle Vision to encourage a modal shift toward cycling by specifically improving the safety and connectivity of the bicycle network (*22*). Nine kilometers of new bicycle paths and 60 more bicycle parking racks were added in 2016 (*23*). Presently, Québec City has around 424 km of bicycle

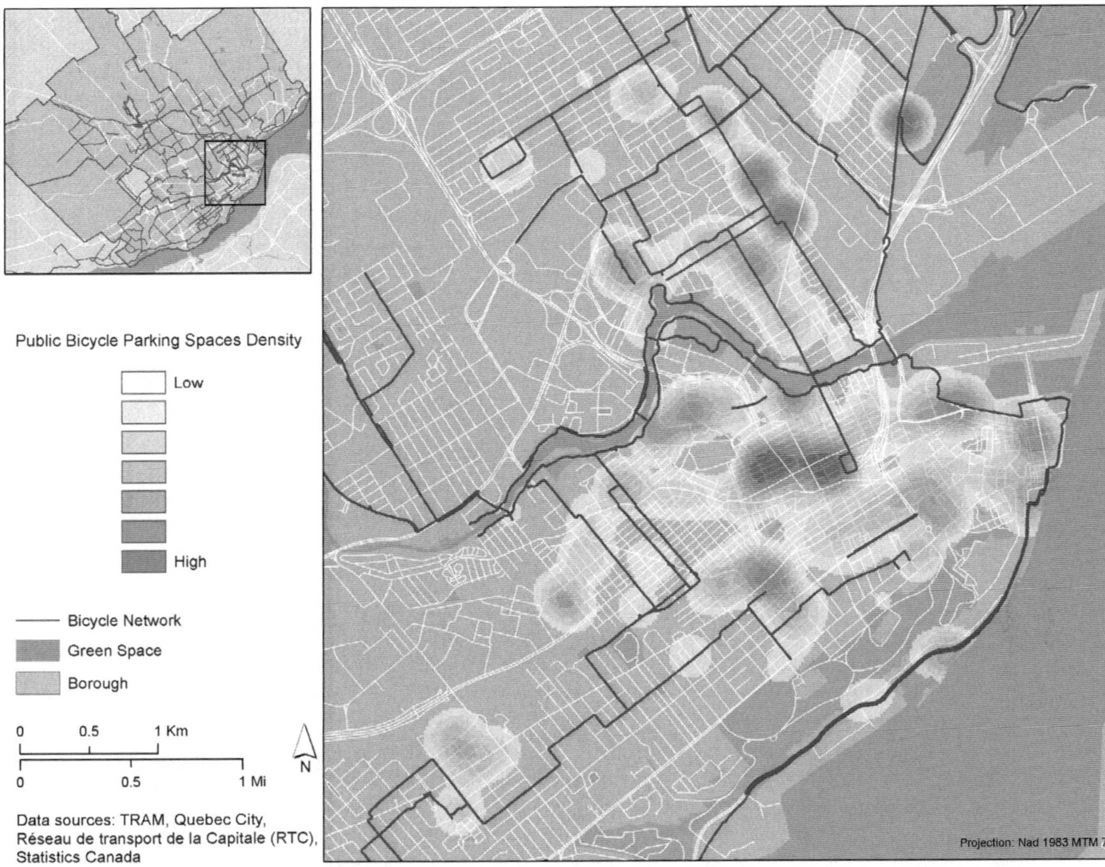

Figure 1. Location and density of municipal bicycle parking spaces in 2017.

network and 480 short-term bicycle parking (bicycle racks) of either three, five, or seven spaces each, for a total of nearly 3,784 public bicycle parking spaces available for cyclists. The distribution of these parking spaces is presented in Figure 1.

Methodology, Data, and Results

Our method consists of a multi-criteria approach to prioritize the location of new long-term and short-term bicycle parking. A flow chart illustrating the main steps of our analysis is displayed in Figure 2. Using data of the destinations of existing and potential cyclists, as well as the proximity of the bicycle network to high-frequency bus stops, we developed a priority index for locations with the highest need for new bicycle parking. A second priority index was then developed that identifies where long-term bicycle parking is needed among the locations identified in the previous step. Finally, the number of recommended bicycle parking spaces for these locations was calculated, taking into account the current supply of bicycle parking. We generated two priority maps, one for long-term and one for short-term bicycle parking needs in the region. The data employed in our analysis will be described in further detail in the following section.

Data Sources

The first data source used in our analysis was the 2011 Québec Origin–Destination Survey (OD), which collects information regarding trip purposes, modes, sociodemographic characteristics and the origin and destination of each trip from respondents. OD data is collected every 5 years by phone, and samples between 7 and 20% of all households living in the city between September and December (*24*). The second source of data used was the location of bus stops served by high-frequency bus routes called Metrobus. Metrobuses operate every 10 min during weekday peak hours. High-frequency bus stops were

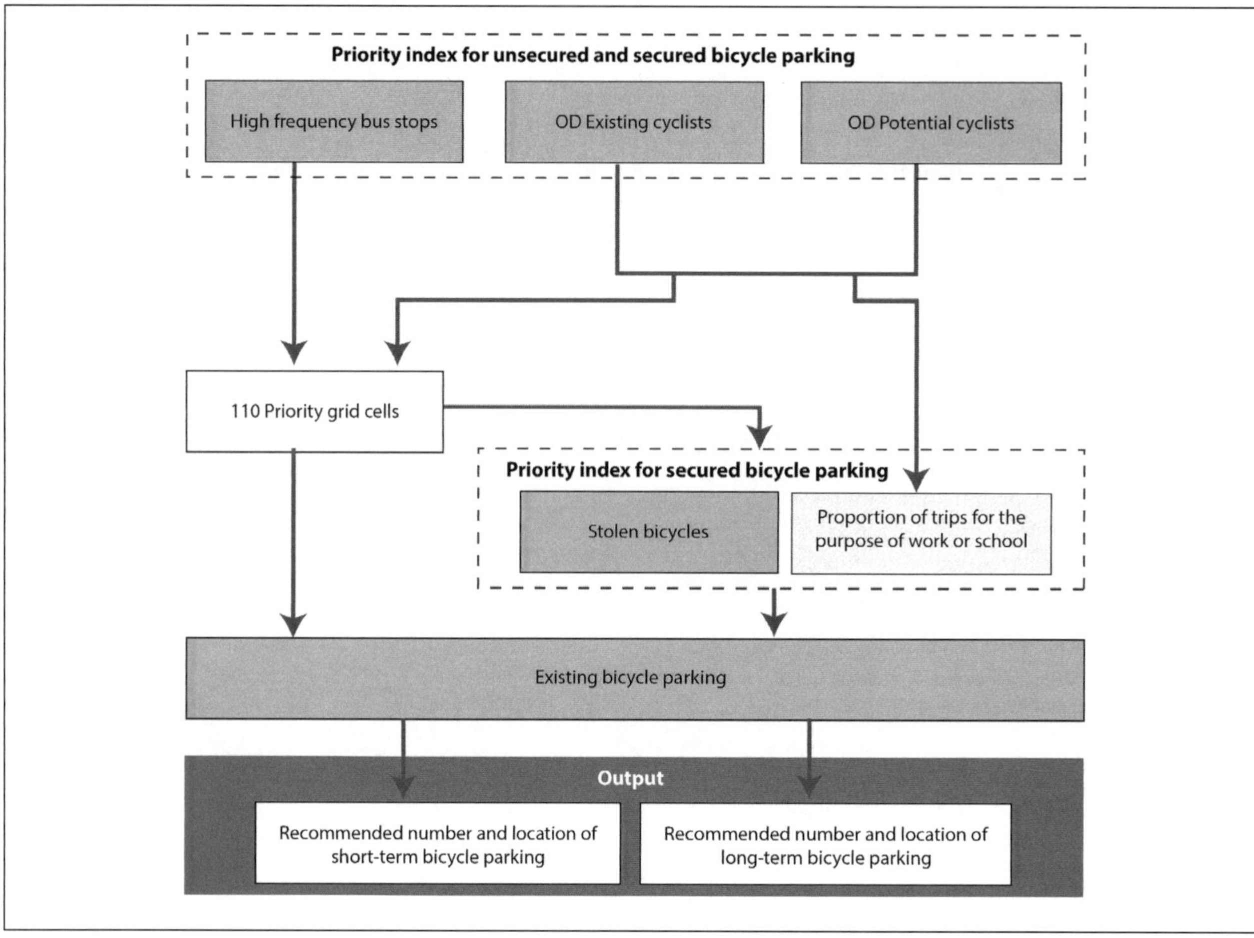

Figure 2. Flow chart of main steps in methodology.

used in this analysis, as Québec City does not have a subway or light-rail service. The data was extracted from the transit agency serving the City of Québec's territory, called Réseau de transport de la Capitale (RTC). The third and fourth data were the number of bicycle parking racks and spaces owned by Québec City and the current bicycle network obtained from Québec City. Our analysis only considered the number of public bicycle racks, since data on privately owned bicycle parking was unavailable. Due to data unavailability, and because the City has a limited influence on the decision to add bicycle parking on university campuses, we excluded Laval University from our analysis, although it is a major cycling destination and requires its own specific analysis. The last data used in this analysis was the 2015 Bicycle Travel Survey conducted by the Transportation Research at McGill (TRAM) group. We extracted two questions from this survey. First, cyclists were asked if they had had a bicycle stolen in the past, and if so, respondents were asked to identify the location where it was stolen.

Prioritization Index for New Bicycle Parking

Using the fishnet tool in ArcGIS, the first step was to generate a grid that covered the extent of our study area, which was adapted from a previous study (*5*). The selected size of each grid cell was 300 × 300 m. This size was found to be optimal to aggregate multiple criteria, such as the proximity of the bicycle network and the destinations of existing and potential cyclists. The same grid cell size was employed by Larsen, Patterson, and El-Geneidy (*5*), however practitioners interested in reproducing our methodology can choose a grid cell size that corresponds to their own city structure. We removed all cells that had their centroid outside the boroughs of Québec City and within Laval University Sainte-Foy campus, which resulted in a working grid composed of 5,185 cells.

To identify and prioritize the best locations to install new bicycle parking, the first step consisted of generating a prioritization index. The index is composed of three indicators: 1) the destinations of existing cyclists, 2) the destinations of potential cyclists, and 3) the proximity to high-frequency bus stops served by the Société de Transport de la Capitale. The description of each indicator is presented below.

Existing Cyclist's Destinations. The first indicator used is the number of trips made by cyclists ending in each of the 5,185 grids. We used only home-based trips from the 2011 OD for all trip purposes. A total of 415 destinations were geocoded using x–y coordinates from OD. Next, we applied the expansion factor of each of these trips, to account for the expected magnitude of individuals in Québec City making these trips. The expansion factor is described as a "sample weight" given to each trip according to their sample strata, and has the ability to adjust for potential bias introduced by the time and day that interviews took place or by underrepresented trips (*24*). After applying the expansion factor, a total of 5,918 trips made by cyclists were spatially joined to our working grid. Finally, we standardized the number of existing cyclists' trips ending in each grid cell with a *z*-score to enable the combination of this measure with others generated later in the study. The distribution of the destinations of existing cyclists is displayed in Figure 3*a*, which shows that their destinations are highly concentrated in downtown Québec.

Potential Cyclists' Destinations. The second indicator is the number of potential cyclists ending their trip in each of the grid cells, using the 2011 OD. Within the context of our study, a potential cyclist trip is defined as a short-distance non-cycling trip, which could be converted into a bicycle trip. Specifically, we used a threshold distance of 5.8 km, which corresponds to the 75th percentile distance of all commuting bicycle trips evaluated from the 2011 OD . These short-distance trips by bicycle would on average take 22 min at a pace of 16 km/h (*25*). We calculated the trip length of these potential cyclists using a straight-line distance for simplicity between each origin and destination.

We adjusted the number of potential trips ending in each grid cell according to the expansion factor provided in the 2011 OD. A total of 23,844 potential cyclist trips were extracted from the OD, which corresponds to 337,928 trips after applying the expansion factor. These trips were then spatially joined to the grid cells. Finally, we standardized the number of potential cyclists' trips ending in each grid cell with a *z*-score. The results are shown in Figure 3*b*. Compared with the destinations of existing cyclists, the distribution of potential cyclists is more dispersed across the Québec City, however a similar high-priority corridor in downtown Québec is observed.

Number of Bus Stops. The last indicator used in our priority index is the number of bus stops served by high-frequency bus routes, named Metrobus. We incorporated proximity to these bus stops to improve the integration of bicycling and public transit. The Metrobus service consists of six high-frequency bus routes that are equipped with bicycle racks holding a maximum of two bicycles, and are the only bus routes operated by RTC equipped with bicycle racks (*26*). These bus routes are the most efficient public transit routes to reach the main employment and activity nodes in Québec City. However, adaptation of our proposed method based on the local supply of transit service is recommended, for example, by using subway stations instead of bus stops or using both high-frequency bus routes and rail stations.

Using the RTC data, we identified 231 bus stops served by high-frequency bus routes. To avoid duplicating the number of bus stops, we only considered bus stops serving one direction. We then generated a buffer of 100 m around each of them, and spatially joined these buffers with the grid cells, which enabled us to systematically sum the number of bus stop buffers intersecting each grid cell. The results were standardized using a *z*-score and are shown in Figure 3*c*.

Combining and Weighting Indicators into an Index. We combined and weighted the three standardized indicators into one bicycle parking index. To arrive at our final index, we used a weighting scheme, in which we applied a higher weighting to the destinations of existing cyclists, namely a weighting of 3, and a weighting of 1.5 to potential cyclists. Although all three indicators are important, we decided to give more importance to existing cyclists' destinations to prioritize the current needs of bicycle parking. However, the application of a weighting scheme should vary according to a region's specific priorities or planning goals. The potential cyclists' trips were given a weight of 1.5 as we also wished to plan for the potential needs of cyclists in a medium-term perspective. Finally, the number of bus stops was integrated without any weighting. The combined and weighted priority index result is displayed in Figure 3*d*.

After determining the priority areas for new bicycle parking, we identified grid cells falling within the top decile of the combined and weighted index, and selected the high-priority areas that were within 100 m of the existing bicycle network. This resulted in 110 grid cells that are within proximity of a bicycle network. After locating these high-priority areas for the installation of new bicycle parking, we first needed to determine the recommended number of spaces that would need to be installed in these grid cells, and second we considered the optimal locations to install new long-term bicycle parking within these grid cells.

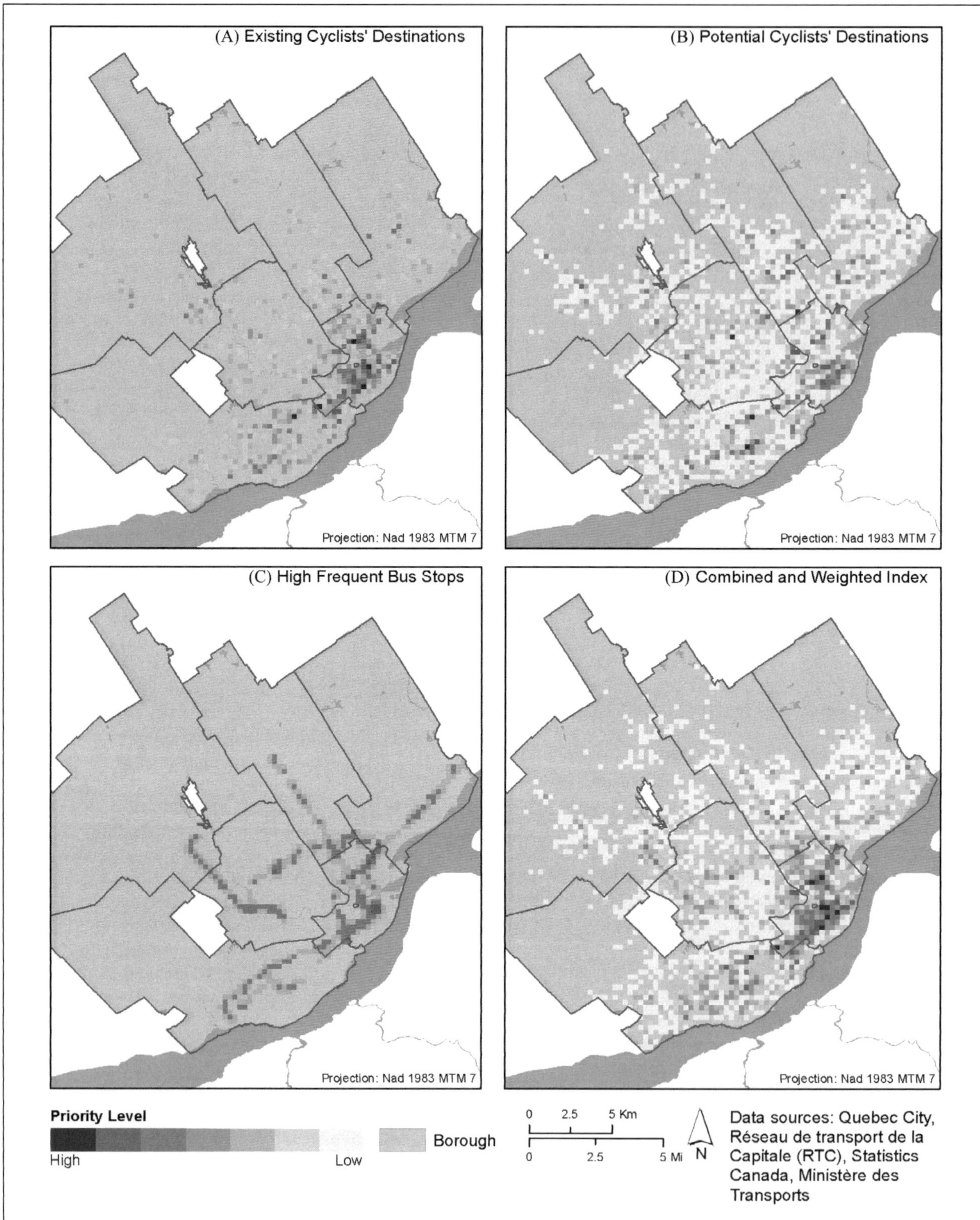

Figure 3. Standardized indicators and combined and weighted index: (*A*) existing cyclists' destinations, (*B*) potential cyclists' destinations, (*C*) high-frequency bus stops, and (*D*) combined and weighted index.

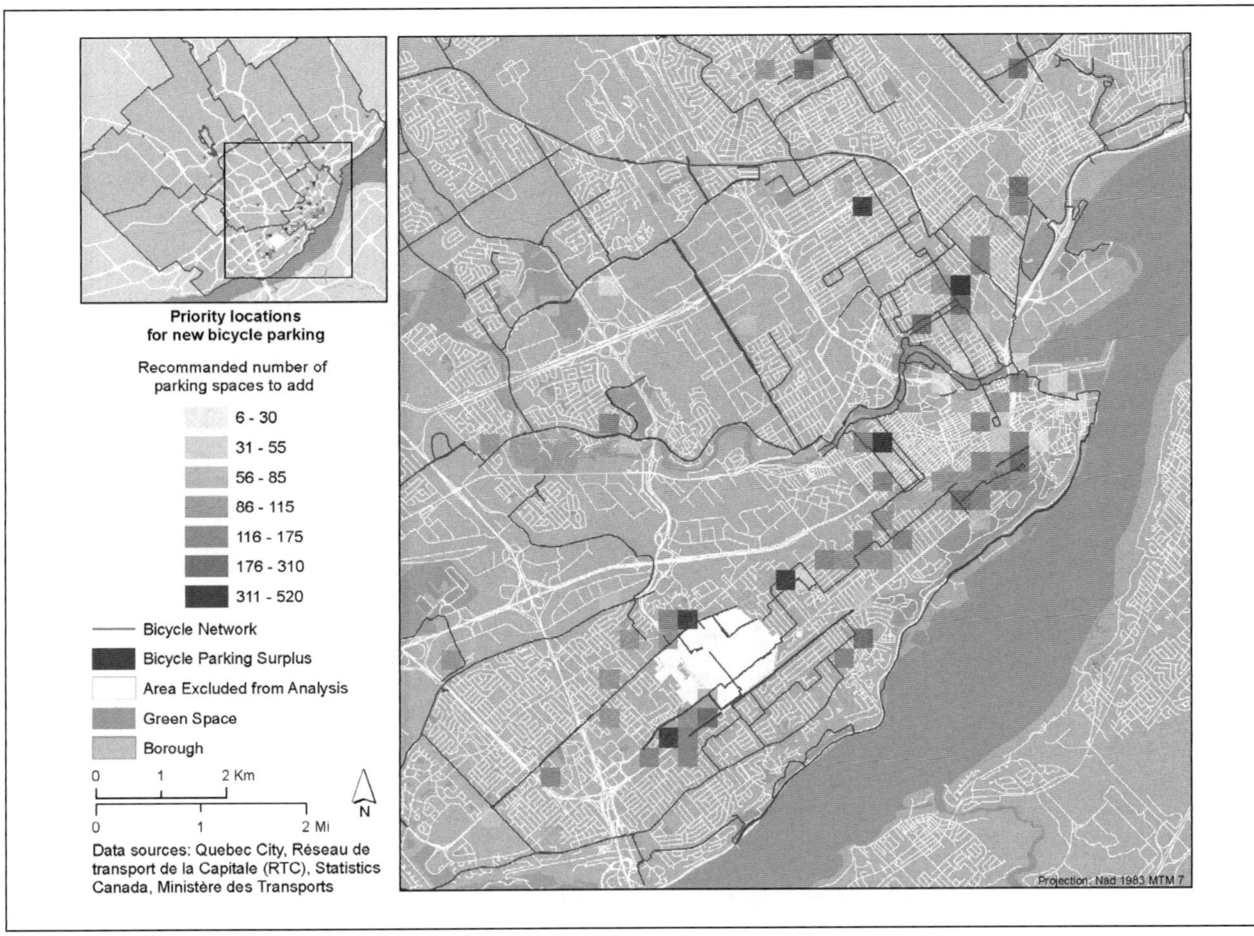

Figure 4. Priority locations to add new bicycle parking and recommended number of bicycle spaces to add.

To estimate the number of bicycle spaces needed in these high-priority grid cells, we summed the number of trips made by existing cyclists and 10% of the trips made by potential cyclists that end in each grid cell. Ten percent of potential trips assumes a moderate mode shift from potential cyclists in the medium future. We then subtracted the bicycle parking demand, identified in the above step, from the existing public parking spaces. Figure 4 presents the final 110 grid cells of high-priority for bicycle parking and the recommended number of new bicycle parking spaces in each grid. These locations could be mainly short-term (bicycle racks), if a region wants to propose long-term bicycle parking more analysis is needed and this will be explained in the following section.

Long-term Bicycle Parking Index

Using the high-priority bicycle parking locations identified above (Figure 4), we next developed a priority index to determine the best locations to invest in long-term bicycle parking. The indicators considered for the prioritization of secured bicycle parking included the location of stolen bicycles and the proportion of existing and potential trips ending in each priority grid cells that were work or school trips (please refer to Figure 2 methodology overview).

Using the 2015 Bicycle Travel Survey conducted by the TRAM group, we geocoded the locations where respondents reported a stolen bicycle. We then spatially joined each location of a reported stolen bicycle to the 110 priority grid cells and summed the total number of bicycles stolen per grid cell. Finally, we standardized the number of stolen bicycles in each grid cell using a *z*-score. The results are shown in Figure 5*a*.

We then evaluated trips made by existing and potential cyclists for the purpose of commuting to work or school. Specifically, we calculated a ratio of trips to work or school compared with trips for all other purposes (e.g., shopping, grocery, health). Cyclists who commute to work or school require access to bicycle parking for a longer time, and the availability of long-term bicycle parking for this type of trip has been found to be an

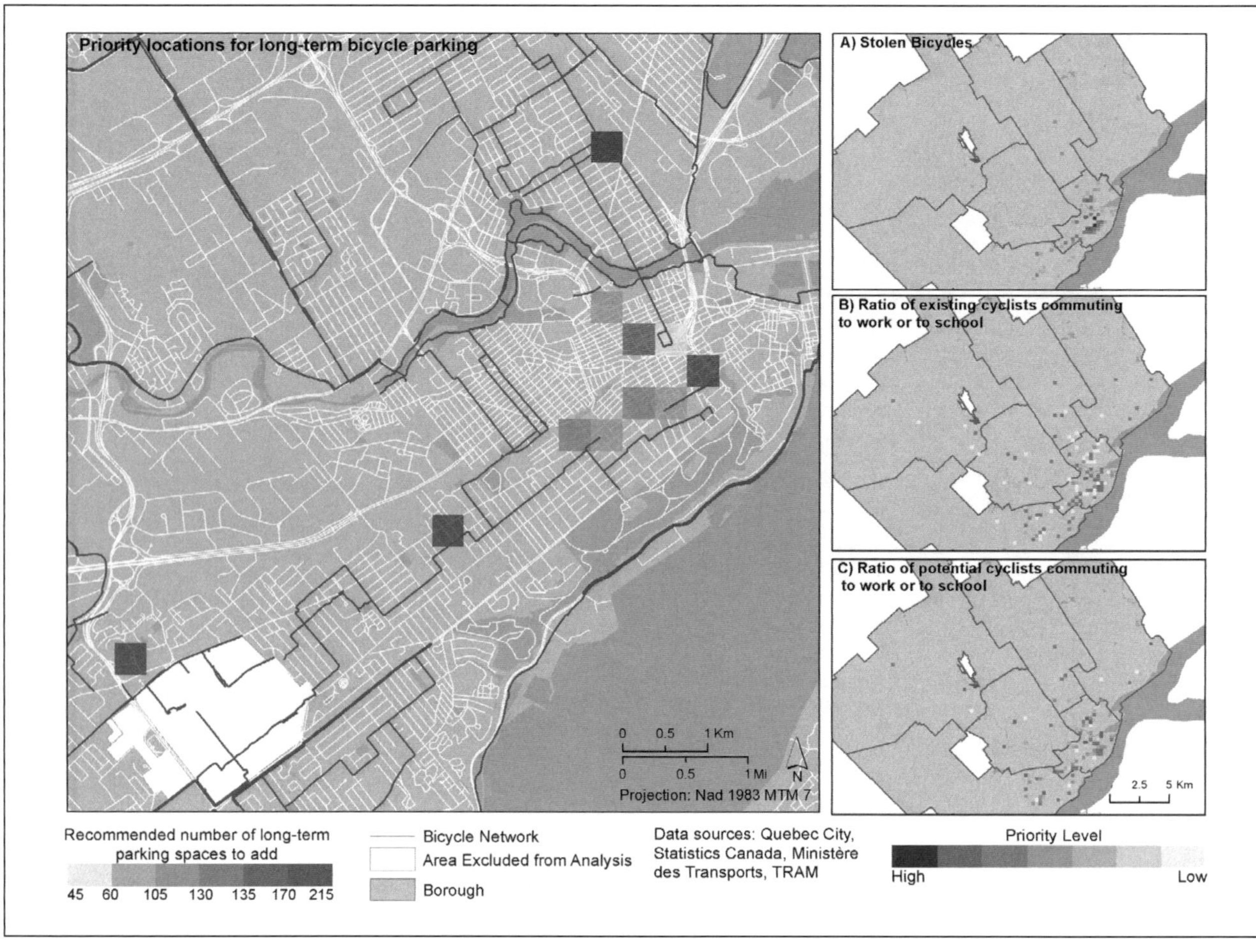

Figure 5. Standardized indicators for the long-term bicycle parking index: (*A*) stolen bicycles, (*B*) ratio of existing cyclists commuting to work or to school, (*C*) ratio of potential cyclists commuting to work or to school.

important determinant for cycling to work (*9*, *13*, *27*, *28*). Accordingly, we want to prioritize locations with higher ratios of work or school trips, as a strategy to encourage more individuals to cycle to work or school. We standardized the ratios of the high-priority grid cells separately for existing and potential cyclists, and the results are shown in Figure 5, *b* and *c*.

Combining and Weighting Indicators into an Index. To generate our secured parking index, we combined and weighted the three standardized indicators. Although all three indicators were selected because of their importance for planning new long-term bicycle parking, we decided to give more weight to the number of stolen bicycles, knowing that the perceived risk of bicycle theft can discourage individuals from using their bicycle (*9*, *10*). Accordingly, a weighting of 3 was attributed to this indicator. The ratio of existing cyclists commuting to work or school was assigned a weight of 2 and the ratio of potential cyclists commuting to work or school was assigned a weight of 1.5. With such criteria, we prioritize locations for investments in long-term bicycle parking in areas where we know there are cyclists currently commuting, however the weighting attributed to potential cyclists was not considerably lower, since we would hope that access to secure bicycle parking would lead to a major mode shift toward cycling.

The cells falling in the top decile of the combined and weighted index were identified as priority locations for investment in long-term bicycle parking. To recommend the number of secured bicycle spaces in each cell, we calculated the optimal number of spaces using the same method as the short-term bicycle parking. We calculated the difference between the number of existing bicycle parking spaces and the current and potential cycling demand in each grid cell, in which we estimated the demand to be the number of existing cyclists whose trip ends in each grid cell plus 10% of all potential cyclists whose trips end in each grid cell. The recommended locations and number of long-term bicycle parking spaces to be installed are displayed in Figure 5. It is important to note that this step could be adjusted if there is an existing supply of long-term bicycle parking in a region.

Discussion and Conclusion

The objective of this paper was to develop a practice-ready GIS-based method to identify optimal locations to add both short-term and long-term bicycle parking, which aims to be flexible and easily adaptable to different contexts. Using Québec City as a case study, we first divided the city into 300- × 300-m grid cells. The use of grid cells is recommended in this kind of analysis as it guides planners to areas according to need, providing them with latitude to closely evaluate this zone and find the appropriate location based on existing land use. This work must be followed by more detailed analysis to locate bicycle parking spaces within the identified grid cells, which requires more local knowledge that would only exist among local planning authorities. We generated the prioritization index for new bicycle parking according to what factors were identified as likely to contribute to the need for new long-term or short-term parking, namely the destinations of existing and potential cyclists and proximity to a high-frequency transit service. Though we did not consider linked trips of existing and potential cyclists, practitioners and researchers could include them to tailor our method according to their needs. These indicators were then combined and a weighting scheme was applied – the devising of which is recommended according to the objectives or local planning goals and priorities of a region. Finally, we only selected grid cells within 100 m of the existing bicycle network, to prioritize locations that contribute the most to the development of a complete cycling network and to ensure high usage levels of these newly proposed facilities. The recommended number of parking spaces for each grid cell was then calculated according to the existing supply of parking and the existing and potential demand for bicycle parking. For the recommended priority locations of long-term bicycle parking, we developed a priority index that considered the location of stolen bicycles and grid cells with high proportions of existing and potential cyclists commuting into each grid cell. The number and the type of indicators utilized could be tailored to other regions according to data availability and policy goals. Moreover, the weighting scheme applied to each indicator could also be modified according to different contexts and local planning goals and priorities.

The strength of our method lies in its flexibility and ability to account for long-term demand for bicycle parking as we specifically considered work and school trip purposes, as access to these secured bicycle facilities have been shown to encourage cyclists to commute to their workplace or school (*10*, *13*, *28*). Long-term bicycle parking is expected to be an integral part of the improvement of Québec City's bicycle network, and is expected to help Québec reach its goal of providing residents with a safe and connected bicycle network to attract and encourage cycling for everyday purposes, such as commuting.

It is important to note that our methodology to locate long-term bicycle parking, proceeded under the assumption that such facilities will be provided free of charge to users. As many regions provide these facilities to users for a fee, demand will need to be adjusted according to the willingness of cyclists to pay to access this service. Knowledge of cyclists' willingness to pay for long-term bicycle parking could be the subject of a future study.

Acknowledgments

This study was made possible with the collaboration of transportation planner Mr. Jean-François Martel of Québec City, who provided the municipal bicycle parking data. We also thank Genevieve Boisjoly for her helpful inputs. This study was funded by the Social Sciences and Humanities Research Council of Canada and the Natural Sciences and Engineering Research Council of Canada.

Author Contributions

The authors confirm contribution to the paper as follows: study conception and design: M.P.V., E.G., A.E-G.; data collection: M.P.V.; analysis and interpretation of results:M.P.V., E.G., A.E-G.; draft manuscript preparation: M.P.V., E.G., A.E-G. All authors reviewed the results and approved the final version of the manuscript.

References

1. Pucher, J., J. Dill, and S. Handy. Infrastructure, Programs, and Policies to Increase Bicycling: An International Review. *Preventive Medicine*, Vol. 50, 2010, pp. S106–S125.
2. Pucher, J., and R. Buehler. Making Cycling Irresistible: Lessons from The Netherlands, Denmark and Germany. *Transport Reviews*, Vol. 28, No. 4, 2008, pp. 495–528.
3. Dill, J., and T. Carr. Bicycle Commuting and Facilities in Major U.S. cities. If You Build Them, Commuters Will Use Them. *Transportation Research Record: Journal of the Transportation Research Board*, 2003. 1828: 116–123.
4. Buelhler, R., and J. Pucher. Cycling to Work in 90 Large American Cities: New Evidence on the Role of Bike Paths and Lanes. *Transportation*, Vol. 39, 2011, pp. 409–432.
5. Larsen, J., Z. Patterson, and A. El-Geneidy. Build It. But Where? The Use of Geographic Information Systems in Identifying Locations for New Cycling Infrastructure. *International Journal of Sustainable Transportation*, Vol. 7, No. 4, 2013, pp. 299–317.

6. Buehler, R. Determinants of Bicycle Commuting in the Washington, DC Region: The Role of Bicycle Parking, Cyclist Showers, and Free Car Parking at Work. *Transportation Research Part D: Transport and Environment*, Vol. 17, 2012, pp. 525–531.
7. Wardman, M., M. Tight, and M. Page. Factors Influencing the Propensity to Cycle to Work. *Transportation Research Part A: Policy and Practice*, Vol. 41, 2007, pp. 339–350.
8. Pucher, J., R. Buehler, and M. Seinen. Bicycling Renaissance in North America? An Update and Re-Appraisal of Cycling Trends and Policies. *Transportation Research Part A: Policy and Practice*, Vol. 45, 2011, pp. 451-475.
9. Rietveld, P., and V. Daniel. Determinants of Bicycle Use: Do Municipal Policies Matter? *Transportation Research Part A: Policy and Practice*, Vol. 38, 2004, pp. 531–550.
10. Rietveld, P. The Accessibility of Railway Stations: The Role of the Bicycle in The Netherlands. *Transportation Research Part D: Transport and Environment*, Vol. 5, 2000, pp. 71–75.
11. van Lierop, D., M. Grimsrud, and A. El-Geneidy. Breaking into Bicycle Theft: Insights from Montreal, Canada. *International Journal of Sustainable Transportation*, Vol. 9, No. 7, 2015, pp. 490–501.
12. Gamman, L., A. Thorpe, and M. Willcocks. Bike Off! Tracking the Design Terrains of Cycle Parking: Reviewing Use, Misuse and Abuse. *Crime Prevention and Community Safety: An International Journal*, Vol. 6, No. 4, 2004, pp. 19–36.
13. Hunt, J. D., and J. E. Abraham. Influences on Bicycle Use. *Transportation*, Vol. 34, 2007, pp. 453–470.
14. Martens, K. Promoting Bike-and-Ride: The Dutch Experience. *Transportation Research Part A: Policy and Practice*, Vol. 41, 2007, pp. 326–338.
15. Van der Spek, S., and N. Scheltema. The Importance of Bicycle Parking Management. *Research in Transportation Business & Management*, Vol. 15, 2015, pp. 39–49.
16. Krizek, K., and E. Stonebraker. Bicycle and Transit: A Marriage Unrealized. *Transportation Research Record: Journal of the Transportation Research Board*, 2010. 2144: 161–167.
17. Bachand-Marleau, J., J. Larsen, and A. El-Geneidy. Much-Anticipated Marriage of Cycling and Transit. How Will It Work? *Transportation Research Record: Journal of the Transportation Research Board*, 2011. 2247: 109–117.
18. Rybarczyk, G., and C. Wu. Bicycle Facility Planning Using GIS and Multi-Criteria Decision Analysis. *Applied Geography*, Vol. 30, 2010, pp. 282–293.
19. García-Palomares, J. C., J. Gutiérrez, and M. Latorre. Optimizing the Location of Stations in Bike-Sharing Programs: A GIS Approach. *Applied Geography*, Vol. 35, 2012, pp. 235–346.
20. Statistics Canada. *Census Profile, 2016 Census, Québec, Ville [Census Subdivision]*. Statistics Canada. http://www12.statcan.gc.ca/census-recensement/2016/dp-pd/prof/details/page.cfm?Lang = E&Geo1 = CSD&Code1 = 2423027&Geo2 = PR&Code2 = 01&Data = Count&SearchText = Quebec&SearchType = Begins&SearchPR = 01&TABID = 1&B1 = All. Accessed October 15, 2017.
21. Vincens, M., M. Vandersmissen, and M. Thériault. Impacts de la restructuration du réseau d'autobus de la ville de Québec sur l'accessibilité aux emplois des femmes et sur la mobilité professionelle. *Cahiers de géographie du Québec*, Vol. 51, No. 144, 2007, pp. 419–446.
22. Ville de Québec. *Vision des déplacements à vélo*. Quebec City, 2016. https://www.ville.quebec.qc.ca/apropos/planification-orientations/transport/docs/vision_deplacements_velo.pdf.
23. —. *Bilan vélo 2016. Détail des projets 2016 et des comptages*, Quebec City, 2017. https://www.ville.quebec.qc.ca/apropos/planification-orientations/transport/velo/docs/bilan_2016_projets_comptages.pdf
24. Ministère des Transports du Québec. *Enquête Origine-Destination 2011. La mobilité des personnes dans la région de Québec. Volet Enquête-ménages. Sommaire des résultats*, 2015.
25. El-Geneidy, A., K. Krizek, and M. Iacono. Predicting Bicycle Travel Speeds Along Different Facilities Using GPS Data: A Proof of Concept Model. Presented at 86th Annual Meeting of the Transportation Research Board, Washington, D.C., 2007.
26. Réseau de transport de la Capitale. *À chacun son bus. Trouvez le service qui convient à vos besoins!* https://www.rtcquebec.ca/Default.aspx?tabid = 104&language = fr-CA.
27. Yan, K., and J. Zheng. Study of Bicycle Parking in Central Business District of Shanghai. *Transportation Research Record: Journal of the Transportation Research Board*, 1994. 1994: 27–35.
28. Taylor, D., and H. Mahmassani. Analysis of Stated Preferences for Intermodal Bicycle-Transit Interfaces. *Transportation Research Record: Journal of the Transportation Research Board*, 1996. 1556: 86–96.

The Standing Committee on Bicycle Transportation (ANF20) peer-reviewed this paper (18-01458).

Article

TRR
JOURNAL OF THE TRANSPORTATION RESEARCH BOARD

Transportation Research Record
2018, Vol. 2672(36) 83–91

DOI: 10.1177/0361198118776812
journals.sagepub.com/home/trr
SAGE

Generation of "Biking Schedules" for Bicycle Travel Analysis

Amr Mohamed[1] and Alexander Y. Bigazzi[2]

Abstract

With an increasing focus on bicycling as a mode of urban transportation, there is a pressing need for improved tools for bicycle travel analysis and modeling. This paper introduces "biking schedules" to represent archetypal urban cycling dynamics, analogous to driving schedules used in vehicle emissions analysis. Three different methods of constructing biking schedules with both speed and road grade attributes are developed from the driving schedule literature. The methods are applied and compared using a demonstration data set of 55 h of 1-Hz on-road GPS data from three cyclists. Biking schedules are evaluated based on their ability to represent the speed dynamics, power output, and breathing rates of a calibration data set and then validated for different riders. The impact of using coarser 3, 5, and 10 s GPS logging intervals on the accuracy of the schedules is also evaluated. Results indicate that the best biking schedule construction method depends on the volume and resolution of the calibration data set. Overall, the biking schedules successfully represent most of the assessed characteristics of cycling dynamics in the calibration data set (speed, acceleration, grade, power, and breathing) within 5%. Future work will examine the precision of biking schedules constructed from larger data sets in more diverse cycling conditions and explore additional refinements to the construction methods. This research is considered a first step toward adopting biking schedules in bicycle travel analysis and modeling, and potential applications are discussed.

Cycling can be a low-cost, healthy, and fun form of transportation with diverse personal and social benefits. Promoting cycling is an increasingly common part of environmental initiatives in the transportation sector to decrease pollution emissions (*1*). As a result of these and other initiatives, bicycle facilities are increasing in many North American cities, accompanied by an increase in cycling for travel. These changes intensify a need for better understanding of real-world cycling behavior and for improved tools for bicycle travel analysis and modeling.

Bicycle speed dynamics are important for both physical and behavioral aspects of bicycle travel analysis. The human power requirements of cycling vary greatly with speed, acceleration, and road grade (*2*). Human power and energy expenditure, in turn, are linked to physical activity levels, breathing rates and pollution inhalation, and many facets of travel behavior such as speed, route, and mode choices (*3–6*). Unfortunately, detailed microsimulation and operational models of on-road urban bicycle dynamics are still in their infancy (*7–9*).

Knowledge of typical urban bicycling dynamics would be useful for a number of applications, including estimating energy expenditure and related impacts from aggregate data such as average trip speed, identifying different bicycle travel regimes such as cautious versus risky riding, and simulating cycling dynamics for vehicle development and testing such as design of electric-assist bicycle motors. Driving schedules were developed to address some similar issues for motor vehicles (*10–13*). Driving schedules, also referred to as driving cycles, are speed profiles designed to represent typical driving patterns, used in fuel and emissions modeling and in vehicle simulation and testing (*14*).

A main application of driving schedules is to enable estimation of realistic vehicle fuel consumption and emissions (which depend on driving dynamics) from aggregate travel model outputs such as average speed (*15*). Lab dynamometer testing of real-world vehicles operating on driving schedules captures the fuel and emissions effects of typical driving dynamics associated with a given average speed. Thus, the driving schedules can essentially be used to disaggregate travel model output. Similarly,

[1]Department of Civil Engineering, University of British Columbia, BC, Canada
[2]Department of Civil Engineering and School of Community and Regional Planning, University of British Columbia, Vancouver, BC, Canada

Corresponding Author:
Address correspondence to Alexander Y. Bigazzi: alex.bigazzi@ubc.ca

bicycle travel analysis could benefit from the ability to estimate realistic cycling dynamics and related impacts from aggregate travel data such as average trip speed.

Like driving schedules, biking schedules can be expected to vary with travelers, topography, traffic, facility type, and more. Road grade significantly influences the speed and energy of cyclists (*4*, *16*, *17*), as well as motor vehicle emissions (*18*, *19*). However, driving schedules are typically created without synchronous grade profiles (*20*, *21*). Due to the strong influence of road grade on cycling speeds, the interdependence of speed and road grade should be included in an accurate representation of bicycle dynamics.

The objective of this paper is to introduce the concept of biking schedules along with methods for their construction. Methodologies of constructing biking schedules that integrate speed and grade dynamics are developed by building on the established driving schedule literature. The methods are evaluated using an on-road bicycling data set. Because biking schedules will most likely be generated from GPS data, the issue of required data resolution is also explored, facilitating application of the method to larger naturalistic data sets in future work.

Proposed Biking Schedule Construction Methods

Schedule construction methodologies can be classified into microtrip based, segment based, pattern based, and stochastic modal approaches (*22*). This study uses the microtrip based approach to develop biking schedules. Observed travel data are divided into small snippets, called microtrips, which form the elementary components of the biking schedules. A schedule is then constructed by appending microtrips together until a desired length is reached. Candidate schedules are evaluated for their similarity to the original travel data, based on a set of predetermined aggregate assessment parameters. Departing from driving schedule methods, the microtrips and schedules in this research include synchronous speed and grade data.

In past research on driving schedules, microtrips have been selected randomly (*21*, *23*, *24*), with sophisticated statistical methods such as Markov chain transition matrices (*12*), or with hybrid methods (*25*). In this paper, three methods of microtrip selection are presented and compared. The following subsections present detailed steps for the proposed biking schedule construction methods.

Microtrip Extraction

Approaches to defining microtrips vary across the literature. One common approach is to define microtrips between two consecutive stops (*11*, *26*), but this method poorly represents travel data with long uninterrupted segments (*25*). Other approaches involve temporal or spatial segmentations of predetermined sizes. These methods have the advantage of producing microtrips of a consistent desired length regardless of speed dynamics, but with the drawback of requiring a speed continuity criterion to produce realistically smooth speed profiles.

Microtrips in this study are delineated at fixed spatial intervals of 250 m—an approach which was demonstrated in recent research to yield the most accurate driving schedules (*27*). It is acknowledged that results from driving schedule studies are limitedly transferable to biking schedules due to differences in speed and acceleration characteristics. Investigation of alternative microtrip definitions for constructing biking schedules is left for future work.

Assessment Criteria

The role of the assessment criteria is to ensure that the developed schedules represent the important characteristics of the calibration data set. Thus, assessment criteria should be selected that are relevant to the purposes of the schedule. Target parameter values for the assessment criteria are calculated from the calibration data set, and then schedules are constructed to reproduce those parameter values as closely as possible. Table 1 lists the 12 assessment criteria proposed for developing biking schedules. These parameters were adopted from the driving schedule literature to represent speed and acceleration dynamics, with new parameters added for road grade.

The performance value (PV) is a single aggregate indicator of the set of assessment criteria, used to evaluate the biking schedules. The PV for each parameter in Table 1 is the absolute percent difference between the value calculated from the calibration data and the value calculated from the biking schedule data, with the exception of SAGPD for which the PV is the root mean square error (RMSE) between the target and schedule time fractions in each cell of the distribution matrix. A lower PV means that the schedule is closer to the target parameters and more representative of the calibration data. The total schedule PV is then the weighted average of PVs for individual parameters, PV_x. Weights are distributed equally among four sets of parameters that represent speed, acceleration, grade, and SAGPD:

$$\begin{aligned} PV = {} & 0.25(PV_{ATS} + PV_{ARS} + PV_{PTI} + PV_{PTC})/4 \\ & + 0.25(PV_{AAA} + PV_{PTA} + PV_{PTD} + PV_{APW})/4 \\ & + 0.25(PV_{AAG} + PV_{PTPG} + PV_{PTNG})/3 + 0.25(PV_{SAGPD}). \end{aligned}$$

Table 1. Assessment Criteria for Biking Schedules

ID	Parameter[1]	Abbreviation	Units
1	Average trip speed	ATS	km/h
2	Average running speed ($v>0$)	ARS	km/h
3	Percentage time idling ($v = 0$)	PTI	%
4	Percentage time cruising ($v>1$, $-0.1<a<0.1$)	PTC	%
5	Average absolute acceleration	AAA	km/h/s
6	Percentage time accelerating ($a>0$)	PTA	%
7	Percentage time decelerating ($a<0$)	PTD	%
8	Average positive work per mass-distance[2] ($a>0$)	APW	m/s^2
9	Average absolute grade	AAG	%
10	Percentage time positive grade ($G > 0.5$)	PTPG	%
11	Percentage time negative grade ($G < -0.5$)	PTNG	%
12	Speed acceleration grade probability distribution[3]	SAGPD	%

Note: [1]v = speed in km/h; a = acceleration in km/h/s; G = grade in %. [2]Expressed as "positive kinetic energy" in (*21*), calculated as $\frac{\sum\left(v_{i+1}^2 - v_i^2\right)\forall v_{i+1} > v_i}{total\ distance}$.
[3]Percentage time in each cell of a 3-dimensional speed-acceleration-grade probability distribution matrix with speed intervals of 5 km/h, acceleration intervals of 0.2 km/h/s (*28*), and grade intervals of 1%.

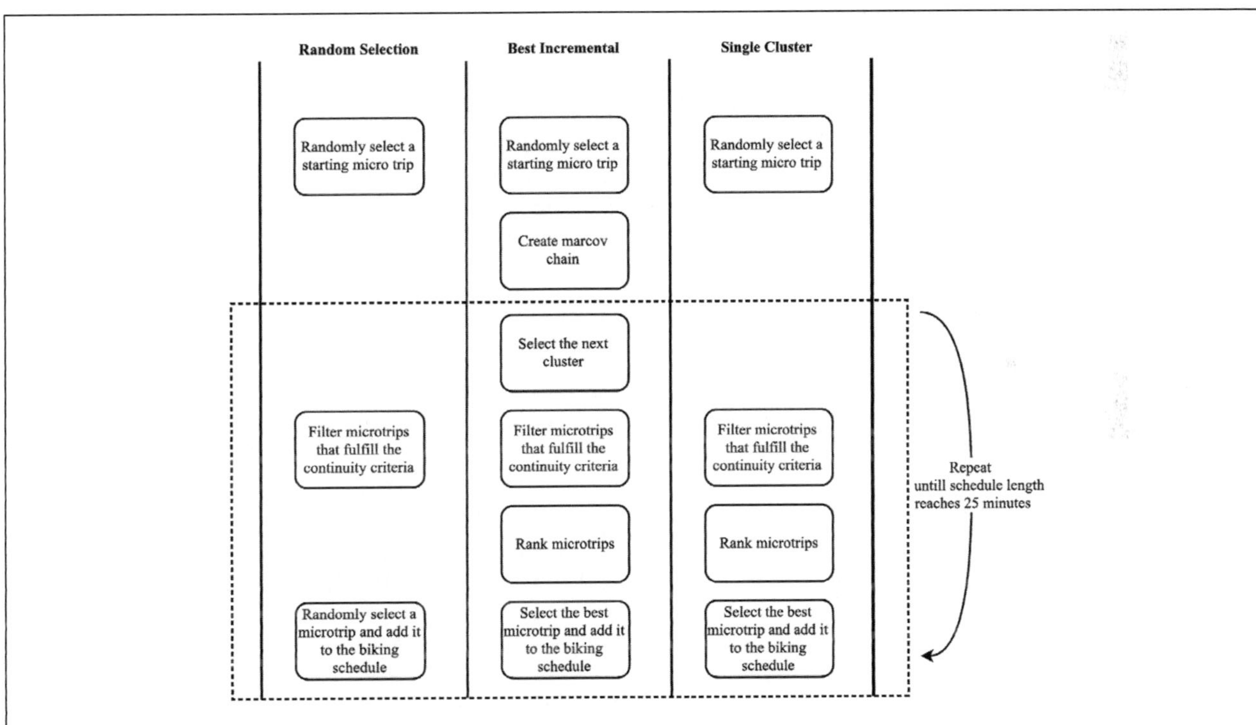

Figure 1. Framework for constructing biking schedules with three different methods.

Biking Schedule Construction

Biking schedules were constructed from the microtrips using three different methods: random selection, best incremental, and single cluster. The details of each method are described in the following subsections. Figure 1 summarizes the overall framework for constructing biking schedules.

For all three methods, the first microtrip in the biking schedule is randomly selected from a subset of trip-starting microtrips. Then, additional microtrips are appended, without repetition, until a schedule duration of 25 min is reached, consistent with the 10–30 min driving schedules common in the literature (*24*). Each subsequent microtrip must meet continuity criteria of an initial speed within 2 km/h and initial grade within 2%

of the end of the previous microtrip. The methods differ primarily in how they select the next microtrip from among those meeting the continuity criteria. For each method, several candidate biking schedules are constructed and the best schedule selected based on lowest PV.

Random Selection Method. In this method, microtrips are randomly selected and appended, restricted only by the continuity criteria, until the target duration (25 min) is reached. Consistent with previous studies (*23*), new biking schedules are repeatedly constructed until 20 candidate schedules with PV $< 15\%$ have been generated, and then the preferred schedule selected from those candidates based on lowest PV.

Best Incremental Method. In this method, also adapted from previous studies (*12*, *29*), microtrips are first clustered by average speed, average acceleration, and average grade using a K-means clustering algorithm. The number of clusters is selected based on the sum of squared errors (SSE) within clusters, as illustrated in Figure 2. The SSE decreases with more clusters (i.e. they become more similar), but with the trade-off of fewer microtrips in each cluster which can create problems for meeting the continuity criteria. The optimal number of clusters is likely context-dependent. In this study, microtrips are grouped into nine clusters.

After grouping the microtrips into clusters, a transition matrix is generated to represent the probability of transitioning between clusters, based on observed sequences of microtrip clusters in the calibration data. A stochastic Markov chain process then generates a sequence of clusters starting from the cluster of the (randomly selected) initial microtrip (*30*). Then, for each successive cluster

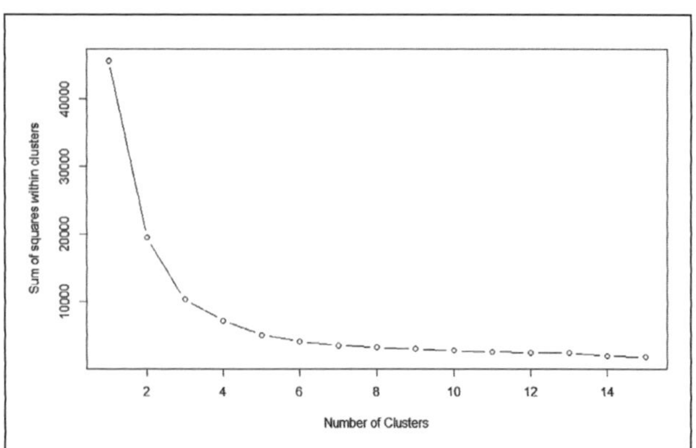

Figure 2. Relationship between number of clusters and SSE within clusters.

1. Microtrips in the cluster are filtered to identify candidate microtrips that meet the continuity criteria;
2. Candidate microtrips are individually appended to the schedule and a provisional PV calculated for each;
3. Candidate microtrips are ranked by lowest provisional PV; and
4. The best microtrip is appended to the schedule.

This process is repeated for each successive cluster in the sequence until the desired schedule length is reached. Consistent with the random selection method, 20 biking schedules are generated using this method and the best schedule selected based on lowest PV.

Single Cluster Method. This method is similar to the best incremental method, but discards the Markov process by combining all microtrips into a single cluster. After selecting the first microtrip randomly, successive microtrips are appended based on fulfilling continuity and lowest provisional PV. Because this is no longer a stochastic process, each starting microtrip generates a single deterministic schedule. Hence, the number of unique candidate biking schedules from this method depends on the number of starting microtrips in the pool. As with the other methods, the best biking schedule is selected from among the candidates based on lowest PV.

Evaluation of Methods

An existing on-road cycling GPS data set is used to demonstrate and evaluate the proposed biking schedule methods. The data set includes 55 h of 1-Hz speed and grade data from three cyclists (A, B, and C) in Portland, Oregon (*3*). Speed and grade data were processed using a kernel smoothing algorithm with bandwidths of 3 and 10, respectively (*28*). Biking schedules are generated for cyclist A (for which the most data are available) and then validated in two ways: by testing application to estimates of cyclist power and breathing rate, and by testing transferability to the other two riders and lower data resolutions. A total of 1,530 microtrips were extracted from the cyclist A data; partial microtrips were discarded.

Power output and breathing rate are calculated using speed and grade from the biking schedules applied to the equations and fixed parameters (mass, resistance factors, etc.) given in (*3*). Transferability to other riders is tested by generating biking schedules for cyclists B and C based on their aggregate travel characteristics (assessment parameters), but using the microtrips from cyclist A. The rationale for this approach is to test the possibility of developing biking schedules for other riders and conditions knowing only aggregate riding characteristics. To

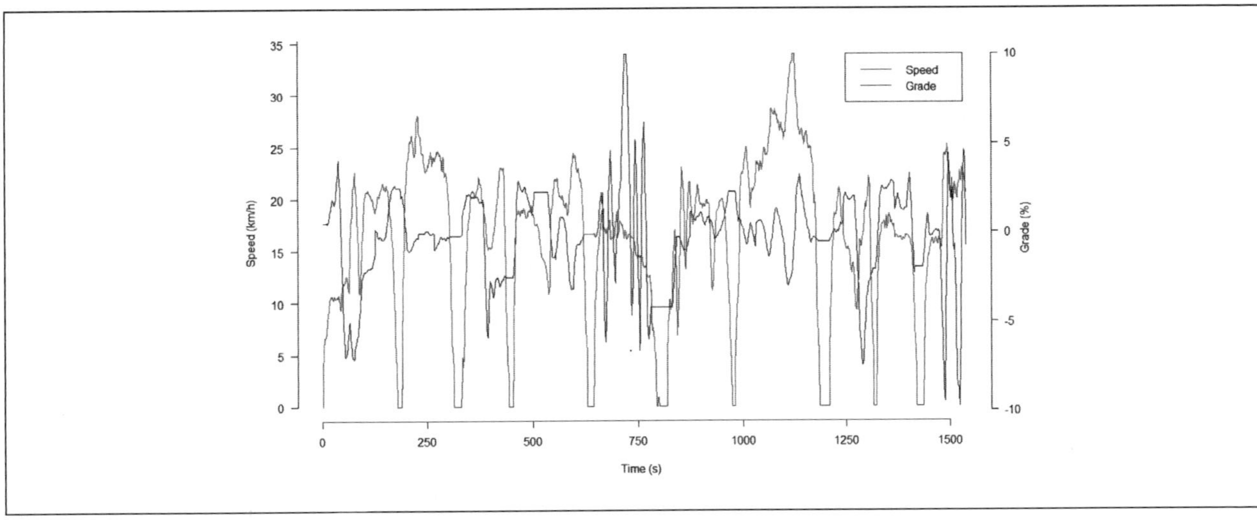

Figure 3. Biking schedule with the best overall PV, generated from the single cluster method.

Table 2. Overall and Parameter PV of Biking Schedules from Each Construction Method

		Construction method		
		Random selection	Best incremental	Single cluster
Individual-parameter PV_x (%)	ATS	8.91	1.40	0.39
	ARS	8.70	0.89	0.08
	PTI	1.94	4.99	3.03
	PTC	6.04	1.19	0.86
	AAA	7.26	0.42	1.57
	PTA	0.96	1.21	1.75
	PTD	3.21	2.60	2.66
	APW	3.04	1.06	2.26
	AAG	2.52	1.03	1.05
	PTPG	4.77	1.15	0.06
	PTNG	21.4	2.78	1.29
	SAGPD	0.15	0.13	0.13
Overall PV (%)		4.93	1.31	1.02

evaluate the effect of GPS data resolution on biking schedule generation, coarser logging intervals of 3, 5, and 10 s are simulated from the original data set by removing observations. Acceleration is recalculated as the difference between consecutive velocity observations, and new biking schedules generated from the revised microtrips and assessment parameters.

Evaluation Results

Processing time to generate the biking schedules varied between methods but was similar among data resolutions. The random selection method took the longest: up to 3 h to identify 20 candidate schedules with PV $<$ 15%. On average, 70 schedules were created to reach 20 candidate schedules. The processing time for this method would increase with more microtrips, more desired candidate schedules, or a lower PV threshold. In contrast, the best incremental method required the least processing time, 10–20 min. The processing time for the single cluster method is mainly determined by the number of microtrips in the trip-starting pool. This method took the most time to develop each schedule, but was still more time efficient than the random selection method because it required fewer iterations to identify 20 candidate schedules. Failure to find microtrips that met the continuity criteria was an issue for all methods, but predominantly a problem for the best incremental method which is restricted to microtrips within clusters rather than the entire pool.

Table 2 gives the overall and individual-parameter PV for the biking schedules generated from each construction method. Based on the PV, the single cluster method yielded the best biking schedule (illustrated in Figure 3), followed closely by the best incremental method. The random selection method substantially underperformed the other two methods according to the PV. Inspecting all generated schedules, there was no clear pattern of certain parameters having consistently higher PV than others.

Figure 4 shows the evolution of PV for each method with biking schedules of increasing length (measured by the number of microtrips). The last data point in each series is the target 25-min schedule. In general, PV improves (decreases) with increasing schedule length, but not monotonically, and the optimum length depends on the construction method. Longer schedules can in some cases degrade the PV due to the constraints of the continuity criteria when selecting microtrips to append.

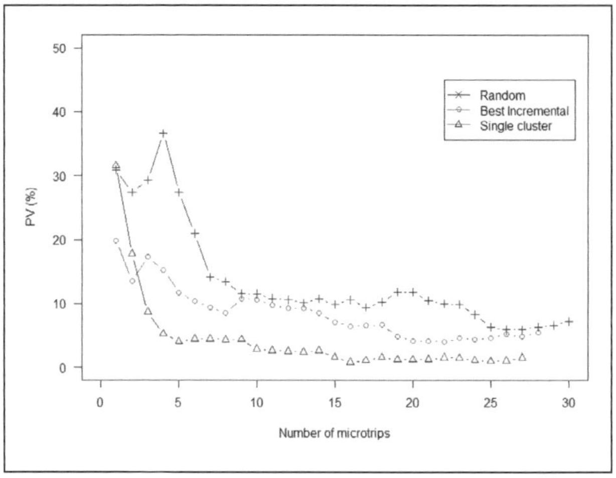

Figure 4. Relationship between PV and drive schedule length for each method.

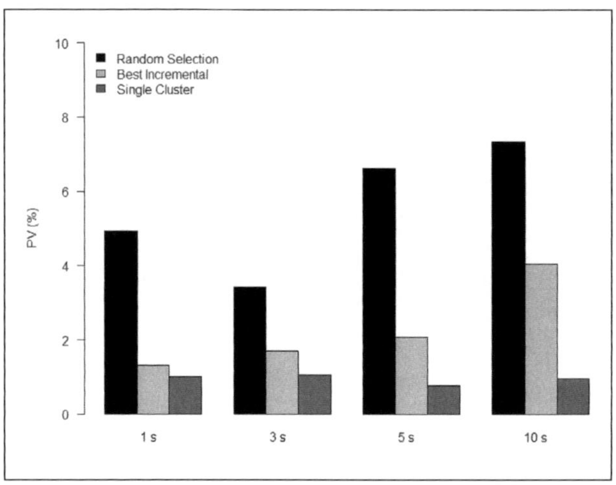

Figure 5. PV across methods using data resolutions of 1, 3, 5, and 10 s.

Substantially shorter schedules generate similar PV for the best incremental and single cluster methods, but the random selection method is less efficient at attaining a low PV. The question of optimal schedule length for biking schedules requires further investigation.

Figure 5 gives the PV results for all three methods using data resolutions of 1, 3, 5, and 10 s (the 1-s results are the same as in Table 2). The ordering among the three methods is consistent across all four data resolutions, with single cluster performing best (lowest PV), followed by best incremental and random selection. The accuracy of the best incremental method degrades at coarser data resolutions. The single cluster method has the most consistently good performance, and no clear relationship with data resolution.

Table 3 gives the cyclist power output and breathing rate results. Biking schedules generated by all methods and data resolutions provide power output and breathing rate estimates within 10% of the estimates from raw data, and most are within 5%, suggesting that biking schedules can plausibly be used for these applications. Somewhat surprisingly, there is no clear relationship between the accuracy of the power and breathing estimates and the PV (shown in Figure 5). If desired, the assessment criteria could potentially be refined to more specifically reflect the determinants of power and breathing, or power and breathing estimates could even be used directly as assessment parameters.

The PV for the biking schedules generated for cyclists B and C (using cyclist A's microtrips) were 2% and 6%, respectively, by the best incremental method, and 1% and 2%, respectively, by the single cluster method. The biking schedules are less precise, as expected, when constructed from a different cyclist's GPS data. Still, the biking schedules are able to represent the cycling dynamics reasonably well based on the assessment criteria. The single cluster method was markedly better for this application, likely because it draws from a larger pool of microtrips and does not rely on a transition matrix that was generated from a different cyclist's data.

Calculated power outputs from these single cluster biking schedules are 11% and 18% higher than from the raw data for cyclists B and C, respectively, and breathing rates are 4% and 8% higher. These differences are likely larger for cyclist C than cyclist B because the dynamics of cyclist C were more distinct from cyclist A. For example, cyclists A, B, and C had mean power outputs of 114, 77, and 57 W, respectively, and mean breathing rates of 21, 20, and 16 L/min. The observation of similar PV but markedly different power/breathing rate accuracy for cyclists B and C supports the idea that the assessment criteria might not well reflect the determinants of cyclist power, and further refinements should be explored.

Conclusion

This paper presents and compares three methods for constructing biking schedules to be used in bicycle travel analysis. The selection of the best method depends on several factors. First, data set size is crucial to the process. A larger data set is preferable for all methods, but particularly for the best incremental method which further segments the microtrips into clusters. For smaller data sets the single cluster method would be preferred. The single cluster method also performed better than other methods with coarser data resolutions. The random selection method is unlikely to be optimal because it generates relatively low-accuracy schedules.

Table 3. Cyclist Power Output and Breathing Rate Calculated from Raw Data and Biking Schedules

		Power output (W)		Breathing rate (L/min)	
		Mean	Difference from raw data (%)	Mean	Difference from raw data (%)
1 sec data	Raw data	114	–	20.8	–
	Random selection	114	0.2	20.8	0.2
	Best incremental	103	9.8	19.1	8.0
	Single cluster	116	1.7	21.1	1.5
3 sec data	Raw data	115	–	20.9	–
	Random selection	113	1.9	20.6	1.1
	Best incremental	112	2.7	20.4	1.7
	Single cluster	123	7.2	22.3	7.0
5 sec data	Raw data	115	–	20.9	–
	Random selection	120	3.9	21.7	3.4
	Best incremental	111	3.6	20.4	3.0
	Single cluster	118	2.6	21.5	2.2
10 sec data	Raw data	116	–	21.1	–
	Random selection	108	7.4	19.8	6.2
	Best incremental	121	4.1	21.9	3.6
	Single cluster	116	<0.1	21.1	<0.1

The accuracy of the method depends on the proximity of the calibration data to the application conditions. In this study, GPS data from one cyclist was able to generate reasonable biking schedules based on the aggregate travel data of two other cyclists in the same 55-hour data set, albeit with less accuracy. Transferability to more remote conditions (other cities, seasons, etc.) requires further investigation. It is expected that larger calibration data sets encompassing more variability (of riders, terrain, trip purposes, bicycle types, weather conditions, facility types, etc.) would have greater utility for generating realistic biking schedules for other contexts. The transferability of biking schedules, however, depends on the consistency of cycling dynamics across contexts, for which we still have little evidence in the literature.

The proposed biking schedules can have a variety of applications. Biking schedules can be used to estimate cyclist power, energy, and breathing rate from aggregate travel data (collected through travel surveys, bikeshare systems, smartphone applications, etc.), and thus improve health effects estimates including physical activity and pollution inhalation. Biking schedules could be segmented by rider type (age, experience), equipment (e-bike, cargo bike, road bike), season or weather conditions, or facility type (bike lane, cycle track, multi-use path) to explore and represent systematic, archetypal differences in urban cycling styles and dynamics among population segments, cities, or facilities. In addition to modeling cycling outcomes, biking schedules could potentially be used to represent the typical energy "costs" of network links and thus applied as inputs to route choice models.

Biking schedules can also be used by bicycle designers and manufacturers, particularly of e-bikes and other human or electric hybrid vehicles, similar to the way driving schedules are used in motor vehicle modeling and design. Representative schedules could be implemented in simulation models and laboratory testing to investigate power consumption and battery life, for example. Segmented biking schedules could provide more customized performance information for specific market segments (sport vs. leisure riders, for example), similar to the city and highway fuel economy information supplied to motor vehicle shoppers. Biking schedules could also be used in research and clinical laboratories to investigate human performance under more realistic cycling conditions than traditional tiered-workload exercise tests.

The proposed biking schedules are a promising new tool for bicycle travel analysis, but further work is needed to develop robust construction methods. Much of the proposed approach was drawn from driving schedule methods, which likely have limited transferability to cycling. The selection of assessment criteria requires further investigation, including the most appropriate parameters to represent outcomes of interest such as energy expenditure, and alternative weighting schemes to adjust the influence of individual assessment parameters on the overall PV. As this is the first known attempt to develop biking schedules, the 12 assessment measures used in this study should be viewed as preliminary. Different micro-trip definitions and methods of determining schedule length should also be explored. Delineating microtrips between consecutive stops would eliminate the need for continuity criteria, which could improve the performance of the best incremental method. Lastly, this study used a limited demonstration data set, and the methods should be further validated on larger data sets encompassing more diverse cycling conditions. Transferability should also be tested by comparing biking schedules and assessment criteria from cycling data in multiple cities.

Author Contributions

The authors confirm contribution to the paper as follows: study conception and design: AB, AM; data collection: AB; analysis and interpretation of results: AM, AB; draft manuscript preparation: AM, AB. All authors reviewed the results and approved the final version of the manuscript.

References

1. Intergovernmental Panel on Climate Change, and O. Edenhofer. (eds.). *Climate Change 2014: Mitigation of Climate Change: Working Group III Contribution to the Fifth Assessment Report of the Intergovernmental Panel on Climate Change.* Cambridge University Press, New York, NY, 2014.
2. Wilson, D. G. *Bicycling Science.* MIT Press, Cambridge, Mass., 2004.
3. Bigazzi, A. Y., and M. A. Figliozzi. Dynamic Ventilation and Power Output of Urban Bicyclists. *Transportation Research Record: Journal of the Transportation Research Board*, 2015. 2520: 52–60.
4. Bigazzi, A. Y., and R. Lindsey. A Utility-Based Bicycle Speed Choice Model with Time and Energy Factors. *Transportation*, (in press).
5. Heinen, E., B. van Wee, and K. Maat. Commuting by Bicycle: An Overview of the Literature. *Transport Reviews*, Vol. 30, No. 1, 2010, pp. 59–96. https://doi.org/10.1080/01441640903187001.
6. Willis, D. P., K. Manaugh, and A. El-Geneidy. Cycling Under Influence: Summarizing the Influence of Perceptions, Attitudes, Habits, and Social Environments on Cycling for Transportation. *International Journal of Sustainable Transportation*, Vol. 9, No. 8, 2015, pp. 565–579. https://doi.org/10.1080/15568318.2013.827285.
7. Ma, X., and D. Luo. Modeling Cyclist Acceleration Process for Bicycle Traffic Simulation Using Naturalistic Data. *Transportation Research Part F: Traffic Psychology and Behaviour*, Vol. 40, 2016, pp. 130–144. https://doi.org/10.1016/j.trf.2016.04.009.
8. Twaddle, H., and G. Grigoropoulos. Modeling the Speed, Acceleration, and Deceleration of Bicyclists for Microscopic Traffic Simulation. *Transportation Research Record: Journal of the Transportation Research Board*, 2016. 2587: 8–16.
9. Twaddle, H., T. Schendzielorz, and O. Fakler. Bicycles in Urban Areas. *Transportation Research Record: Journal of the Transportation Research Board*, 2014. 2434: 140–146. https://doi.org/10.3141/2434-17.
10. Alessandrini, A., and F. Orecchini. A Driving Cycle for Electrically-Driven Vehicles in Rome. *Proceedings of the Institution of Mechanical Engineers, Part D: Journal of Automobile Engineering*, Vol. 217, No. 9, 2003, pp. 781–789.
11. André, M. The ARTEMIS European Driving Cycles for Measuring Car Pollutant Emissions. *Science of The Total Environment*, Vol. 334–335, 2004, pp. 73–84. https://doi.org/10.1016/j.scitotenv.2004.04.070.
12. Lin, J., and D. A. Niemeier. Regional Driving Characteristics, Regional Driving Cycles. *Transportation Research Part D: Transport and Environment*, Vol. 8, No. 5, 2003, pp. 361–381. https://doi.org/10.1016/S1361-9209(03)00022-1.
13. Newman, P. W. G., and J. R. Kenworthy. Use and Abuse of Driving Cycle Research: Clarifying the Relationship Between Traffic Congestion, Energy, and Emissions. *Transportation Quarterly*, Vol. 38, No. 4, 1984, pp. 615–635.
14. Wang, Q., H. Huo, K. He, Z. Yao, and Q. Zhang. Characterization of Vehicle Driving Patterns and Development of Driving Cycles in Chinese Cities. *Transportation Research Part D: Transport and Environment*, Vol. 13, No. 5, 2008, pp. 289–297. https://doi.org/10.1016/j.trd.2008.03.003.
15. Smit, R., A. L. Brown, and Y. C. Chan. Do Air Pollution Emissions and Fuel Consumption Models for Roadways Include the Effects of Congestion in the Roadway Traffic Flow? *Environmental Modelling and Software*, Vol. 23, No. 10–11, 2008, pp. 1262–1270.
16. Berry, M. J., T. R. Koves, and J. J. Benedetto. The Influence of Speed, Grade and Mass during Simulated off Road Bicycling. *Applied Ergonomics*, Vol. 31, No. 5, 2000, pp. 531–536. https://doi.org/10.1016/S0003-6870(00)00022-3.
17. Parkin, J., and J. Rotheram. Design Speeds and Acceleration Characteristics of Bicycle Traffic for Use in Planning, Design and Appraisal. *Transport Policy*, Vol. 17, No. 5, 2010, pp. 335–341. https://doi.org/10.1016/j.tranpol.2010.03.001.
18. Prati, M. V., G. Meccariello, L. Della Ragione, and M. A. Costagliola. *Real Driving Emissions of a Light-Duty Vehicle in Naples. Influence of Road Grade.* SAE Technical Paper 2015-24-2509. Society of Automotive Engineers, Warrendale, Pennsylvania, 2015.
19. Wyatt, D. W., H. Li, and J. Tate. *Examining the Influence of Road Grade on Vehicle Specific Power (VSP) and Carbon Dioxide (CO_2) Emission over a Real-World Driving Cycle.* SAE Technical Paper 2013-01-1518. Society of Automotive Engineers, Warrendale, Pennsylvania, 2013.
20. Kamble, S. H., T. V. Mathew, and G. K. Sharma. Development of Real-World Driving Cycle: Case Study of Pune, India. *Transportation Research Part D: Transport and Environment*, Vol. 14, No. 2, 2009, pp. 132–140. https://doi.org/10.1016/j.trd.2008.11.008.
21. Seers, P., G. Nachin, and M. Glaus. Development of Two Driving Cycles for Utility Vehicles. *Transportation Research Part D: Transport and Environment*, Vol. 41, 2015, pp. 377–385. https://doi.org/10.1016/j.trd.2015.10.013.
22. Dai, Z., D. Niemeier, and D. Eisinger. *Driving Cycles: A New Cycle-Building Method That Better Represents Real-World Emissions.* Department of Civil and Environmental Engineering, University of California, Davis, 2008.
23. Amirjamshidi, G., and M. J. Roorda. Development of Simulated Driving Cycles for Light, Medium, and Heavy Duty Trucks: Case of the Toronto Waterfront Area. *Transportation Research Part D: Transport and Environment*, Vol. 34, 2015, pp. 255–266. https://doi.org/10.1016/j.trd.2014.11.010.
24. Hung, W. T., H. Y. Tong, C. P. Lee, K. Ha, and L. Y. Pao. Development of a Practical Driving Cycle

Construction Methodology: A Case Study in Hong Kong. *Transportation Research Part D: Transport and Environment*, Vol. 12, No. 2, 2007, pp. 115–128. https://doi.org/10.1016/j.trd.2007.01.002.
25. Giakoumis, E. G. *Driving and Engine Cycles*. Springer International Publishing, Cham, Switzerland, 2017.
26. Berzi, L., M. Delogu, and M. Pierini. Development of Driving Cycles for Electric Vehicles in the Context of the City of Florence. *Transportation Research Part D: Transport and Environment*, Vol. 47, 2016, pp. 299–322. https://doi.org/10.1016/j.trd.2016.05.010.
27. Nouri, P., and C. Morency. Evaluating Microtrip Definitions for Developing Driving Cycles. *Transportation Research Record: Journal of the Transportation Research Board*, 2017. 2627: 86–92. https://doi.org/10.3141/2627-10.
28. Brady, J., and M. O'Mahony. Development of a Driving Cycle to Evaluate the Energy Economy of Electric Vehicles in Urban Areas. *Applied Energy*, Vol. 177, 2016, pp. 165–178. https://doi.org/10.1016/j.apenergy.2016.05.094.
29. Ashtari, A., E. Bibeau, and S. Shahidinejad. Using Large Driving Record Samples and a Stochastic Approach for Real-World Driving Cycle Construction: Winnipeg Driving Cycle. *Transportation Science*, Vol. 48, No. 2, 2014, pp. 170–183. https://doi.org/10.1287/trsc.1120.0447.
30. Lin, J., and D. A. Niemeier. Estimating Regional Air Quality Vehicle Emission Inventories: Constructing Robust Driving Cycles. *Transportation Science*, Vol. 37, No. 3, 2003, pp. 330–346.

The Standing Committee on Bicycle Transportation (ANF20) peer-reviewed this paper (18-02134).

Article

TRR
JOURNAL OF THE TRANSPORTATION RESEARCH BOARD

Transportation Research Record
2018, Vol. 2672(36) 92–102

DOI: 10.1177/0361198118801354
journals.sagepub.com/home/trr
SAGE

Use of Deep Learning to Predict Daily Usage of Bike Sharing Systems

Hong Yang[1], Kun Xie[2], Kaan Ozbay[3], Yifang Ma[4], and Zhenyu Wang[5]

Abstract

The use of bikes among stations is often spatiotemporally imbalanced, causing many problems in daily operations. Predictively knowing how the system demand evolves in advance helps improve the preparedness of operational schemes. This paper aims to present a predictive modeling approach to analyze the use of bicycles in bike sharing systems. Specifically, a deep learning (DL) approach using the convolutional neural networks (CNNs) was proposed to predict the daily bicycle pickups at both city and station levels. A numerical study using data from the Citi Bike system in New York City (NYC) was performed to assess the performance of the proposed approach. Other than the historical records, relevant information like weather was also incorporated in the modeling process. The modeling results show that the proposed approach can achieve improved predictive performance in both city- and station-level analyses, confirming the merits of the proposed method against other baseline approaches. In addition, including information from neighboring stations into the models can help improve the performance of station-level prediction. The predictive performance of the CNN was also found to be related to parameters such as temporal window, number of neighboring stations, learning ratio, patch size, and the inclusion of additional data such as drop-offs. Thus, the implementation of the proposed models requires necessary calibration to determine appropriate parameters for a given bike sharing system.

Public bike sharing is becoming increasingly prevalent in many cities worldwide. Many existing bike sharing schemes (BSSs) are equipped with automatic rental systems to facilitate accessing and returning bicycles (*1*). Through the rental systems, real-time records of each trip become available. In addition, real-time data on available docks and bicycles at each station are also available. These data not only offer riders a convenient way to query bicycle information, but also help operators to judge their system performance responsively. Given the easy access to the system data, researchers have been leveraging the data to help operators understand their BSS usage predictively. Over the years, tremendous efforts have been made to develop models for bike usage prediction. In general, these models have attempted to use a set of determinants to explain BSS usage (e.g., pickups and drop-offs). The success of such models largely lies in the simplified assumptions on the statistical causal relationships between the variables. Despite the simplicity, these tractable mathematical models hardly capture the complexity of the changes in daily usage of BSSs. Thus, this motivates us to seek more reliable approaches that can work in the context of unclear causal relationships and limited explanatory variables.

Recently, deep neural networks (DNNs) have shown some impressive results in a variety of challenging tasks such as speech recognition. The capability of DNNs to automatically learn complex and nonlinear patterns from observed data makes them applicable to a wide range of problems, including those related to transportation systems. Meanwhile, given the accessibility of massive BSS records and other limited data (e.g., weather information), deep learning (DL) methods such as convolution neural networks (CNNs) and deep belief networks (DBNs) hold promise for performance improvements in

[1]Department of Modeling, Simulation & Visualization Engineering, Transportation Research Institute, Old Dominion University, Norfolk, VA
[2]Department of Civil and Natural Resources Engineering, University of Canterbury, Christchurch, New Zealand
[3]C2SMART Center (A Tier 1 USDOT UTC), Department of Civil and Urban Engineering & Center for Urban Science and Progress (CUSP), Tandon School of Engineering, New York University (NYU), Brooklyn, NY
[4]Department of Information Technology and Decision Sciences, Old Dominion University, Norfolk, VA
[5]Department of Modeling, Simulation & Visualization Engineering, Old Dominion University, Norfolk, VA

Corresponding Author:
Address correspondence to Hong Yang: hyang@odu.edu

analyzing the complex scenarios of BSS usage that may have nonlinear and heterogeneous patterns.

Thus, this paper aims to examine the potential of using DL techniques to forecast the daily usage of BSSs. It is believed that the DL approach explored in this paper is a useful and insightful way of fundamentally rethinking the prediction problem of bicycle usage. To be more specific, the predicted "usage" in this paper is considered to be the number of pickups. Depending on the scope of analysis, analysts may consider other definitions of usage, for example, the number of trips between a certain pair of stations or the total number of pickups and drop-offs at a station.

Literature Review

There are a few studies that have focused on the prediction of BSS usage. These studies have investigated factors that may affect bike usage. Typical determinants relateing to weather, socio-demographics, economic attributes, transportation facilities, and so forth have been frequently identified (*2–4*). For example, Hampshire and Marla (*5*) studied the impact of built environment on hourly pickups and drop-offs at the sub-city district level in two cities in Spain. These studies were mainly conducted using daily (*4, 6–8*) or monthly (*2, 9*) aggregated data in modeling. Some others focused on the hourly usage of BSSs (*3, 10–12*).

Existing studies were often focused on the activity (pickups and/or drop-offs) of the stations or more globally the state of the system. They modeled the BSS usage at city (system) level (*10*), sub-city district level (*5*), neighborhood level (*13*), and station clusters (*14*). For example, Wang (*7*) predicted the city-level hourly bike rentals of the Citi Bike system in New York City (NYC). The aggregation on a larger spatial scale in these studies can improve the model performance (*13*). However, the bike usage aggregated at these scales was less pertinent than using bike activity data at a station level. The spatial aggregation may group together stations with significant demand profiles but fail to capture fine-grained spatial effects, which in turn results in less-reliable models for prediction (*11*).

To facilitate operational applications, a few studies have thus attempted to statistically predict the bike usage at a station level. For example, Buck and Buehler (*6*) built the multivariate regression model to identify determinants of the average daily checkouts per station for the Capital BSS in Washington, D.C. In the study (*8*), it was found that regression models based on ordinary least squares estimators are not appropriate because the bicycle usage is usually not normally distributed. Alternatively, previous studies have developed negative binomial (NB) models to forecast BSS usage (*4, 8*). Considering excessive zero counts, Rudloff and Lackner (*11*) introduced hurdle models to predict the hourly bike usage at a station level with data from the Citybike Wien in Vienna. Despite its promising performance, the hurdle model is relatively complex and difficult to implement without automated variable selection procedures. Other than the statistical modeling, several studies also sought to predict BSS usage using data mining techniques (*7, 12, 14*).

Despite the success in modeling bike usage with identified determinants, many issues are still associated with current modeling practices. One important issue is the subjective extraction of explanatory variables. Many studies used different spatial criteria to obtain the independent variables. For example, Buck and Buehler (*6*) linked each station to independent variables (e.g., population and bike lane supply) measured within a 0.5-mile buffer zone. In contrast, others used smaller buffer zones such as 400-meter zones (*9*) and 300-meter zones (*15*) to obtain similar variables (e.g., bus stops and length of bike infrastructure). However, as shown in (*15*), modeling results will often be sensitive to the range of the buffer zone. Unfortunately, there was no clear guidance to define the buffer range for data extraction with better quality. Another major issue is temporal inconsistency between the dependent variable and the explanatory variables. For example, bike usage statistics in 2014 were modeled in (*13*). However, taxi usage in 2013, population, and housing data from the 2010 US Census were used as covariates. Likewise, variables such as socioeconomic attributes used in (*15*) were collected in 2013 and 2014 whereas the bike flow data were calculated in 2011. The temporal inconsistency makes it difficult to explain bike usage by covariates collected from different periods. Meanwhile, the need to synthesize extensive data sets often makes it difficult to collect them consistently over several years (*8*). For example, many variables such as population are dynamically changing over time. From an operational perspective, including historical survey data (e.g., historical population and jobs) cannot reflect effectively actual temporal variations at present. Last but not least, modeling issues related to multi-collinearity, variable selection, determination of clusters, model transferability, etc., also limit the use of existing approaches.

The present paper builds upon the lessons learned from early empirical studies and contributes to the research community by developing a new approach for predicting bike usage of BSSs.

Proposed Methodology

Earlier studies have made great efforts to establish statistical models for predicting BSS usage. However, many issues discussed above have limited their use, especially for operational applications. This study seeks instead to take advantage of recent developments in machine learning techniques to tackle the prediction problem.

Neural Network Model

As discussed in the previous section, many statistical models cannot quantify well the bike usage of BSSs. One notable issue is that bike usage data are not normally distributed. In addition, the assumption on the linearity between bike usage and relevant variables is often violated. On the other hand, neural networks will be able to avoid such issues even when some data are skewed and a nonlinear relationship exists. Neural networks consist of multiple layers of interconnected nodes (neurons) for learning tasks. These nodes are called perceptrons and mathematically resemble different linear regression (LR) models. The difference between an LR model and a perceptron is that the latter feeds the signal obtained from an LR model into an activation function that may or may not be linear. In general, given a neural network with N layers, the final output p of layer n could be represented as follows:

$$p = f(W_n x_{n-1} + b_n) \quad (1)$$

where f is the activation function (e.g., sigmoidal function); W_n is the weighted coefficient of layer n; x_{n-1} is the output of layer $n-1$ and serves as the input of layer n; and b_n is the bias for layer n. By quantifying the prediction error using an error measure e, the parameters of all layers of the network are jointly optimized to make the estimates approximate the desired output within the predefined error threshold ε or the maximum iteration I. The gradient descent algorithm can be used to determine how the parameters should be updated to reduce the prediction error. Assume there are I iterations, the parameters of each layer in the ith iteration are modified:

$$\mathrm{W}_n^i = \mathrm{W}_n^{i-1} - \eta \frac{\partial e(p^i, y)}{\partial \mathrm{W}_n^{i-1}} \quad (2)$$

$$\mathrm{b}_n^i = \mathrm{b}_n^{i-1} - \eta \frac{\partial e(p^i, y)}{\partial \mathrm{b}_n^{i-1}} \quad (3)$$

where η is the learning ratio that controls the step size and $i = 1, 2, ..., I$. By repeatedly taking small steps in the direction opposite to the gradient, the hidden layers will learn to capture the relationship between the input data and the target output data, and the output layer will predict the desired results based on the learned relationship.

However, in case of high-dimensional input, just using the activation function to approximate the relationship between the raw input and output can be time-consuming and inefficient. To better capture the nonlinearity characteristics, the idea of DL is initiated to build models that represent data at multiple levels of abstraction and can discover accurate representations autonomously from the data themselves via various methods such as auto encoder, belief propagation, and convolution.

CNN Model

Recently, the DL concept that originated from artificial neural network research has received increasing attention. Many research efforts have promoted its application in various areas such as signal processing, artificial intelligence, and so forth, owing to its achievable high performance. As such, various DL methods such as DBNs, deep Boltzmann machines (DBMs), and DNNs, have been introduced. Among them, the CNN under the DNN category has achieved state-of-the-art performance in various applications (*16*). It is a subclass of neural networks with constrained connectivity patterns between some layers (*17*). Typically, they are very useful if input data exhibit certain patterns, for example, the ordering of image pixels in a grid and the temporal structures of an audio signal. Thus, considering the complex spatio-temporal patterns of bike usage in a BSS, this paper explores the possibility of using the CNNs for its predictive analysis.

As shown in Figure 1, a CNN contains three types of layers: convolutional layers, pooling layers, and hidden layers. The convolutional layer takes a stack of feature maps as input and convolves each of these with a set of learnable filters to produce a stack of output feature maps to form a deep model. In the ith iteration, this is efficiently implemented by replacing the matrix–vector product $W_n^{(i,j)} \otimes X_{n-1}^{(i)}$ in Equation 4 with a sum of convolutions. The output feature map $X_n^{(i,j)}$ is represented as follows:

$$\mathrm{X}_n^{(i,j)} = f(W_n^{(i,j)} \otimes X_{n-1}^{(i)} + b_n^{(i,j)}) \quad (4)$$

where $j = 1, 2, ..., J; i = 1, 2, ..., I; n = 1, 2, ..., N$; $\otimes$ represents convolution operation; the matrix $W_n^{(i,j)}$ represents the filters of layer n; and $b_n^{(i,j)}$ represents the bias for the feature map. Note that a feature map $X_n^{(i,j)}$ is calculated by computing a sum of convolutions with the feature maps of the previous layer that has a predefined patch size C.

The convolution operation takes advantage of the input structure and reduces the number of parameters needed during the training process. During the convolution, each perceptron is connected to the patched subset of perceptrons in the previous layer using the following equation:

$$W_n^{(i,j)} \otimes X_{n-1}^{(i)} = \sum_{\mathrm{k}=1}^{K} W_n^{(i,j,k)} \times X_{n-1}^{(i)} \quad (5)$$

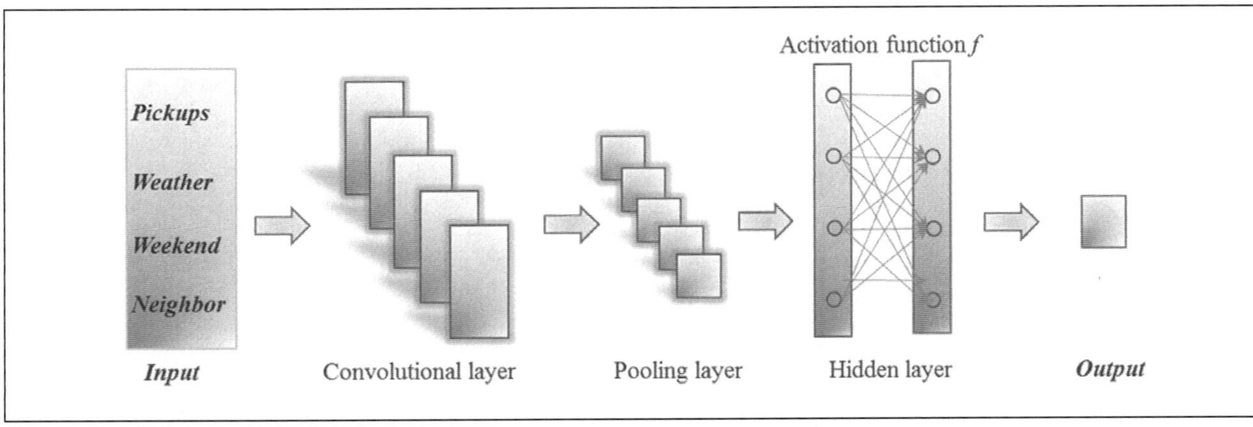

Figure 1. A schematic overview of a convolutional neural network.

where k denotes the number of convolution patches. This means that each perceptron detects features across the input. With predefined patch size C, applying feature detectors across the entire input enables us to capture the pattern of the input data, for example, the archived bike usage data of a BSS.

After modeling local relationships in input using convolutional layers, the dimensionality of the feature maps is further reduced by inserting a pooling layer. This allows the hidden layer to model correlations across a larger part of the input with a lower resolution. The pooling layer reduces the dimensionality of a feature map by averaging its outcomes across local regions of the input (*18*). Thus, the model is more invariant to small translations of the input, which provides the robustness to outliers. Unlike convolutional layers, pooling layers typically do not have any trainable parameters. The outputs of pooling layers are used as the input for hidden layers consisting of neural perceptrons, relevant connections, and activation functions. Using the updating mechanism in Equations 2 and 3, parameters are trained to fit the model.

Convolutional and pooling layers enable higher layers in the network to obtain a coarser representation of the input. Thus, the complicated relationships between variables such as weather and historical data and current daily usage can be modeled.

Numerical Study

Data Sources

This paper uses the data from the Citi Bike System in NYC as a case study. As a privately owned BSS in NYC, Citi Bike was launched in May 2013 with 6,000 bikes at 330 docking stations in Manhattan and parts of Brooklyn and will be expanded to over 700 stations by the end of 2017. It is one of the largest BSSs in the world. This paper uses it as a case study and tests the proposed prediction approach.

The trip data archived from January 1, 2015 to September 30, 2017 were used in following analysis. The data were downloaded from the Citi Bike official webpage (https://www.citibikenyc.com/system-data). The total number of bike pickups in this period was 36,030,073. The only external data sought in this study include weather and holiday information because they have been found to tightly relate to bike usage. We obtained weather data from the Weather Underground (https://www.wunderground.com/) and the date information from the calendar. The daily maximum and minimum temperature, precipitation—snow and rainfall—measured by the weather station in the city, and weekend information are assembled with the daily pickups of each station as well as the entire system to test the model's performance.

Descriptive Analysis

The collected data were analyzed to explore the characteristics of the system usage. The daily pickups of the study period are shown in Figure 2. Figure 2*a* illustrates the patterns of bike pickups for the entire system. On average, the daily usage was 36,174 times. It should be noted that there was no data from January 23–26, 2016 owing to the deadly Winter Storm Jonas passing over NYC. The black line in Figure 2*a* denotes the total daily pickup, and the red and green ones are rentals by male and female riders, respectively. The blue line in Figure 2*a* shows the usage of riders without gender information. These curves indicate that there was more bike usage in summer and fall but much less in winter.

Figure 2*b* shows daily pickups with respective to the daily maximum temperature. It can be seen that pickups did not change linearly with the change of temperature.

Overall, the bike usage shows an increasing trend when temperature gradually increases up to 77°F. The peak value of 54,399 times occurs at 77°F. However, when the temperature is relatively high, the daily pickups tend to decrease. Such a trend suggests that a LR model using temperature as a regressor will not work appropriately. Figure 2*c* shows the spatial distribution of the average daily pickups at different stations. It should be noted that the distribution of pickups among stations is not uniform (e.g., darker (red) dots denote more frequently used stations). Heavily used stations are concentrated in the downtown and midtown of Manhattan. The most frequently used station (ID = 519) on average had 370 pickups a day. The imbalanced distribution of the bike usages among stations should be attributable to the specific location, built environment, and social activities in the area. However, many of these factors (e.g., number of employees near each station; number of tourists) that quantify these attributes are not easy to (continuously) obtain. On the other hand, including some outdated survey data in the analysis will lead to a fundamentally flawed prediction. Thus, it would be more practical to use models with less data needs that can capture spatial variations of daily BSS usage.

Rather than relying on the untraceable factors (e.g., daily visitors/population in a city), we examined bike usage under different weather conditions. Figure 2*d* shows the change of the average daily trips with respect to precipitation. The usage was noticeably higher for the days without rain (39,128 times). Nevertheless, there were still many pickups in the presence of high precipitation, that is, (2, 2.5] inches. The average pickups with precipitation of (2, 2.5] inches greatly decreased to less than 10,000 times a day. Likewise, Figure 2*e* shows a clear relationship between the average daily pickups and snowfall. The pickups dropped drastically for the days

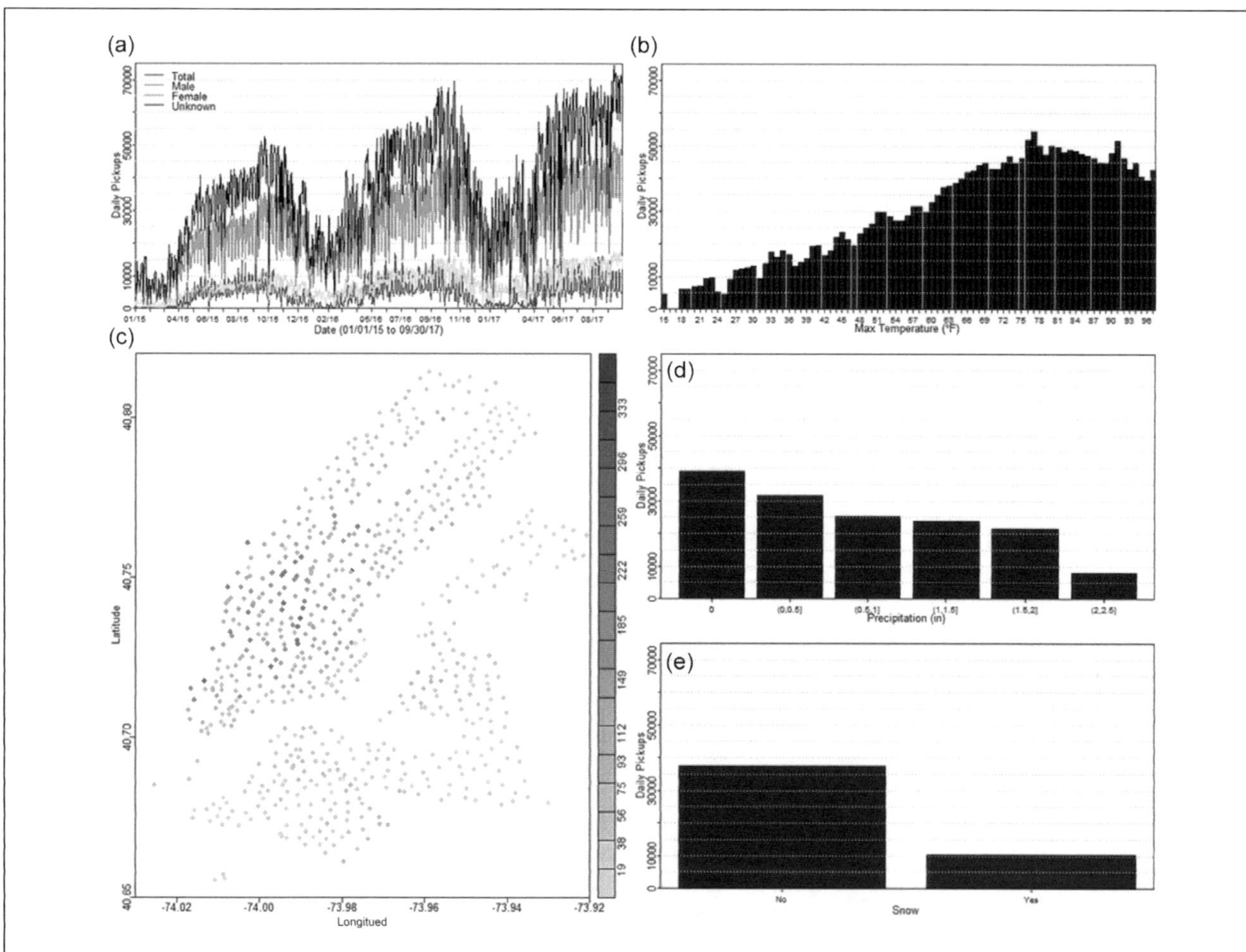

Figure 2. Descriptive analysis of Citi Bike pickups (01/01/2015 to 09/30/2017): (*a*) daily pickups; (*b*) average daily pickups based on temperature; (*c*) average daily pickups at each station: (*d*) average daily pickups based on precipitation; and (*e*) average daily pickups based on snow.

with snow. Specifically, the average pickups on snow days were less than one third of those on days without snow. Both the rainfall and snowfall have shown a notable impact on the bike usage. Thus, it would be helpful to include these objective measurements in predicting bike usage.

Results and Discussion

In this paper, we assembled the Citi Bike usage data along with the weather and holiday data to create the final data set for analysis. By applying the proposed CNN method, the data set was used to train the models and predict the usage of the bike sharing system. Specifically, data from January 1, 2015 to December 31, 2016 were used to train the CNN models so that the information of different periods and factors could be incorporated. The remaining nine months of data were then used for testing. The data structure can be represented as $(x_1, x_2, x_3, x_4, x_5, x_6, x_7, y_d, p_d)$, where x_1 represents historical daily trips and it is a vector consisting of the daily pickups observed in a given time window (e.g., one week); x_2 and x_3 are maximum and minimum temperature (°F) of each day, respectively; x_4 is daily precipitation (inches); x_5 denotes whether it was a rainy day; x_6 denotes whether it was a snowy day; x_7 is whether it was a holiday (weekend) or not; y_d is the actual pickups of dth target day; and p_d is predicted pickups of dth target day. Other factors such the availability of bicycle corridors (*19*) that may affect the usage were not included. Both city- and station-level models were developed, which will be detailed below. When developing the city-level models, the daily pickups of each station were aggregated to obtain x_1 and y_d for the entire system.

To evaluate the performance of the proposed models, we choose three metrics that are frequently used in studies on BSS usage prediction. These metrics include mean absolute error (MAE), root-mean-squared-error (RMSE), and R^2 defined as follows:

$$MAE = \frac{1}{D}\sum_{i=1}^{D}|p_i - y_i| \tag{6}$$

$$RMSE = \sqrt{\frac{1}{D}\sum_{i=1}^{D}(p_i - y_i)^2} \tag{7}$$

$$R^2 = 1 - \frac{\sum_{i=1}^{D}(y_i - p_i)^2}{\sum_{i=1}^{D}(y_i - \bar{y})^2} \tag{8}$$

where D is the total number of observations in the test data set; p_i is the predicted value; and y_i is the actual observed number of pickups. Different metrics focus on different aspects: (a) MAE directly explains by how many pickups the predictions are off; (b) RMSE focuses on the penalty of predictions far away from the actual observations; and (c) R^2 is the coefficient of determination.

City-Level Prediction

The city-level CNN model was developed with an epoch of 40, patch size of 2, and a learning ratio of 1×10^{-5}. A detailed discussion on the choices of the parameters is presented in a later section. For comparison, the Neural Network (NN) and Autoregressive Integrated Moving Average (ARIMA) time series models were also developed. As mentioned earlier, the data from 2015 and 2016 were used to train each of the models and the 2017 data were used for testing the models. The modeling results based on the three approaches are shown in Figure 3*a*: (i) the red line is based on NN model; (ii) the green line is based on the ARIMA model; (iii) the blue line represents the CNN modeling result; and (iv) the black line denotes actual pickups for reference. The proposed models were implemented in Python3.5 with Intel® Core™ i5-6300U CPU at 2.40 GHz. The training process using up to 2 years of data needs less than 5 minutes for each epoch, depending on data size and model parameters. Once a model is trained, the prediction for the testing data only takes a few seconds.

The modeling results show that the CNN model outperforms the others, with the time series model being worse. The corresponding MAE, RMSE, and R^2 of the NN model are 7,717.614 ± 271.599 (mean ± SD), 9,394.204 ± 209.640, and 0.726 ± 0.012, respectively. Likewise, the three performance measures for the ARIMA model are 8414.234, 11031.790, and 0.622, respectively. In contrast, the performance measures of the CNN-based model are 6613.445 ± 246.165, 8168.797 ± 163.519, and 0.793 ± 0.008, respectively. These small MAE and RMSE, and large R^2 performance measures confirm that the proposed model improved prediction of the daily pickups at the city level.

Station-Level Prediction

Based on the same modeling structure, the station-level modeling results for a station (ID = 79) are illustrated in Figure 3*b*. As in the city-level prediction, CNN performs better than the other models. Its MAE, RMSE, and R^2 are 14.264 ± 0.201, 18.995 ± 0.274, and 0.721 ± 0.008, respectively. The baseline neural network method resulted in slightly worse results for this station, with MAE = 14.887 ± 0.231, RMSE = 19.784 ± 0.403, and R^2 = 0.698 ± 0.012, respectively. Once again, the performance of the ARIMA model was less preferred: MAE = 20.710, RMSE = 26.604, and R^2 = 0.453.

Compared with the city-level data, the pickups associated with individual stations can have larger variations. Explanatory variables may dynamically change or cannot

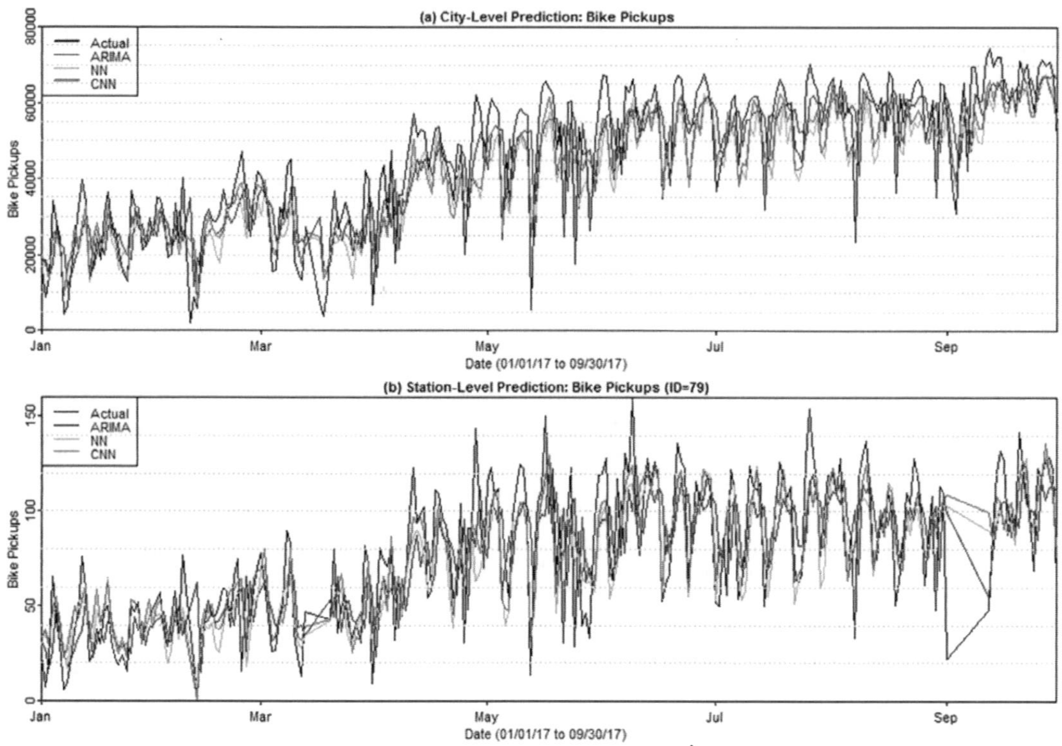

Figure 3. Predictive performance: (*a*) City-level; and (*b*) Station-level.

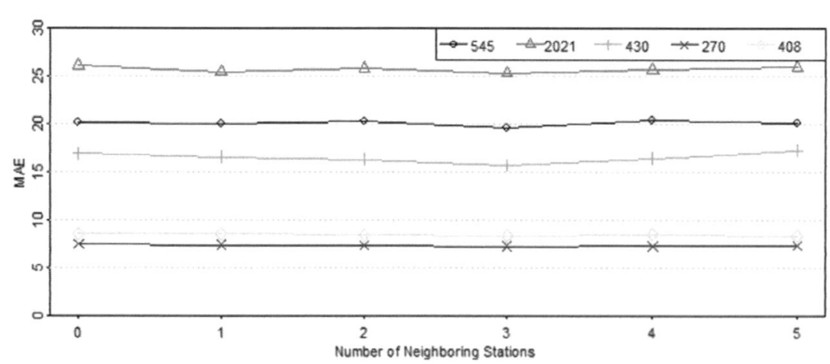

Figure 4. Model performance with different numbers of nearest neighboring stations.

be accurately obtained day to day. Thus, it is difficult to incorporate their reliable measurements in the day-to-day prediction models. As an alternative, this paper proposes to include the spatial relationship between the target station and its neighbors to enhance the performance.

Thus, for station-level prediction, we further assemble the information of neighboring stations to the data set of the target station. As an example, we have used five stations (ID = 545, 2021, 430, 270, and 408) as the target stations to test the proposed approach. For each station, we have tested the effect of including a different number of the nearest neighboring stations in the model. As shown in Figure 4, including additional neighbors in the model improves its performance in terms of the MAE reduction. Nevertheless, the marginal benefit is not obvious when three or more neighbors were

Table 1. Station-Level Predictive Performance

District	Station ID	Include neighbors	MAE	RMSE	R^2
Downtown	79 (High)	Yes	14.015 ± 0.228	18.876 ± 0.213	0.725 ± 0.006
		No	14.264 ± 0.201	18.995 ± 0.274	0.721 ± 0.008
	408 (Low)	Yes	8.411 ± 0.028	10.849 ± 0.044	0.692 ± 0.002
		No	8.604 ± 0.118	10.977 ± 0.177	0.685 ± 0.010
Midtown	545 (High)	Yes	19.564 ± 0.284	25.234 ± 0.345	0.788 ± 0.006
		No	20.200 ± 0.455	25.672 ± 0.531	0.781 ± 0.009
	2021 (Low)	Yes	25.323 ± 0.282	32.265 ± 0.313	0.649 ± 0.005
		No	26.097 ± 0.534	32.769 ± 0.227	0.638 ± 0.005
Uptown	3164 (High)	Yes	28.587 ± 0.702	35.932 ± 0.866	0.609 ± 0.019
		No	28.975 ± 0.710	36.627 ± 0.888	0.594 ± 0.020
	305 (Low)	Yes	18.419 ± 0.157	23.931 ± 0.261	0.725 ± 0.006
		No	19.137 ± 0.224	24.092 ± 0.232	0.722 ± 0.005
Brooklyn	430 (High)	Yes	15.709 ± 0.180	20.359 ± 0.140	0.708 ± 0.004
		No	16.924 ± 0.549	21.686 ± 0.505	0.669 ± 0.015
	270 (Low)	Yes	7.304 ± 0.123	9.309 ± 0.142	0.526 ± 0.015
		No	7.501 ± 0.119	9.506 ± 0.112	0.505 ± 0.012
Queens	3124 (High)	Yes	8.022 ± 0.068	11.222 ± 0.115	0.560 ± 0.009
		No	8.141 ± 0.112	11.426 ± 0.099	0.544 ± 0.008
	3121 (Low)	Yes	5.329 ± 0.080	8.660 ± 0.142	0.418 ± 0.019
		No	5.570 ± 0.344	8.962 ± 0.375	0.376 ± 0.054

included. This may be attributed to the fact that the entropy does not increase much with more redundant information of neighboring stations. In addition, adding too many neighbors has the high risk of including irrelevant information in the model. Thus, for further analysis, we include three nearest neighbors for station-level prediction. With their information, the R^2 values for the five target stations are between 0.5 and 0.8.

Other than the five stations tested above, we also assessed the proposed CNN approach with additional stations that are distributed across different districts in NYC. The five districts include downtown, midtown, and uptown Manhattan, Brooklyn, and Queens. In total, two stations were randomly selected from each district to cover both high and low usage scenarios. Table 1 presents the results when applying the CNN approach to each station with and without (three) neighboring stations. The results were based on the average of five simulations. Overall, the MAE, RMSE, and R^2 (mean ± SD) for the model with the nearest neighboring stations are consistently lower than those without neighbors. Therefore, it can be argued that the CNN approach including neighboring stations helps improve the predictive performance. In addition, the consistent findings among various stations suggest that the proposed method is applicable to different stations.

Sensitivity Analysis

To further verify the performance of the proposed approach under different scenarios, sensitivity analysis was conducted.

Impact of Adding Drop-Off Data in Models

This study examined the impact of including drop-off data in the models. Other than the use of historical pickup data, historical drop-offs were also included in the input to predict the daily pickups of each bike station. The corresponding MAE, RMSE, and R^2 of the extended models are shown in Figure 5. Each symbol in the figures represents the average value of five simulation runs for a selected station. Overall, they suggest that adding additional information on the drop-off records can slightly improve the performance of the proposed CNN models since most points in each figure are close to the reference line. Despite the slight improvement, this will bring extra burdens on data acquisition and processing, and require more computation efforts.

Impact of Historical Data

The analysis presented above used one week of historical data as the input for x_1. To confirm this is an appropriate choice and analyze the impact of temporal window on the modeling results, we test five stations (ID = 545, 2021, 430, 270, and 408) considering different temporal windows. Here x_1 was varied from one week to four weeks. In other words, it used trips of the latest 7 to 28 days as the historical input to predict the target day's pickups. As shown in Figure 6, the predictive performance in terms of MAE, RMSE, and R^2 show that increasing the temporal window led to degraded performance: MAE and RMSE increased with more weeks whereas R^2 decreased. Thus, this suggests that a

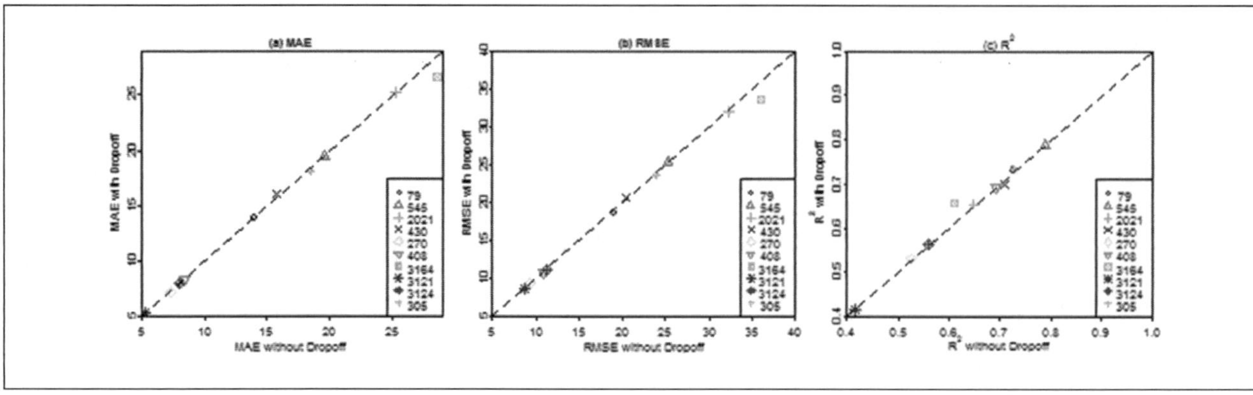

Figure 5. Impact of adding drop-off data as an input in model development.

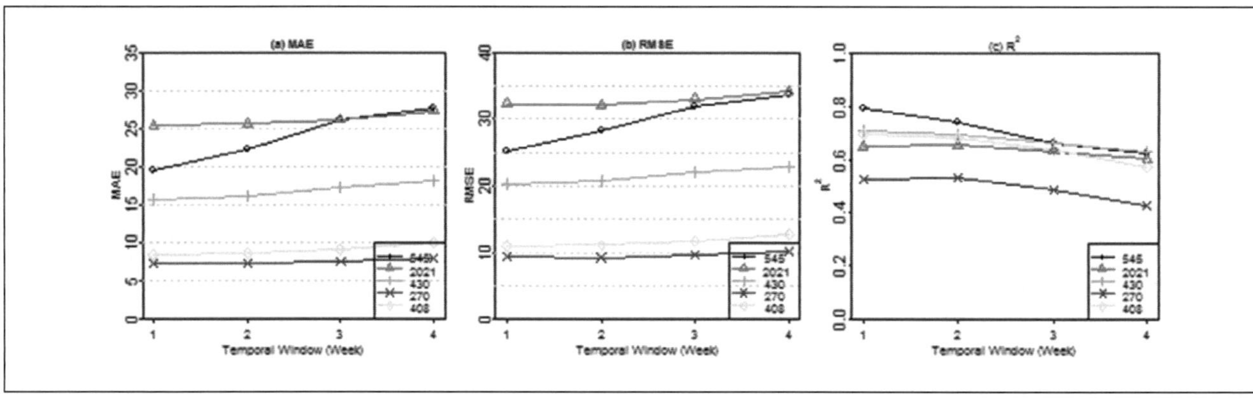

Figure 6. Impact of different temporal windows.

temporal window of one week is sufficient to achieve relatively good performance.

CNN Parameter Analysis

There are two key parameters that may affect the performance of CNN models: learning ratio and patch size. A high learning ratio may lead to quick approximation but degraded performance. A low learning ratio may achieve high performance but sacrifice computational resources. As to the patch size, it determines to which level the pattern is transformed through convolution. A large patch size means that more abstract feature representation of the neural network will be captured from the input. Thus, these two parameters need to be carefully considered to balance the performance and computation complexity.

As shown in Figure 7, different combinations of learning ratio and patch size were tested. The tested learning ratios include 1.0×10^{-5}, 5.0×10^{-5}, 1.0×10^{-4}, 1.5×10^{-4}, and 2.0×10^{-4}. The tested patch sizes include 2, 3, and 4. Note that we used one week of historical data, thus the patch size cannot be very large. Figure 7*a* shows the performance of city-level prediction. When the patch size was two and the learning ratio was 1.0×10^{-5}, the MAE reached the lowest value. Overall, increasing the patch size increased the MAE but reduced the training time due to high-level abstraction. The increase in the learning ratio led to higher MAE. Similar findings were found when testing the station-level prediction for different scenarios shown in Figure 7, *b–d.* The change of modeling performance with respect to these parameters suggests the necessity for calibration when implementing the proposed approach in other BSSs.

Conclusion

The prediction of BSS usage continues to attract growing interest. However, many existing models that use many variables to explain the change of bike usage are not flexible enough to envision the real-world data due to the need for restrictive statistical modeling assumptions and

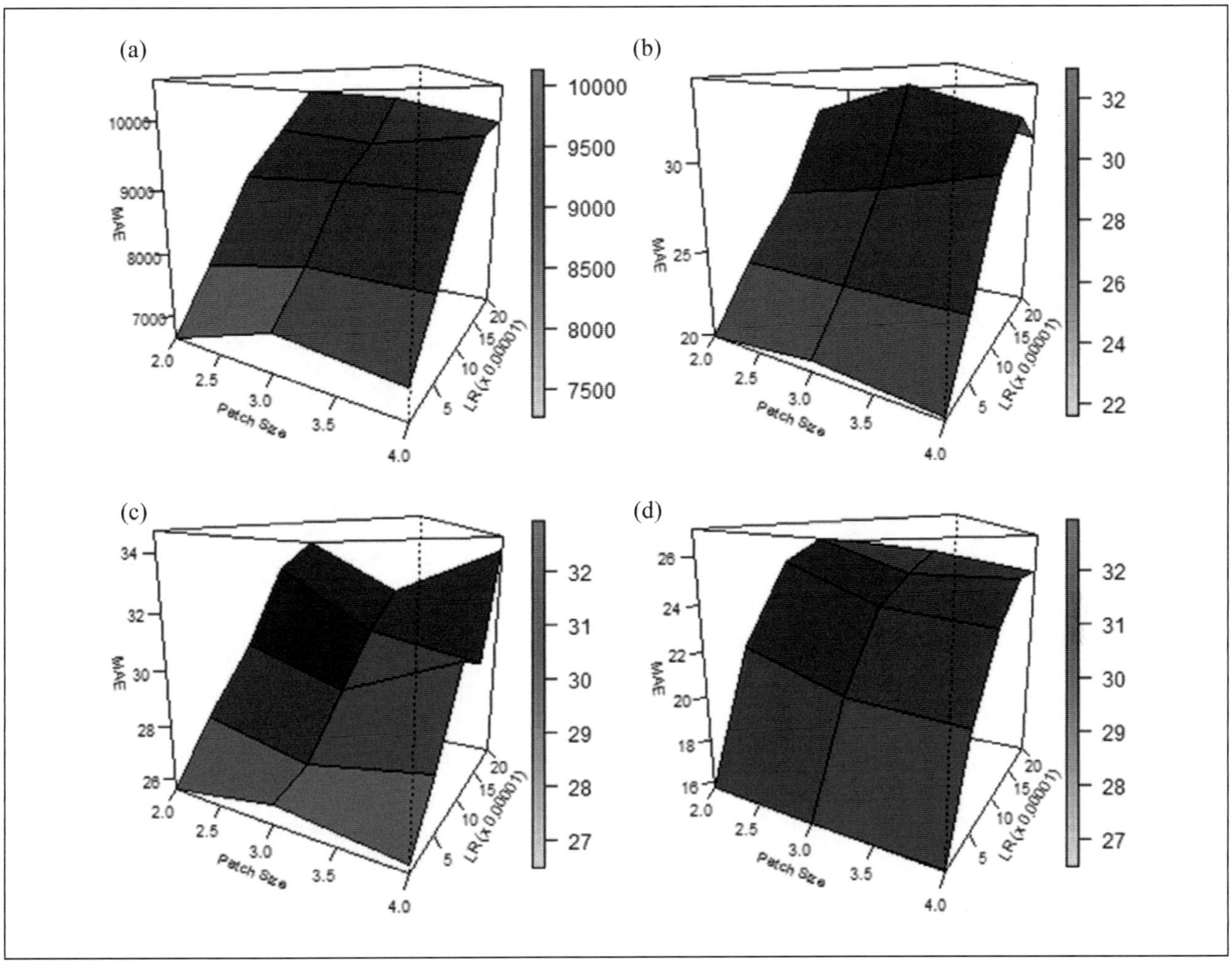

Figure 7. Performance changes with respect to different model parameters; (*a*) City level; (*b*) Station ID=545; (*c*) Station ID=2021; (*d*) Station ID=430.

exhaustive data acquisition among others. This paper proposed a DL approach based on the CNN model. The proposed approach applied historical BSS data and very limited external variables that can be measured objectively to predict daily bike usage of BSSs. We obtained promising results based on the NYC Citi Bike system by demonstrating that both the city- and station-level predictive analyses can achieve better performance if the CNN model was used. The improved performance for models applied to multiple stations confirmed that the proposed approach is also applicable for different stations. The performance of the station-level models can be further improved if information about the nearest neighboring stations was included. In addition, the sensitivity analysis is necessary for the prediction of any BSS because it can help determine the most practical and appropriate model parameters. Overall, the benefit of using the CNN method lies in that it uses the convolution transformation and activation functions in neural networks to account for the complex nonlinear relationship between the target variable and the explanatory variables. Thus, it helps improve the predictive performance for modeling the unknown casual relationships between bike usage and various factors.

Despite the promising performance, this paper has used one typical BSS in a high-density urban area as the case study to evaluate the proposed prediction approach. A useful extension to this problem would be to verify it with data from additional BSSs, especially those of smaller systems. In addition, the comparison of the proposed approach with other DL algorithms is also suggested as future endeavors.

Acknowledgments

The research work presented in this paper was partially supported by the C2SMART Tier 1 University Transportation Center at New York University. The contents of the paper do not necessarily reflect the official views or policies of any sponsoring agencies.

Author Contributions

The authors confirm contribution to the paper as follows: study conception and design: Hong Yang; data collection: Hong Yang, Kun Xie, and Zhenyu Wang; analysis and interpretation of results: Kun Xie, Kaan Ozbay, and Yifang Ma; draft manuscript preparation: Hong Yang, Zhenyu Wang, Kaan Ozbay, and Yifang Ma. All authors reviewed the results and approved the final version of the manuscript.

References

1. Raviv, T., and O. Kolka. Optimal Inventory Management of a Bike-Sharing Station. *IIE Transactions*, Vol. 45, No. 10, 2013, pp. 1077–1093.
2. Rixey, R. Station-Level Forecasting of Bikesharing Ridership: Station Network Effects in Three US Systems. *Transportation Research Record: Journal of the Transportation Research Board*, 2013. 2387: 46–55.
3. Faghih-Imani, A., N. Eluru, A. M. El-Geneidy, M. Rabbat, and U. Haq. How Land-Use and Urban Form Impact Bicycle Flows: Evidence from the Bicycle-Sharing System (BIXI) in Montreal. *Journal of Transport Geography*, Vol. 41, 2014, pp. 306–314.
4. Wang, X., G. Lindsey, J. E. Schoner, and A. Harrison. Modeling Bike Share Station Activity: Effects of Nearby Businesses and Jobs on Trips to and from Stations. *Journal of Urban Planning and Development*, Vol. 142, No. 1, 2016, p. 04015001.
5. Hampshire, R. C., and L. Marla. An Analysis of Bike Sharing Usage: Explaining Trip Generation and Attraction from Observed Demand. Presented at 91st Annual Meeting of the Transportation Research Board, Washington, D.C., 2012.
6. Buck, D., and R. Buehler. Bike Lanes and Other Determinants of Capital Bikeshare Trips. Presented at 91st Annual Meeting of the Transportation Research Board, Washington, D.C., 2012.
7. Wang, W. *Forecasting Bike Rental Demand Using New York Citi Bike Data*. Dublin Institute of Technology, 2016.
8. Morency, C., M. Trépanier, A. Páez, H. Verreault, and J. Faucher. *Modelling Bikesharing Usage in Montreal Over 6 Years*. Working Paper, CIRRELT-2017-33. Interuniversity Reserach Centre on Enterprise Networks Logistics and Transportation, Montreal, Quebec, Canada, 2017.
9. Daddio, D. W. *Maximizing Bicycle Sharing: An Empirical Analysis of Capital Bikeshare Usage*. University of North Carolina at Chapel Hill, Chapel Hill, NC, 2012.
10. Gebhart, K., and R. B. Noland. The Impact of Weather Conditions on Capital Bikeshare Trips. Presented at 92nd Annual Meeting of the Transportation Research Board, Washington, D.C., 2013.
11. Rudloff, C., and B. Lackner. Modeling Demand for Bikesharing Systems: Neighboring Stations as Source for Demand and Reason for Structural Breaks. *Transportation, Research Record: Journal of the Transportation Research Board*, 2014. 2430: 1–11.
12. Liu, J., L. Sun, W. Chen, and H. Xiong. Rebalancing Bike Sharing Systems: A Multi-source Data Smart Optimization. *Proc., 22nd ACM SIGKDD International Conference on Knowledge Discovery and Data Mining*, San Francisco, CA, 2016, pp. 1005–1014.
13. Singhvi, D., S. Singhvi, P. I. Frazier, S. G. Henderson, E. O'Mahony, D. B. Shmoys, and D. B. Woodard. Predicting Bike Usage for New York City's Bike Sharing System. *Proc., AAAI Workshop: Computational Sustainability*, Austin, TX, 2015, pp. 110–114.
14. Li, Y., Y. Zheng, H. Zhang, and L. Chen. Traffic Prediction in a Bike-Sharing System. *Proc., 23rd SIGSPATIAL International Conference on Advances in Geographic Information Systems*, Seattle, WA, ACM, New York, 2015.
15. Tran, T. D., N. Ovtracht, and B. F. d'Arcier. Modeling Bike Sharing System Using Built Environment Factors. *Procedia CIRP*, Vol. 30, 2015, pp. 293–298.
16. Yan, Y., M. Chen, M.-L. Shyu, and S.-C. Chen. Deep Learning for Imbalanced Multimedia Data Classification. *Proc., 2015 IEEE International Symposium on Multimedia (ISM)*, Miami, FL, IEEE, New York, 2015, pp. 483–488.
17. LeCun, Y., L. Bottou, Y. Bengio, and P. Haffner. Gradient-Based Learning Applied to Document Recognition. *Proceedings of the IEEE*, Vol. 86, No. 11, 1998, pp. 2278–2324.
18. Dieleman, S., K. W. Willett, and J. Dambre. Rotation-Invariant Convolutional Neural Networks for Galaxy Morphology Prediction. *Monthly Notices of the Royal Astronomical Society*, Vol. 450, No. 2, 2015, pp. 1441–1459.
19. Chen, X., W. Xiong, Z. Huang, and H. Yang. Planning Bicycle Corridor for Shanghai Central City. Presented at 86th Annual Meeting of the Transportation Research Board, Washington, D.C., 2007.

The Standing Committee on Bicycle Transportation (ANF20) peer-reviewed this paper (18-03033).

Article

TRR
JOURNAL OF THE TRANSPORTATION RESEARCH BOARD

Transportation Research Record
2018, Vol. 2672(36) 103–115

DOI: 10.1177/0361198118799170
journals.sagepub.com/home/trr
SAGE

Introducing a Design Framework for a Multi-Modal Public Transportation System, Focusing on Mixed-Fleet Bike-Sharing Systems

Mehrnaz Ghamami[1] and MohammadHossein (Sam) Shojaei[1]

Abstract

Bike-sharing is increasingly becoming more popular. Electric bikes as an emerging transportation technology have extended range and are less physically demanding, compared with regular bicycles, thus they can be incorporated into regular bike-sharing systems to attract more users. This study aims at capturing the users' preference, while considering investors' limitations and societal cost and benefits of each mode. The problem is defined as a mixed-integer non-liner problem, with non-linear objective function and constraints. Because of the computationally challenging nature of the problem, a metaheuristic algorithm based on simulated annealing algorithm is proposed for its solution. The performance of the algorithm is tested in this study and convergence patterns are observed. The main findings of this study which are derived from the hypothetical numerical example, include but are not limited to: (1) the most popular public modes are bus and pedelec, because these two modes (bus and pedelec) are less expensive and have the ability to traverse longer distances in comparison to similar modes (i.e., e-scooter/car and bike), and (2) for small communities with short travel distances (feasible within the ranges of active modes), users would not choose fuel-consuming modes, and thus their choice is insensitive to fuel cost.

Bike-sharing is an increasingly evolving concept across the globe, offering diverse benefits such as flexible mobility, reduced fuel use, the corollary reductions in emissions, and increased physical activity levels (*1*). The evolution of bike-sharing has been such that, beginning from only a handful of bike-sharing programs in the late 1990s, there are currently above 400 functional bike-sharing systems worldwide (*2*). This drastic growth owes to the underlying factors of raised public and government awareness about the downsides of car use, as well as the affordability of the bike-sharing service (*3*).

Bike-sharing enables the users to enjoy the advantages of biking as an active mode of transportation, without the complications of private bike ownership (i.e., regular maintenance). There are numerous studies in the literature on bike-sharing systems, which predominantly focus on the existing bike-sharing systems and investigate the spatial-temporal performances or mobility patterns of these systems through data mining methods (*4–7*). However, attention to network and facility location design of bike-sharing systems from strategic planning perspectives is not as common. Lin and Yang formulated a mathematical model for a bike-sharing system design problem (*8*). They consider both users' and investors' perspectives, respectively, by considering the level of service in terms of demand coverage and system set up costs. There is a variety of studies focusing on hub location inventory models. However, application of these models to bike-sharing system has not gained enough attention (*9*). Another important concern of planners when designing bike-sharing systems is the locations of bike stations, as a key determinant of its success (*10–12*). In addition to optimum locations of stations, fleet sizes, and distribution of supply, considering the relocation of bikes to maximize the served demand has been an interest to researchers (*13–16*).

Another rapidly growing aspect of today's transportation is represented by electric bicycles, commonly known as e-bikes (*17*). The market share of e-bikes and e-bike-sharing systems depends on a combination of factors such as increasingly bicycle-friendly cities and technological advances. In general, there are two major types of

[1]Department of Civil and Environmental Engineering, Michigan State University, East Lansing, MI

Corresponding Author:
Address correspondence to Mehrnaz Ghamami: ghamamim@msu.edu

e-bikes (*18*): (1) bicycle-style e-bikes (BSEBs), which resemble conventional bicycles in both appearance and operation; one type of BSEB, which is referred to as ped-elec, requires the rider to constantly keep pedaling while assisting the rider through electric assistance to the pedals; and (2) scooter style electric bicycles (SSEBs), which share many similar features to gasoline-fueled scooters while essentially relying on electric power. As opposed to pedelecs, SSEBs do not require pedaling. E-bikes are superior to conventional bicycles in many distinct aspects such as the ability to traverse longer distances and reach higher speeds, and being more convenient, especially over hilly terrains. Also, although less so than conventional bikes, pedelecs would still provide health benefits by engaging the rider in the physical activity of pedaling, despite the electric assistance (*17*). In the United States, an electric bike has a motor power up to 750 W, operable pedals, and a maximum assisted speed of 20 mph (*19*).

Along with the variety of benefits e-bikes have compared with conventional bicycles, they are also more costly. To overcome the cost barrier, inclusion of e-bikes in bike-sharing systems has been proposed (*20*). This solution also has the potential to attract more users from other transportation modes and contribute to the social benefits of conventional bike-sharing systems (*21*). In general, e-bike-sharing systems are anticipated to contribute to reduced single-occupancy car trips, improved air quality through reduced CO_2 emissions, enhanced public health by increasing physical activity levels, improved roadway mobility and safety (especially for cyclists, owing to more cyclists than car users), and to support local economies and tourism, and at a larger scale the societal quality of life (*22*).

Despite the conceptual foundations for e-bike-sharing systems and high hopes for their promising advantages, few studies have investigated the design of such systems. In a pilot study at the campus of University of Tennessee, Knoxville, a small bike-sharing system offering both regular and electric bikes was set up. This campus was considered a good candidate for implementation of an e-bike-sharing system, because of the hilly terrain and the vast spatial expanse of the campus. The cycling system was able to attract more users to cycling and was a success (*23*).

Aware of the shortage of studies on the design of e-bike-sharing systems, this study aims to bridge this gap by introducing a design framework for a public transportation system which offers e-bikes as well as regular bikes and buses. The next section describes the problem statement, followed by the mathematical model and solution method. The numerical analysis includes describing the case study, testing the algorithm performance, and the sensitivity analysis.

Problem Statement

The main contribution of this study is considering the trade-off between user preference, investment cost, and social costs/benefits. The multifaceted objective is to minimize the investment cost, along with other social costs, such as emission cost, while maximizing the revenue and health benefits of the system. Quantification of these costs and benefits in an intended mixed-fleet bike-sharing system offering both conventional bikes, e-bikes of both pedelec and e-scooters, as well as a bus system for a target area, is required. This contributory aspect to the present research helps investors and transportation planners to strike a trade-off between different facets of their decision on initiating a multi-modal transportation system. Moreover, another major contribution of this study is considering user choice through a utility function, added as a constraint to the proposed mathematical model.

Mathematical Model

This study considers a transportation system, including the shared or public modes (i.e., bikes, pedal-assist electric bicycles or pedelecs, e-scooters, and buses), as well as the private modes (i.e., cars and motorcycles), available to the users in the target area. Let I denote the set of the transportation modes, and $|I| = N_I$. Figure 1 depicts a schematic sketch of a target area, divided into several

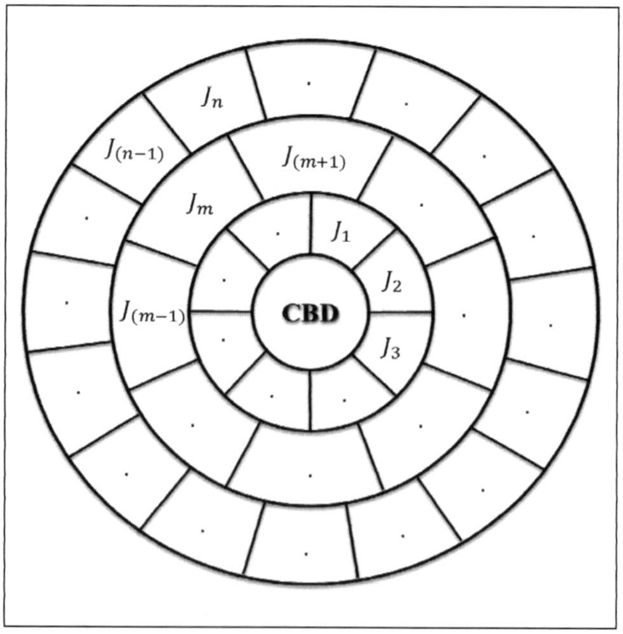

Figure 1. Sketch of the study area; each cell represents a user group (J_j), with a specified financial power and at a specific commuting distance from CBD.

user groups with specific populations, in such a way that each group contains users with the same financial power for their daily transportation, and the same distance from the destination, which is the central business district (CBD). Let J denote the set of the user groups, with $|J| = N_J$. We assume that the transportation modes ($i \in I$) are numbered from 1 to N_I, and the user groups ($j \in J$) are numbered from 1 to N_J.

Each user group is at a certain distance from the destination or CBD, denoted by d_j, with a total demand for transportation denoted by Ψ_j, and the financial power denoted by F_{P_j}.

The characteristics of each mode considered in this study are as follows. The capacity of each mode is denoted by Ω_i. Each mode i has an average speed represented by v_i. The expense of using mode i by a member of user group j is denoted by C_{uij} expressed in dollars. The user cost or C_{uij} would be the fare for public modes denoted by C_{r_i}, and would be the daily cost of ownership (C_{p_i}) and fuel cost for the private modes ($C_{f_{ij}}$). The comfortable traveling distance by mode i is denoted by D_i in miles. The fuel consumption rate of mode i is denoted by F_i which signifies the amount of fuel that mode consumes to traverse unit of distance (i.e., gallons per mile). The purchase and maintenance costs of mode i are, respectively, denoted by C_{p_i} and C_{mi}. This study focuses on CO_2 as the dominant emission; the amount of CO_2 that mode i emits is denoted by G_i expressed in grams of CO_2 produced per unit fuel (i.e., gallon). The monetary value of the health benefit gained from an hour of using mode i, which is associated with the physical activity level of mode i, is denoted by h_i. It is worth noting that only bike and pedelec are considered to involve physical activity and thus other transportation modes are assumed to not provide any health benefits.

The fuel price and fuel tax are, respectively, denoted by C_l and C_t. The objective in this study is multifaceted. Primarily, both the entire system's perspective and the users' perspective are considered. From the entire system's viewpoint, the study intends to minimize the system costs while maximizing the system gains. From the users' standpoint, the study develops a utility function that considers various aspects of users' decision-making. The utility function accounts for the mode choice behavior of the users in each user group.

Having introduced the parameters used in this study, the mathematical model is presented next:

$$min\ Z = \sum_i (\gamma_c C_{s_i} + \sum_j \left((\gamma_e E_{ij} - \gamma_h H_{ij} - \gamma_r C_{vij})\Gamma_{ij} + \rho_{ij}\left(\Psi_j \pi_{ij} - \left(\frac{\pi_{ij}\Psi_j}{\sum_{j \in J}(\pi_{ij}\Psi_j)} \right) N_i \Omega_i \right) \right)) \quad (1)$$

s.t

$$C_{s_i} = \begin{cases} 0 & if\ i \in \{car, motorcycle, walking, private\ bike\} \\ N_i C_{p_i} + N_i C_{mi} & if\ i \in \{shared\ bike, pedelec, e-scooter, bus\} \end{cases} \quad (2)$$

$$E_{ij} = \begin{cases} \frac{d_j F_i G C_f}{\Omega_i} & if\ i \in \{car, motorcycle, bus\} \\ 0 & if\ i \in \{walking, shared/private\ bike, pedelec, e-scooter\} \end{cases} \quad (3)$$

$$H_{ij} = \begin{cases} 0 & if\ i \in \{e-scooter, car, motorcycle, bus\} \\ h_i \frac{d_j}{v_i} & if\ i \in \{walking, shared/private\ bike, pedelec\} \end{cases} \quad (4)$$

$$C_{vij} = \begin{cases} C_{ti} d_j F_i & if\ i \in \{car, motorcycle\} \\ C_{r_i} & if\ i \in \{shared\ bike, pedelec, e-scooter, bus\} \end{cases} \quad (5)$$

$$\Gamma_{ij} = \begin{cases} \Psi_j \pi_{ij}\ if\ N_i\Omega_i > \sum_{j \in J} \Psi_j \pi_{ij} \\ \left(\frac{\pi_{ij}\Psi_j}{\sum_{j \in J}(\pi_{ij}\Psi_j)} \right) N_i\Omega_i\ if\ N_i\Omega_i \leq \sum_{j \in J} \Psi_j \pi_{ij} \end{cases} \forall i \in I, j \in J \quad (6)$$

$$\rho_{ij} = \begin{cases} 0\ if\ N_i\Omega_i > \sum_j \Psi_j \pi_{ij} \\ D_{ui}.d_j\ if\ N_i\Omega_i \leq \sum_j \Psi_j \pi_{ij} \end{cases} \forall i \in I, j \in J \quad (7)$$

$$D_i - d_j \leq \mu_{ij} M \quad \forall i \in I, j \in J \quad (8)$$

$$D_i - d_j > \left(\mu_{ij} - 1 \right) M \quad \forall i \in I, j \in J \quad (9)$$

$$C_{p_i} - F_{p_j} \leq \xi_{ij} M \quad \forall i \in I, j \in J \quad (10)$$

$$C_{p_i} - F_{p_j} > (\xi_{ij} - 1) M \quad \forall i \in I, j \in J \quad (11)$$

$$U_{ij} = \beta_0 + \beta_1 d_j^{\eta_{1i}} + \beta_2 N_i \Omega_i + \beta_3 T^{\eta_{2i}} + \beta_4 \frac{d_j}{v_i} + \beta_5 C_{uij} + \beta_6 \left(1 - \mu_{ij} \right) + \beta_7 (1 - \xi_{ij}) + \beta_8 E_{ij} + \beta_9 H_{ij} \quad \forall i \in I, j \in J \quad (12)$$

$$C_{uij} = \begin{cases} C_{p_i} + C_{f_{ij}} & i \in \{car, motorcycle, private\ bike\} \\ C_{r_i} & i \in \{shared\ bike, pedelec, e-scooter, bus\} \end{cases} \quad (13)$$

$$C_{f_{ij}} = C_l d_j F_i \quad \forall i \in I, j \in J \tag{14}$$

$$\pi_{ij} = \frac{e^{U_{ij}}}{\sum_x e^{U_{xj}}} \quad \forall i \in I, j \in J \tag{15}$$

$$\sum_{i \in I} \pi_{ij} = 1 \quad \forall j \in J \tag{16}$$

$$\sum_{i \in I} C_{p_i} N_i \leq \text{B} \quad \forall i \in \{shared\ \text{bike, pedelec, e} - \text{scooter, bus}\} \tag{17}$$

$$\mu_{ij} \in \{0, 1\} \quad \forall i \in I, j \in J \tag{18}$$

$$\xi_{ij} \in \{0, 1\} \quad \forall i \in I, j \in J \tag{19}$$

$$\pi_{ij} \in [0, 1] \quad \forall i \in I, j \in J \tag{20}$$

$$\text{N}_\text{i} \geq 0 \quad \forall i \in \{shared\ \text{bike, pedelec, e} - \text{scooter, bus}\} \tag{21}$$

The objective of this study is to minimize the investors' costs, while maximizing the system's revenue, minimizing environmental (CO_2 emission) concerns, maximizing societal health benefits, and minimizing the entire system's unserved demand cost. The decision variables in this problem are the numbers of public modes(i.e., $N_i, \forall i \in \{\text{bike, pedelec, e} - \text{scooter, bus}\}$).

C_{s_i} denotes the investment cost for mode i, defined by the normalized purchase cost C_{p_i}, plus the maintenance cost C_{mi} of the mode (Constraint 2). In fact, the investment cost is incurred by the authority implementing the public transportation system. E_{ij} is the emission cost function for mode i adopted by group j. The emission production cost for each user is a function of the group's distance from destination (d_j), the fuel consumption rate of the mode (F_i), the CO_2 production rate of mode i (G_i), the CO_2 burden cost (C_f), divided by the capacity of the mode (Ω_i) (Constraint 3). H_{ij} is the health benefit value function of mode i used by group j, which is a product of hourly value of adopting a mode (h_i) and the time needed to reach the destination by that mode (Constraint 4). It is noteworthy that health benefit values in this study are calculated as inactivity costs forgone as a result of engaging in physical activity. In other words, inactivity costs for American adults are considered to be avoided if individuals are active per physical activity guidelines for Americans (*24*). C_{vij} is the revenue function, which is the fare of a mode if it is a public mode of transportation and the fuel tax if the mode is private (Constraint 5).

Γ_{ij} (Constraint 6) represents the adoption of mode i by user group j. When the demand for mode i in group j is less than the availability, the number of users experiencing the societal costs and generating the revenue equals the demand. However, in the case in which the demand exceeds the availability, the number of users experiencing the societal costs and generating the revenue would be restricted to availability and there would be unserved demand. In the case in which there is unserved demand, there would be penalty, also known as the inconvenience cost of remaining unserved, which is denoted by ρ_{ij} (Constraint 7). In this study, it is assumed that the unserved demand for shared/public modes would have to rely on taxi/Uber to make the commute; the average unit cost is thus calculated as D_u.

Constraints 8–9 stipulate distance feasibility of each mode for each user group. If mode i is feasible for user group j, from the distance comfortability perspective, which means if the group's distance to CBD is less than the comfortable distance of the mode, μ_{ij} would equal 1; otherwise, μ_{ij} would be zero. Similarly, Constraints 10–11 investigate financial feasibility. U_{ij} denotes utility of mode i for group j (Constraint 12). The utility of each mode for each user group depends on the factors such as distance, mode availability, topography of the area, travel time, user expenses, distance and financial feasibilities, emissions costs, and health benefits. The coefficients of utility function are derived from the literature (*25*, *26*).

User expenses, denoted by $C_{u_{ij}}$, would be the fare if i is a public mode. In the case in which i is a private mode (i.e., private bike, motorcycle, car), user expenses would entail normalized daily purchase cost of the relevant mode, and if the private mode is motorcycle or car, the fuel cost ($C_{f_{ij}}$) incurred by user group j to make the the commute to the CBD is also included. A private mode's fuel cost denoted by $C_{f_{ij}}$ is product of fuel price, the vehicle's consumption rate, and the user's distance from destination (CBD) (Constraints 13 and 14). π_{ij}, which is the probability of group j using mode i, is calculated based on logit model and the utility function U_{ij} (Constraint 15). Constraint 16 ensures that the sum of probabilities across the modes for each group j equals unity. Constraint 17 mandates that the cost of purchasing the modes of the shared and public system not be greater than the total system implementation budget (B). Constraints 18–21 are feasibility constraints.

The problem is a mixed integer as the decision variables are a combination of integer values (the mode counts) as well as non-integer, binary values. Moreover, the probability function used for users' mode choice modeling is a nonlinear function of the mode counts. As the probability function is a component of the objective function and one of the constraints as well, the problem is of a nonlinear nature.

Solution Method

Because of the nonlinearity of the utility function, the optimization model is computationally challenging for the available commercial solvers. Thus, a metaheuristic algorithm is proposed and modified to fit the proposed model. Metaheuristics based on simulated annealing (SA) is inspired by annealing in metallurgy. The iterative process resembles the heating and controlled cooling of a solid material to increase the size of its crystals and reduce their defects. At the cooling stage, an equilibrium state should be achieved at each temperature before moving to a lower temperature. The final solution is achieved at the minimum or final temperature.

An SA-based algorithm typically has two main steps. In the first, it searches over the feasible set of the integer solutions, starting from a current feasible solution and then moving to a neighbor feasible solution. The second step compares the objective functions of the current and the new solutions and, based on the difference, replaces the current solution with the new one with a probability. The probability is gradually reduced as the solution process proceeds. SA schemes allow larger objective function values (worse solutions) relative to the current solution to be accepted, which offers a mechanism to avoid getting trapped in local optimum solutions. This feature is very useful when the problem is known to have multiple local optima. There are some transportation-related studies (i.e., facility location models) in the literature which have adopted the SA algorithm to solve flow-capturing mixed-integer programs (*27*, *28*). In these studies, the SA algorithm has proven efficiency.

The initial solution used in this study, which is derived from a series of sensitivity analyses, is set to be one vehicle for each public mode:

Input: Maximum number of temperature changes K_0, Maximum number of inner iterations at each temperature K_1, different costs, vehicle and user characteristics, utility function parameters.
Output: $N_i^* \in \mathrm{N}$
Initialize:
Set the current temperature stage $t = 0$, choose initial temperature u_t
Set a state variable $\xi = 1$, which indicate that a mode count should be added. $\xi = 0$ when a mode count should be removed.
Initialize $N_i^t \in \mathrm{N}$
While $t < K_0$, **do**
  Set inner iteration index $k = 0$. Set $N_i^k = N_i^t$
  While $k < K_1$ **do**
    Calculate U_{ij}
    Calculate Γ^k
    Set Λ be a weighted matrix of different modes (assumed to have the same values).
    Set a random number $\gamma = u[0, 1]$.
    If $\Lambda > \gamma$ **then**
      $\xi = 0$
    else
      $\xi = 1$
    end if
    If $\xi = 1$ **then**
      $N_i^k = N_i^k + 1$
      If $B_c \geqslant B$ **then**
        $N_i^k = N_i^k - 2$
        end if
    else
      $N_i^k = N_i^k - 1$
      If $N_i^k = 0$ **then**
        $N_i^k = N_i^k + 2$
        end if
    end if
    Set Γ^k and Γ^{k-1} be the objective function values associated with the perturbed and current solutions respectively
    Set k = k + 1 and the perturbed solution as N_i^k. Draw a random number $\gamma = u[0, 1]$.
    If $\Gamma^k < \Gamma^*$ **then**
      $\Gamma^* = \Gamma^k, N_i^* = N_i^k$
    end if
    If $\Gamma^k > \Gamma^{k-1}$ and $\exp\left(\frac{\Gamma^{k-1}-\Gamma^k}{u_t}\right) > \gamma$ **then**
      Discard the perturbed solution, i.e., set $N_i^k = N_i^{k-1}$.
    end if
  end while
  Set $t = t + 1$ and and $u_t = \theta u_0$, where $\theta = 0.85$.
end while
$\Gamma^*$and N_i^*

Numerical Example

Case Study

The proposed model has the ability to capture a variety of modes and user groups. However, a case study based on a hypothetical network is contrived which encompasses walking, biking (private and shared), pedelec, scooter, bus, motorcycle, and car, to be used by six user groups. The modes' characteristics used for this case study are displayed in Table 1; these values are based on the average values in the United States. The configuration of the user groups is available in Table 2. Among the public modes, buses have the highest comfortable traveling distance. Thus, it is decided that the farthest user group is at a distance that users can at least use buses to commute to the CBD.

The health benefit hourly values in this study are calculated following an approach proposed by Trubka et al. (*35*), and based on the inactivity costs in the United States (*36*); in this approach it is assumed that the inactivity costs would be avoided if one abides by the physical activity guideline for Americans (*24*). Another noteworthy aspect in the proposed modeling framework

Table 1. Modes' Characteristics

Characteristics	Walking	Bike (shared/private)	Pedelec	E-scooter	Car	Bus	Motorcycle
Comfortable traveling distance (mi) (D_i)	1(29)	3.8(29)	5.01	6.2	12.1 (29)	10.2(29)	12.1(29)
Average speed (mph) (ν_i)	3.5(29)	10(29)	12	13.4	28.9 (29)	11.4(29)	29.5(29)
Average daily cost of ownership ($) ($C_{p_i}$)	-	0.28	1.6	1.03	5.75	68.5	1.52
Average daily maintenance cost ($) ($C_{m_i}$)	-	0.34	1.94	1.25	3.85	1.1	2.07
Health benefit hourly value ($/h)	14.44	15.2	14.06	-	-	-	-
Trip cost of public modes (Fare)	-	$0.22(\frac{\$}{hr})$	$0.22(\frac{\$}{hr})$	$2.58(\frac{\$}{hr})$	-	$1.25(30)	-
Fuel cost ($/gallon) ($C_l$)	-	-	-	-	2.485 (31)	2.485	2.485
Fuel consumption rate (gallon/mi) (F_i)	-	-	-	-	0.043 (32)	0.307 (32)	0.023(32)
Emission production rate (grams of CO_2/gallon) (G)	-	-	-	-	8,887 (33)	8,887	8,887
Financial burden of emission ($/grams of CO_2) (C_f)	-	-	-	-	0.0002	0.0002	0.0002(25)
Fuel tax (cents /gallon) (C_t)	-		-	-	26.3 (34)	26.3	26.3

Table 2. User Specifications

User group	1	2	3	4	5	6
Distance from CBD area (mi)	1	1	5	5	10	10
Financial power ($/commute)	2.6	12	2.6	12	2.6	12

is that the utility function determines the desirability of a mode based on various factors including distance from the CBD, mode availability, topography of the area, travel time, user expenses, distance and financial feasibilities, emissions costs, and health benefits. Among these factors, the only varying component is the mode availability, which is defined in this study as the mode counts multiplied by the capacity of a specific mode. Public mode counts are the decision variables and intended to be optimized in the objective function which considers a trade-off between the implementation cost and unserved demand cost. Considering this trade-off, the system chooses to incur unserved demand cost in lieu of providing more of a specific public mode, if the cost of provision is greater than the unserved demand cost. Therefore, if any of the modes considered in the modeling framework were acceptable to the users and if the investment cost was comparable to the inconvenience cost of unserved demand, the users would have been assigned to one of the existing modes. However, depending on the various characteristics of each mode (mentioned above), it is not always the optimum decision to serve all the users in the system with the public modes discussed in the model. In such instances, a user who is unserved has to opt out for another mode that is not included in the public transit system. Therefore, this mode is assumed as taxi/Uber. The unserved demand cost, D_{ui} can be further adjusted for each specific mode i for improved accuracy.

Algorithm Performance

Considering a budget of $15,000, setting the inner iteration at 150 and the outer iteration at 200, and starting with an initial solution of 1 shared bike, 1 pedelec, 1 e-scooter, and 1 bus, the algorithm successfully converges for this case study after 90 outer iterations. The algorithm performance can be seen in Figure 2. The optimum mode counts are determined as 123 shared bikes, 1928 pedelecs, 26 e-scooter, and 105 buses, having consumed almost 69% of the budget, namely $10,340.

Cost Sensitivity Analysis

This section focuses on investigating the effects of different costs on the users' mode choice and optimum configuration of public modes. Thus, various scenarios are investigated for financial burden of emission, health benefit value, trip cost, and fuel cost.

Financial Burden of Emission. The sensitivity of the model is tested for four scenarios of financial burden of emission: scenario 1 with $\$2 \times 10^{-6}$/grams of CO_2, scenario 2 with

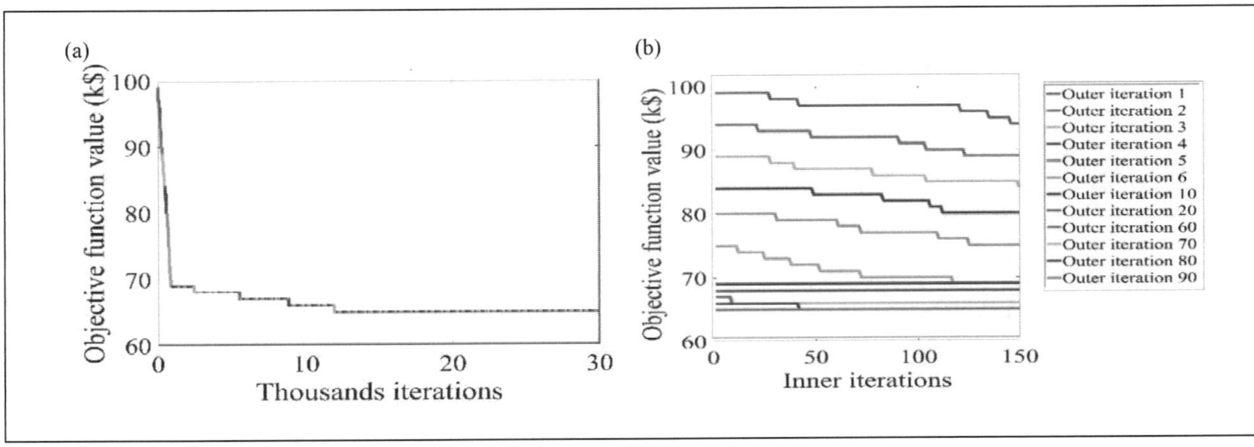

Figure 2. SA algorithm performance: (*a*) general trend of the objective function and (*b*) objective function trend for different inner and main (outer) iterations.

2×10^{-4}/grams of CO_2 (base case), scenario 3 with $ 0.01/grams of CO_2, and scenario 4 with $ 0.02/grams of CO_2. The optimum mode counts and demand percentages for each mode under each of the four scenarios are shown in Figure 3, *a* and *b*, respectively. As the financial burden of emission increases, the system would provide fewer buses, to a point that in the fourth scenario the system does not provide any buses. In return, more pedelecs are provided. Looking into the changes in demands across the scenarios in Figure 3*b*, it can be observed that demand for bus declines with the rise of emission burden cost, and the users turn to pedelec as a primary choice, followed by walking. The ability of pedelec to cover longer trips is the reason users prefer to choose pedelec over walking and biking, when buses are not a utilitarian option as a result of being pollutant.

Health Benefit Hourly Values. Four scenarios were implemented to investigate the model's sensitivity to the value of health benefits. In these four scenarios, the health benefit hourly value for each of the active modes were, respectively, multiplied by factors of 0.1, 1, 2, and 5. The optimum mode counts and demands for each of the modes across the scenarios are shown in Figure 3, *c* and *d*, respectively. As expected, with the increase of health benefit hourly values (i.e., hospital or other illness treatment costs increases), the demand for bus and e-scooter, which are assumed to have no health benefit value, decline. On the other hand, more users would choose pedelec and walking because of their increased health benefit values. It is also observed that demands for both shared and public bikes decrease as well, even though biking has a higher health benefit value than the other active modes, namely pedelec and walking. The reason people would prefer walking over biking in particular is the distance coverage of the two modes and the groups' distance configuration of the first two user groups. These groups would choose to walk more because it is free of charge and would also provide comparable health benefits to biking. On the other hand, users at the distance of 5 mi from CBD would predominantly choose to ride pedelec, as pedelec accounts for longer comfortable traveling distance than bike, and is less costly than e-scooter. The last two groups would mainly depend on bus as it is feasible for them from a distance perspective, while also being more affordable than car and motorcycle. Based on the approach in this study for calculating health benefit hourly values, increase in the health benefit hourly values of the modes results from increase in the inactivity costs, including doctor and hospital fees. As a result of this increase, the public transportation system should provide more pedelecs as they enjoy both reasonable comfortable ranges and health benefit values, while reducing the numbers of shared bikes, e-scooters and buses.

Trip Cost. To investigate the sensitivity of the user groups to the public modes' fares, four scenarios are arranged such that the trip costs/fares of the base case are multiplied by factors of 0.5, 1, 1.5, and 2.5. Figure 3*f* shows that as the trips costs increase, more users would choose pedelec and the demands for the other public modes drop. Also, the number of users who choose to walk increases. The underlying reason is that users from the first two groups whose commuting distances are within the comfortable distance of walking would choose to walk in order to avoid costs of increasingly expensive public modes. The third and fourth groups' users also would prefer pedelec more, as it is a feasible choice which is less costly than e-scooter and bus. Aligned with

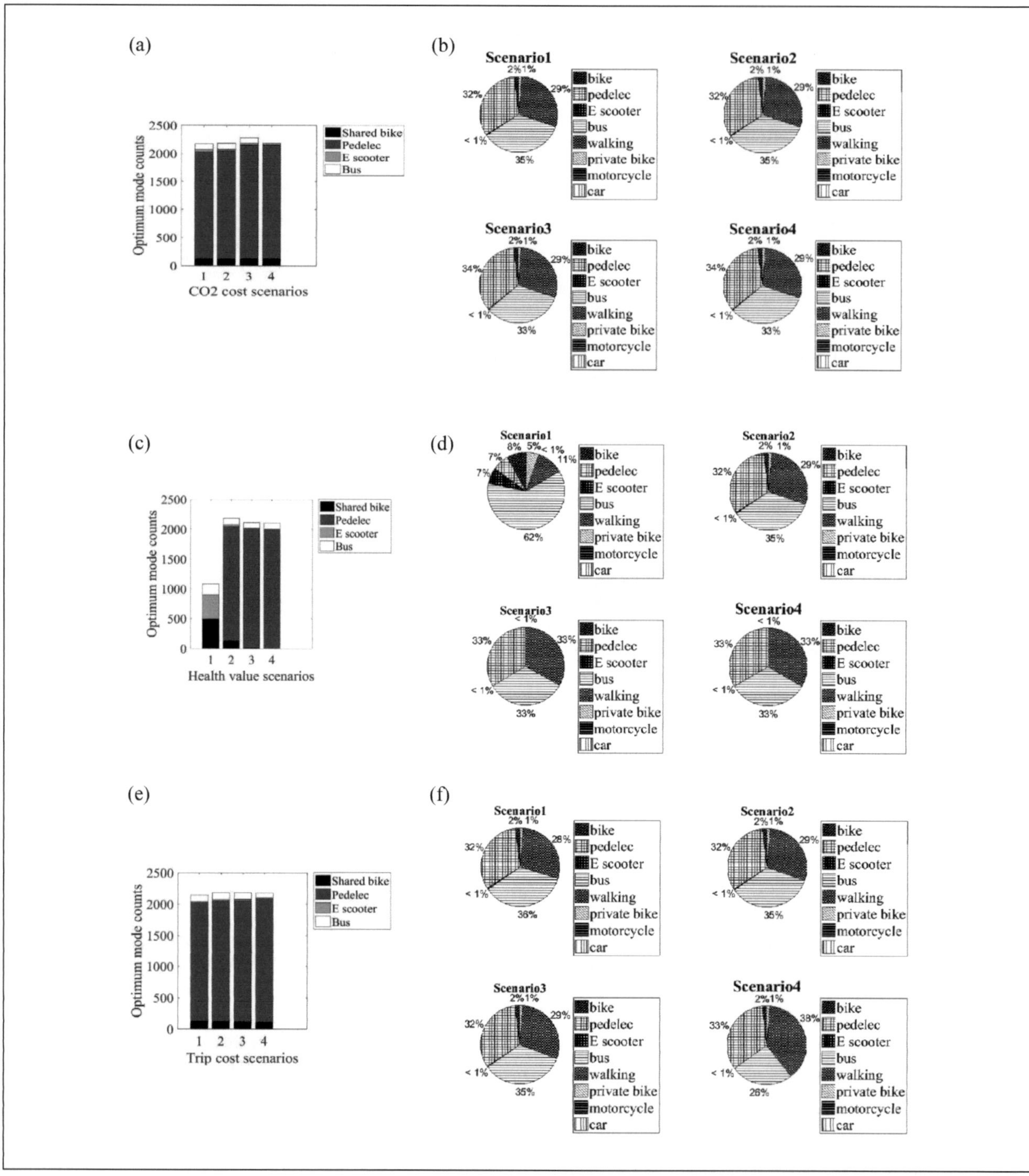

Figure 3. Cost sensitivity analysis: (*a*) optimum mode counts (numbers of each public mode) for CO_2 cost scenarios; (*b*) demand percentages for CO_2 cost scenarios; (*c*) optimum mode counts (numbers of each public mode) for health benefit hourly value scenarios; (*d*) demand percentages for health benefit hourly value scenarios; (*e*) optimum mode counts (numbers of each public mode) for trip cost scenarios; (*f*) demand percentages for trip cost scenarios.

Note: Values of CO_2 burden cost are the four scenarios are as follows: scenario 1 with $\$2\times10^{-6}$/grams of CO_2, scenario 2 with $\$2\times10^{-4}$/grams of CO_2 (base case), scenario 3 with \$ 0.01/grams of CO_2, and scenario 4 with \$ 0.02/grams of CO_2. In the four health benefit scenarios, health benefit hourly values of each of the active modes were multiplied by factors of 0.1, 1, 2 and 5, respectively. In the four trip cost scenarios, trip costs/fares of the base case are multiplied by factors of 0.5, 1, 1.5, and 2.5, respectively.

Table 3. Distance Sensitivity Analysis Scenarios

Scenario 1			Scenario 2			Scenario 3		
User group	Distance from CBD	Financial power	User group	Distance from CBD	Financial power	User group	Distance from CBD	Financial power
1	1	2.6	1	1	2.6	1	2	2.6
2	1	12	2	1	12	2	2	12
3	5	2.6	3	3	2.6	3	2.5	2.6
4	5	12	4	3	12	4	2.5	12
5	10	2.6	5	10	2.6	5	10	2.6
6	10	12	6	10	12	6	10	12

the demand trends, the optimum count of pedelec increases as trip costs increase, but the system reduces the number of buses and shared bikes (Figure 3*e*) while still satisfying the demands. Only in the first scenario is there unserved demand for e-scooter, as the system does not provide e_scooter in the first scenario. The reason here is that the system decides to leave the demand for e-scooter unserved in this case, rather than incurring the cost of providing the demanded e-scooters, because of the large investment costs. Another important finding by trip cost sensitivity analysis was that increasing public modes' fares by above 150% would prevent users from choosing public modes.

Fuel Cost. Four fuel-cost scenarios are considered to analyze the sensitivity of the model to fuel cost. In the four scenarios, the current fuel cost is multiplied by 0.1, 1, 2, and 10, respectively. Under these changes, the optimum mode counts and demand percentages remain unvarying. This suggests the insensitivity of the model to fuel cost. The reason is that because of the limited travel distances of user groups, the fuel-dependent modes are not very popular even in the base case condition of fuel cost. In other words, for example, the first two groups would normally prefer walking as it is both free of charge and also beneficial for health, the third and fourth groups tend to opt for pedelec as it would be less costly than other modes while also providing health benefits, and the last two groups would be inclined to use bus as the comfortable range of bus accommodates the distances of the last two groups from the CBD and is more affordable than the other modes. Under these circumstances, further changes in fuel cost only affect a relatively small percentage of the system's demand, so the changes are not visible.

Budget Sensitivity Analysis

Changes in the budget considerably affect the investment power in public modes. To explore the sensitivity of the model to the initial budget, this section considers four scenarios which change the initial budget of the numerical experiment, respectively, by factors of 0.5, 0.75, 1, and 1.5. As can be seen in Figure 4, the optimum mode counts of scenarios 2–4 are identical as the budget constraint is not binding. The percentages of used budget versus the initial budget are 99.99%, 91.89%, 68.92%, and 45.95% for scenarios 1–4, respectively. It can be observed the best investment strategy, irrespective of the available budget, is to circumvent investing in e-scooters, unless the cost and characteristics of this mode changes.

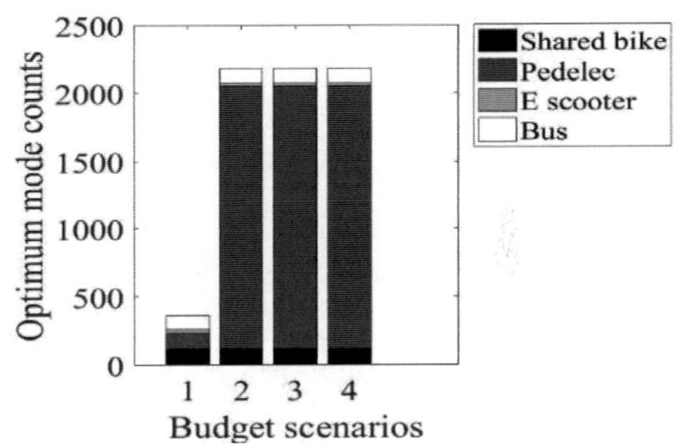

Figure 4. Budget sensitivity analysis.

Distance Sensitivity Analysis

To analyze the sensitivity of the model to user groups' travel distance distribution, three different scenarios were tested as shown in Table 3.

Optimum mode counts and demand percentages for the three scenarios are presented in Figure 5. In the first scenario, which is the base case scenario in this study, the most popular public modes are bus and pedelec. The reason is that these two modes (bus and pedelec) are less

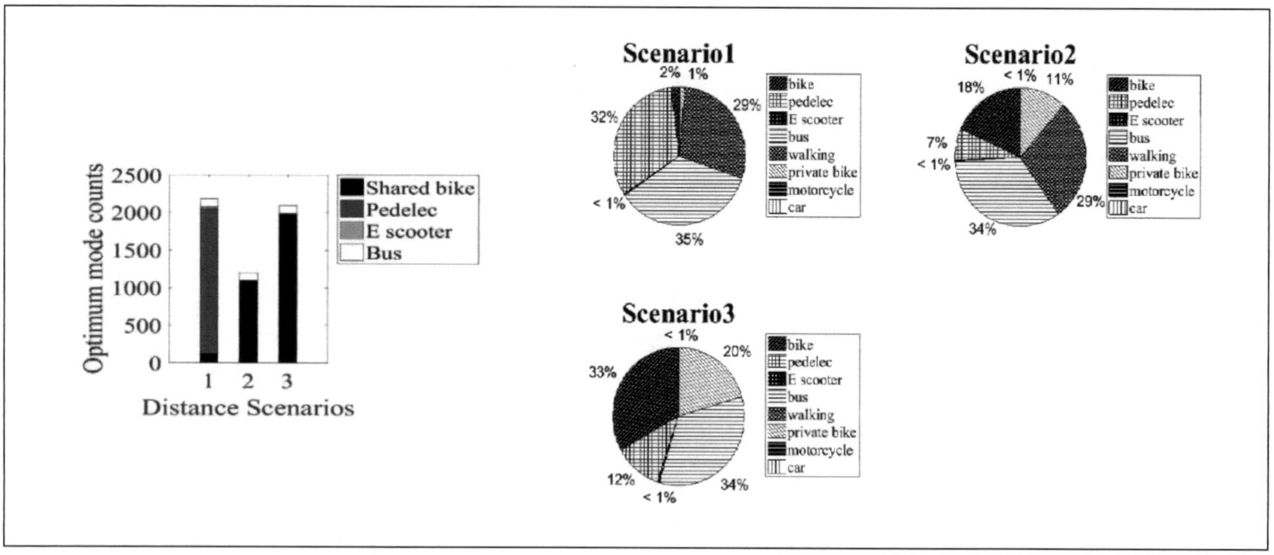

Figure 5. Distance sensitivity analysis: (*a*) optimum mode counts (numbers of each public mode) for different distance scenarios; (*b*) demand percentages for different distance scenarios.

expensive and have the ability to traverse longer distances in comparison to similar modes (i.e., e-scooter/car and bike). The other popular mode in the first scenario is walking for the first two user groups that are located closer to CBD area. In scenario 2, the percentage of the users who choose to walk remains almost invariant, as walking remains a feasible mode for the first two user groups. Bus would still be the most popular public mode, and shared bike competes to take pedelec's place. The reason is that the users in this scenario are distributed closer to CBD. In particular, distances from the CBD of the users in the third and fourth group fall within the comfortable range of bike in the second scenario. Therefore, as shared bike is less expensive than pedelec, users reasonably switch from pedelec to shared bike. In scenario 3, walking becomes an infeasible mode for all the groups, as the distances from CBD has increased. As a result, the demand for shared bike significantly increases, as shared bike becomes the feasible and affordable mode for the first four groups. The demand for private bike also significantly increases, as private bike is also feasible for the first four groups as well as being more affordable than the other modes.

Demand Sensitivity Analysis

In this section, four scenarios are arranged to explore the sensitivity of the model to the demand. In the base case scenario, each user group has a population of 1000. For scenarios 1–4 for the demand sensitivity analysis, demand variations by factors of 0.1, 1, 2, and 10 were considered. In addition, the initial budget is also increased to a high value ($150,000) to ensure that budget constraint would not be violated. The optimum mode counts and demand percentages are shown in Figure 6. As the modal demand distribution is a function of mode characteristics and not the total demand, demand percentages/distributions for different modes remain unchanged across the scenarios as expected (Figure 6*b*). According to Figure 6*a*, the optimum mode counts seem to have high correlation with demand except for pedelec in scenario 4. This phenomenon occurs because increasing pedelec up to 10 times of the base case optimum value would not justify the associated benefits, and thus at this scale the system decides to have unserved demand for pedelec, because the unserved demand cost is less than the cost that would be incurred to satisfy the demand.

Conclusion

Bike-sharing is an increasingly thriving concept across the globe. As a contributory approach, this study accounts for users' perspective through a utility function and considers the investors and societal concerns/costs and benefits in the objective function. The costs of the system entail implementation and maintenance costs as well as emission costs, whereas the benefits include quantified health benefits and system revenue. The problem is formulated as a mixed-integer nonlinear problem. The proposed algorithm for solving this problem is based on the SA algorithm, which showed successful convergence through a hypothetical numerical example. The main findings derived from the hypothetical numerical example are as follows:

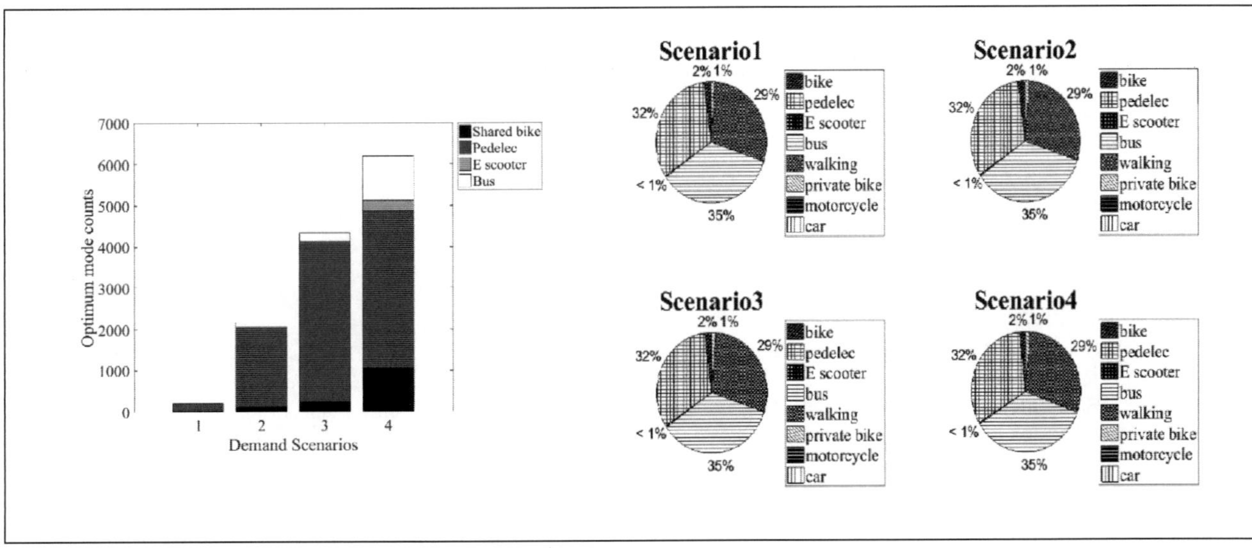

Figure 6. Demand sensitivity analysis: (*a*) optimum mode counts for demand scenarios; (*b*) demand percentages for demand scenarios.

- As the value of health benefit increases, the system requires more pedelecs, while reducing the numbers of shared bikes, e-scooter, and buses.
- The most popular public modes are bus and pedelec, because these two modes (bus and pedelec) can traverse longer distances in comparison to walking and biking which are free of charge and more affordable, respectively. On the other hand, bus and pedelec are less expensive than e-scooter, motorcycle, and car.
- Increasing financial burden of emission was found to reduce the demand and the optimum mode counts for bus while increasing the demand and optimum mode counts for pedelec.
- For small communities with short travel distances, as users would not choose fuel-consuming modes, the model is insensitive to fuel cost.
- Similarly, when trip costs/fares of the public modes are raised, demands for bus and e-scooter would decline in general, and users would choose to walk or ride pedelec instead.

It is noteworthy that the findings are reliant on the hypothetical case study and the parameters are estimated based on the average values in the United States. In fact, the flexible modeling framework and successful solution algorithm suggest that the proposed framework can be deployed by related authorities in decision-making organizations such as metropolitan planning organizations, municipality departments, and so forth.

The utility function in this study is premised on the existing literature and the coefficients are estimated. One of the next steps of this study would be to derive the parameters of the utility function from a detailed survey study. Eventually, the performance of the proposed metaheuristic algorithm can be tested using an analytical solution method. In other words, as metaheuristics are anticipated to yield near optimum solutions, analytical methods can be tested to investigate possible improvement of the results by the proposed metaheuristic, that is, SA algorithm.

One of the limitations of this study is disregarding the last mile of travel, which affects the health benefits provided by walking to bike-sharing stations or public transit mode. However, as it has been consistently disregarded for both modes this should not affect the main findings of the study. This can simply be addressed by adjusting the parameter values, based on the planners' intentions.

Another extension to this study can be considering the rebalancing of the mixed fleets of the proposed bike-sharing system. This process is normally executed using motorized vehicles which lead to emission generation, and disregarding it could overrate bike-sharing as a clean and emission-less transportation mode. Similarly, emissions from electricity production can also be considered. Another concern regarding biking and e-bike riding that has not been addressed in this study is their safety. In fact, there are safety concerns in using these modes on roadways, as they are not as protected as motorized transportation modes. However, these modes are known to be safer for the other users in the system.

Author Contributions

The authors confirm contribution to the paper as follows: study conception and design: MG and MHS; data collection: MG and MHS; analysis and interpretation of results: MG and MHS; draft manuscript preparation: MG and MHS. All authors reviewed the results and approved the final version of the manuscript.

References

1. Shaheen, S., S. Guzman, and H. Zhang. Bike-Sharing in Europe, the Americas, and Asia. *Transportation Research Record: Journal of the Transportation Research Board*, 2010. 2143: 159–167.
2. The Bike-Sharing Blog. *International Bike-Share Database*. 2015. http://bike-sharing.blogspot.com/search?q = background. Accessed October 28, 2017.
3. Fishman, E. Bikeshare: A Review of Recent Literature. *Transport Reviews*, Vol. 36, No. 1, 2016, pp. 92–113.
4. Vogel, P., T. Greiser, and D. C. Mattfeld. Understanding Bike-Sharing Systems using Data Mining: Exploring Activity Patterns. *Procedia - Social and Behavioral Sciences*, Vol. 20, 2011, pp. 514–523.
5. Jäppinen, S., T. Toivonen, and M. Salonen. Modelling the Potential Effect of Shared Bicycles on Public Transport Travel Times in Greater Helsinki: An Open Data Approach. *Applied Geography*, Vol. 43, 2013, pp. 13–24.
6. Borgnat, P., P. Abry, P. Flandrin, C. Robardet, J.-B. Pouquier, and E. Fleury. Shared Bicycles in a City: A Signal Processing and Data Analysis Perspective. *Advances in Complex Systems*, Vol. 14, No. 3, 2011, pp. 415–438.
7. Saberi, M., M. Ghamami, Y. Gu, M. H. Shojaei, and E. Fishman. *Graph Analysis of the Effects of a Public Transportation Disruption on Bike-Sharing Usage in London*. Transportation Research Board of the National Academies, Washington, D.C., 2017. https://trid.trb.org/view.aspx?id = 1437473.
8. Lin, J.-R., and T.-H. Yang. Strategic Design of Public Bicycle Sharing Systems with Service Level Constraints. *Transption Research Part E: Logistics and Transportation Review*, Vol. 47, No. 2, 2011, pp. 284–294.
9. Lin, J.-R., T.-H. Yang, and Y.-C. Chang. A Hub Location Inventory Model for Bicycle Sharing System Design: Formulation and Solution. *Computers and Industrial Engineering*, Vol. 65, No. 1, 2013, pp. 77–86.
10. García-Palomares, J. C., J. Gutiérrez, and M. Latorre. Optimizing the Location of Stations in Bike-Sharing Programs: A GIS Approach. *Applied Geography*, Vol. 35, No. 1–2, 2012, pp. 235–246.
11. Romero, J. P., A. Ibeas, J. L. Moura, J. Benavente, and B. Alonso. A Simulation-Optimization Approach to Design Efficient Systems of Bike-Sharing. *Procedia - Social and Behavioral Sciences*, Vol. 54, 2012, pp. 646–655.
12. Angelopoulos, A., D. Gavalas, C. Konstantopoulos, D. Kypriadis, and G. Pantziou. An Optimization Model for the Strategic Design of a Bicycle Sharing System. *Proc., 20th Pan-Hellenic Conference on Informatics*, Patras, Greece, November 2016, ACM Press, New York, 2016, pp. 1–6.
13. Schuijbroek, J., R. C. Hampshire, and W. J. van Hoeve. Inventory Rebalancing and Vehicle Routing in Bike-Sharing Systems. *European Journal of Operational Research*, Vol. 257, No. 3, 2017, pp. 992–1004.
14. Kaspi, M., T. Raviv, and M. Tzur. Bike-Sharing Systems: User Dissatisfaction in the Presence of Unusable Bicycles. *IISE Transactions*, Vol. 49, No. 2, 2017, pp. 144–158.
15. Brinkmann, J., M. W. Ulmer, and D. C. Mattfeld. Inventory Routing for Bike-sharing Systems. *Transportation Research Procedia*, Vol. 19, 2016, pp. 316–327.
16. Kadri, A. A., I. Kacem, and K. Labadi. A Branch-and-Bound Algorithm for Solving the Static Rebalancing Problem in Bicycle-Sharing Systems. *Computers and Industrial Engineering*, Vol. 95, 2016, pp. 41–52.
17. Fishman, E., and C. Cherry. E-Bikes in the Mainstream: Reviewing a Decade of Research. *Transport Reviews*. http://www.tandfonline.com/doi/pdf/10.1080/01441647.2015.1069907?needAccess = true. Accessed July 3, 2017.
18. Weinert, J., C. Ma, and C. Cherry. The Transition to Electric Bikes in China: History and Key Reasons for Rapid Growth. *Transportation (Amst)*, Vol. 34, No. 3, 2007, pp. 301–318.
19. Macarthur, J., and N. Kobel. *Regulations of E-Bikes in North America*. http://pdxscholar.library.pdx.edu/trec_reports. Accessed July 3, 2017.
20. Ji, S., C. R. Cherry, L. D. Han, and D. A. Jordan. Electric Bike-Sharing: Simulation of User Demand and System Availability. *Journal of Cleaner Production*, Vol. 85, 2014, pp. 250–257.
21. Silvester, S., S. K. Beella, A. van Timmeren, P. Bauer, J. Quist, and S. van Dijk. Exploring Design Scenarios for Large-Scale Implementation of Electric Vehicles: The Amsterdam Airport Schiphol Case. *Journal of Cleaner Production*, Vol. 48, 2013, pp. 211–219.
22. Dill, J., and G. Rose. Electric Bikes and Transportation Policy. *Transportation Research Record: Journal of the Transportation Research Board*, 2012. 2314: 1–6.
23. Langford, B., C. Cherry, T. Yoon, S. Worley, and D. Smith. North America's First E-Bikeshare. *Transportation Research Record: Journal of the Transportation Research Board*, 2013. 2387: 120–128.
24. Office of Disease Prevention and Health Promotion. *Summary - 2008 Physical Activity Guidelines*. health.gov, 2008. https://health.gov/paguidelines/guidelines/summary.aspx. Accessed October 31, 2017.
25. Nie, Y. M., M. Ghamami, A. Zockaie, and F. Xiao. Optimization of Incentive Polices for Plug-in Electric Vehicles. *Transptation Research Part B Methodological*, Vol. 84, 2016, pp. 103–123.
26. Halldórsdóttir, K., L. Christensen, T. C. Jensen, and C. G. Prato. *Modelling Mode Choice in Short Trips - Shifting from Car to Bicycle*. http://orbit.dtu.dk/fedora/objects/orbit:110709/datastreams/file_7742744/content. Accessed July 30, 2017.
27. Zockaie, A., H. Z. Aashtiani, M. Ghamami, and Y. Marco Nie. Solving Detour-Based Fuel Stations Location Problems. *Computer-Aided Civil and Infrastruct Engineering*, Vol. 31, No. 2, 2016, pp. 132–144.
28. Ghamami, M., A. Zockaie, and Y. M. Nie. A General Corridor Model for Designing Plug-in Electric Vehicle

Charging Infrastructure to Support Intercity Travel. *Transportation Research Part C: Emerging Technologies*, Vol. 68, 2016, pp. 389–402.
29. AASHTO. *Commuting in America 2013: The National Report on Commuting Patterns and Trends.* 2015. http://traveltrends.transportation.org/Documents/B14_Bicycling and Walk Commuting_CA14-4_web.pdf. Accessed July 24, 2017.
30. *Fares.* CATA. https://www.cata.org/Fares. Accessed August 1, 2017.
31. *Average Prices By State.* Michigan Gas Prices. http://www.michigangasprices.com/Prices_Nationally.aspx. Accessed August 1, 2017.
32. Alternative Fuels Data Center. *Maps and Data - Average Fuel Economy of Major Vehicle Categories.* https://www.afdc.energy.gov/data/10310. Accessed August 1, 2017.
33. United States Environmental Protection Agency. *Greenhouse Gas Emissions from a Typical Passenger Vehicle.* https://www.epa.gov/greenvehicles/greenhouse-gas-emissions-typical-passenger-vehicle-0. Accessed August 1, 2017.
34. *Michigan's Gas Tax Will Now Be One of the Highest in the Nation - Story.* WJBK. http://www.fox2detroit.com/news/local-news/226262779-story. Accessed August 1, 2017.
35. Trubka, R., P. Newman, and D. Bilsborough. The Costs of Urban Sprawl – Physical Activity Links to Healthcare Costs and Productivity. *Environment Design Guide*, 2010, pp. 1–13. http://www.crcsi.com.au/assets/Resources/fa19593b-16ff-45b8-9fee-0a3a71aac5cf.pdf. Accessed March 18, 2017.
36. Carlson, S. A., J. E. Fulton, M. Pratt, Z. Yang, and E. K. Adams. Inadequate Physical Activity and Health Care Expenditures in the United States. *Progress in Cardiovascular Diseases*, Vol. 57, No. 4, 2015, pp. 315–323.

The Standing Committee on Bicycle Transportation (ANF20) peer-reviewed this paper (18-06683).

Article

TRR
JOURNAL OF THE TRANSPORTATION RESEARCH BOARD

The Relationship between Experienced and Imagined Bicycling Comfort and Safety

Transportation Research Record
2018, Vol. 2672(36) 116–124

DOI: 10.1177/0361198118787635
journals.sagepub.com/home/trr
SAGE

Dillon T. Fitch[1] and Susan L. Handy[2]

Abstract

A common method used to evaluate road designs for bicycling is a survey of stated opinions based on imagined bicycling experiences that are used to represent real experiences. However, we know little about the connection between imagined and real bicycling experiences. In this study, we examine the relationship between bicyclists' (first-person experienced) and video survey participants' (imagined) ratings of bicycling comfort and safety. We do this with real-world bike rides/surveys (n = 20) and a blocked experiment conducted through a web-based video survey of female undergraduates (n = 1,203). We show that imagined ratings resemble first-person experienced ratings, but that they have a systematic negative bias (approximately 10% to 15%). This suggests that imagined bicycling experiences seem less comfortable and safe compared with real experiences. We compare both methodological and psychological explanations for this result and the consequent bicycling implications for each.

Understanding what environments are comfortable and safe for bicyclists is essential for planners aiming to increase bicycling, particularly for non-experienced riders. Surveys probing people's qualitative perceptions (or feelings) about bicycling environments thus have an important role in bicycle planning. Surveys help planners judge the perspectives of bicyclists and would-be bicyclists, and help determine bicyclists' preferences for different road designs. For example, the New York City Bicycle Survey (*1*) included a series of questions about preferences for riding on different bicycle facilities. The results indicated that "riders are most interested in demarcated, differentiated, or otherwise dedicated bicycle lanes and greenways" (p. 11). Researchers also employ surveys in their efforts to better understand the connection between bicycling environments and bicycling behavior, and the subjective value that bicyclists place on different elements of the environment.

Surveys about the road environment for bicyclists tend to fall into three main categories: descriptive, comparative, and behavioral. Descriptive surveys ask people about their feelings, emotions, and attitudes toward different bicycling environments. These types of surveys require participants to score a qualitative concept (e.g., comfort) on a quantitative scale (e.g., Likert items, anchored 10 point scales). Comparative (i.e., stated preference/choice) bicycle surveys focus on tradeoffs between different bicycling environments. These surveys are perhaps more demanding of the imagination in that they require the participant to choose between hypothetical alternative bicycling environments (often with added information about time and cost). However, they have the benefit of providing a basis for estimating the economic value of each bicycling environment (e.g., [*2*]). Finally, behavioral bicycling surveys focus on how people behave in different bicycling environments. These surveys usually ask people where they will ride or where they think bicyclists should ride given new road designs; planners use the results to help improve safety. Each of these types of surveys has utilized textual or visual cues to prompt survey participants. For example, behavioral evaluations of the environment have been measured using text (*3*), abstract drawings (*4*), images (*5*), video (*2*), and most recently virtual reality (*4*). Although there is no agreement on the most appropriate medium for surveying hypothetical bicycling experiences, in general, more realistic survey methods may offer improved behavioral insights (*6*).

The goal of this study is to shed some light on the performance of these different survey methods. We focus on descriptive video surveys of bicyclist comfort and safety. These types of survey have two primary challenges. First, descriptive scales are commonly designed in absolute terms (e.g., the response to one Likert item should not be dependent on a response to another). However, participants may

[1]Institute of Transportation Studies, University of California at Davis, CA
[2]Department of Environmental Science and Policy & Institute of Transportation Studies, University of California at Davis, CA

Corresponding Author:
Address correspondence to Dillon T. Fitch: dtfitch@ucdavis.edu

report scores in a relative sense by basing later scores on previous scores, consciously or unconsciously. For example, if a participant first reports to "strongly agree" with the statement *I felt comfortable* they may feel inclined to later report "strongly agree" to the statement *I felt safe* because they might feel the need for these responses to be consistent. Whereas had the participant received the statement *I felt safe* first, they may have reported "somewhat agree." The degree to which participants use the scale as intended (as absolute) is unknown. Second, video is a unique medium in that it can represent an environment over time (even if only a very short time) in contrast to still graphics, images, and text. This means video has a more realistic appearance when surveying people about hypothetical bicycling experiences. However, the selection of video clips to match the hypothetical experience intended by the researcher makes this method more challenging. For example, if a researcher is interested in how comfortable a person is bicycling on a four-lane road with no bike lane, they can easily ask that directly using text. This leaves the context entirely up to the imagination of the participant. When representing that four-lane road with no bike lane in video, researchers have control of the context, but it comes with the cost of making many additional choices (e.g., time of day, weather conditions, amount/type/speed of vehicular traffic, width of road lanes, with or without on street parking, tree shade, etc.). No matter the choice of medium, all methods aimed at measuring hypothetical experiences have the challenge of ensuring that the participant response is based on the researcher-intended experimental manipulation.

Although our focus is on descriptive video surveys, all of these methods (and mediums) have an imagined component that makes them susceptible to misrepresenting real experiences and associated feelings. For many human behaviors, biases have been reported when comparing surveys of real and hypothetical experiences (e.g., product purchasing [*7*], moral decisions [*8*], travel behavior [*9,10*]). These biases can be problematic if planners make policy or design decisions based on surveys. For stated choice experiments, these biases (known as hypothetical bias) can range from anywhere between 25% and 300% (*9*). Studies of stated and revealed preferences about bicycling yield some evidence that these biases may be small for objective variables (e.g., expected travel time) (*11*). However, evidence on bias for subjective variables such as perceived comfort and safety is not currently available. In the context of judging comfortable and safe bicycling environments, it is unclear how representative surveys are of real bicycling experiences or real-life travel decisions. The uncertainty of the magnitude of potential misrepresentation motivates our primary research question: in surveys of bicycling comfort and safety, what is the relationship between imagined and experienced bicycling comfort and safety? In addition, because we represent qualitative feelings (comfort and safety) as quantities on an absolute scale, we pose a secondary methodological question: to what extent do survey participants use this scale in a relative way?

Methods

Study Design

We addressed our research questions through a sample of female undergraduates at the University of California, Davis who bicycled or participated in a video survey (hereafter referred to as the *bicyclists* and the *video participants*, respectively). The bicyclists' ratings we refer to as *first-person* because they rate their real personal experience in the survey, whereas the video participants' ratings we refer to as *imagined* because they are imagining a bicycling experience through the video. Importantly, this imagined experience is tied to the bicyclists' experience because the video clips are taken from the bicyclists' rides. We chose female undergraduates because they are a conservative target population when planning roads for bicycling (*12*), both because women are more sensitive to traffic safety (*13,14*), and because undergraduates do not have a lot of bicycling experience (i.e., none of the participants regularly bicycled for travel until coming to Davis). As an incentive, we awarded the bicyclists ($n = 20$) a \$35 gift certificate for their participation, which included questionnaires, a mental stress test, an interview, and physiological measurements while bicycling. Because of the cost and duration of this experiment (3 to 4 h per participant) the bicyclist sample size is small. The bicyclists range from 2nd- to 4th-year students, all of whom had previously learned to bicycle when they were a child.

As a part of the bicyclists' participation, they rode in three distinct road environments (Figure 1) while recording video. They rode in these environments in assigned orders designed to counterbalance first-order carry-over effects. An example of a first-order carry-over effect in this study would be if a bicyclist's rating of a collector road was influenced by their prior rating of a major arterial. By balancing the order of roads the bicyclists experienced, we ensured there was no systematic effect of order on the aggregated ratings as a group. Bicyclists rated their experience on a 10-point comfort/safety scale by providing their most common level (mode) and range of levels (minimum and maximum) of comfort/safety. Note that we only asked participants about comfort/safety with regards to interacting with vehicular traffic (we would not expect bicyclists to consider, for example, pavement roughness as a *direct* factor in this measure; the expectation that bicyclists focused on interactions with vehicular traffic was confirmed in post-experiment interviews). The comfort/safety scale was a measure designed to encompass both feelings of comfort and safety. As these were difficult for the bicyclists to separate (e.g., responses to 5-point Likert statements about comfort and safety separately were nearly perfectly correlated) the combined term was used.

Figure 1. Experimental road environments: collector (*a*), minor arterial (*b*), major arterial (*c*). Images from Google Street View.

The video participants ($n = 1{,}203$) were assigned five short video clips (extracted from the recordings of the bicyclists' rides) in a web survey and were asked to rate their comfort/safety on the same 10-point scale. This web survey was an add-on to the 2016–2017 University of California, Davis annual campus travel survey (see Gudz et al. for the 2015–2016 survey [*15*]). Although the video experiment had to be kept short to minimize overall survey burden, it allowed us access to a large sample of female undergraduates. The web survey consisted of a blocked experimental design. Each experimental block consisted of an "anchor" video (explained below), followed by four videos of one road type with four levels of traffic presented in geographic order of the bicyclists' rides (e.g., in Block 1, A_2 (light traffic on the collector) is presented first because it occurred at the beginning part of that road) (see Table 1). Experimental blocks were balanced between anchor and road type, but not between road type and traffic, nor between person and road type. Each participant received one experimental block, which presents five video clips. This design ensured we could make between-subject inferences about all experimental environments.

Table 1. Experimental Design

		Experimental environments[a]			
Block	Anchor	Road type by traffic condition			
1	A_A	A_2	A_1	A_3	A_4
2	A_A	B_1	B_2	B_3	B_4
3	A_A	C_2	C_4	C_3	C_1
4	B_A	A_2	A_1	A_3	A_4
5	B_A	B_1	B_2	B_3	B_4
6	B_A	C_2	C_4	C_3	C_1
7	C_A	A_2	A_1	A_3	A_4
8	C_A	B_1	B_2	B_3	B_4
9	C_A	C_2	C_4	C_3	C_1

[a]Treatment consists of an anchor video of "typical traffic," followed by four sequential road types (A = collector, B = minor arterial, C = major arterial) with unique levels of traffic (e.g., numeric subscripts 1 = none, 2 = light, 3 = moderate, 4 = heavy or potential interaction).

What is An Anchor and Why Do We Use One?

We are loosely borrowing the concept of anchoring from Tversky and Kahneman (*16*). This concept is based on the observation that when asked to give a number, people will give different numbers depending on their exposure to an initial value or reference point. In situations where a numerically correct answer exists, this heuristic means that people tend to systematically favor deductions near the anchoring value. In our case, there is no correct value (we are measuring a subjective feeling), but we have a 10-point scale that is interpreted based on each video participants' judgment of the endpoints ("uncomfortable/unsafe" and "safe/comfortable"). The concern is that the initial video a participant watches skews their ratings of subsequent videos: an initial video that rates as highly safe/comfortable could lead to overall higher ratings, whereas an initial video that rates as uncomfortable/unsafe could lead to overall lower ratings. Although our scale is intended to be absolute (not relative), by explicitly including a common anchoring video for each of the experimental blocks (one video for each road type with typical traffic conditions), we can examine the influence of the anchor on the remaining scores in case participants are using the scale in a relative way.

We used a different process to correct for anchoring with the bicyclists. Each bicyclist rated the roads immediately following their riding experience. Because of anchoring, these ratings could be biased by the bicyclist's rating on the first road she rode (because she had nothing to compare it with). For example, if she rated the first road a 7, then found the next two roads to be more comfortable/safe, she would tend to give ratings of 8 or higher. But if that first road had instead come last, she might have given it a far lower rating in comparison with the other two. To correct for this, we gave the bicyclists the opportunity to adjust their ratings for any road in an interview at the end of the completed experiment. They did not have a predefined "anchor" the way the video participants did, but they were given the opportunity to adjust scores once they had experienced all of the road types.

Video Selection

Video clips of approximately 10 s were selected based on video quality, representativeness of the road environment, and representativeness of traffic conditions. The video camera was secured to the bicyclists' helmets and thus the camera viewing angle closely matches the view angle of the bicyclists. Because head movements caused considerable shakiness in the image, we chose video clips where there was minimal head movement. We also avoided video clips where the sun angle was low enough to cause glare. To ensure representativeness of the road environment, we selected two video clips on the outward leg of each road type and two on the return leg (bicyclists rode out and back). We presented the four video clips for each road type in sequential (geographic) order as the bicyclist would have experienced them. However, the video clips for a road type were taken from up to four different bicyclists' recordings, thus they are not representative of any individual experience. Finally, we selected video clips so as to balance traffic conditions across the roads: each road type includes a video clip that has no passing vehicles, one with a few slow-passing vehicles, one with fast passing or large vehicles, and finally one with either large passing vehicles or situations of potential vehicle conflict (see Table 1). Anchor videos were selected to illustrate a typical environment with typical traffic for each road type. For example, the collector anchor has only a slow-passing vehicle, whereas the major arterial anchor has many fast passing vehicles. Furthermore, anchor videos are distinct from non-anchor videos (e.g., $A_A \notin \{A_1, A_2, A_3, A_4\}$ in Table 1).

Analysis

We estimated overall video participants' comfort/safety by counting the number of responses at each point of the scale for each road type. We divided these counts by the number of responses to produce a normalized histogram. We estimated the bicyclists' comfort/safety from the scores reported by the bicyclists after their rides (including any adjustments based on the interviews). To characterize their comfort/safety during their entire experience on a road type, we used each bicyclists' mode, minimum, and maximum survey responses of comfort/safety to form a triangular distribution (min, mode, max). We combined each triangular distribution (one for every bicyclist) into an aggregated distribution for each road type by summing 10 bin histograms (one for each level in the 10-point scale) and dividing by the sum in each bin. This creates a normalized histogram that can be compared to the video participants' comfort/safety histogram.

We conducted three different comparisons between imagined and first-person comfort/safety. First was a simple comparison of aggregated ratings representing the overall difference. Second, we repeatedly (100 iterations) sampled 20 individuals from an attitude-matched subset of the video participant population and compared the sample ratings to the first-person ratings. We did this to address two potential confounds in our sampling. First, our small bicyclist sample size could have randomly favored one side of the 10-point scale and had we had a larger sample we might have found that the bicyclists' responses would have been less extreme. By repeatedly sampling 20 video participants, we could estimate a 20-person video sample variance. If the bicyclists' ratings were within that variance, it strengthened the argument that sample size is the reason for disagreement between bicyclists and video participants. Second, we used an attitude-matched (self-reported bicycling ability and comfort on a four-lane road with no bike lane, hereafter referred to as *ability* and *general comfort*) subset to guard against the fact that our bicyclist sample was likely to favor women with greater ability and general comfort than the norm (because they volunteered for an on-road bicycling experiment). Finally, we compared the bicyclists' ratings with the most able and comfortable video participants (those who responded in the highest category of both ability and general comfort). This comparison allowed us to examine the relationship between these attitudes and the comfort/safety rating to ensure they were appropriate variables to create a bicyclist-matched sample. In other words, we expect comfort/safety scores to be greatest for this biased sample. In all comparisons, we calculated three statistics to represent the differences. These were:

1. Kullback–Leibler (KL) divergence: This is the directional "distance" between two distributions. This measure has an intuitive interpretation in terms of information theory such that a divergence close to 0 would indicate both distributions have the same information, and a divergence close to 1 would indicate that knowing one distribution does not help in gaining information about the second distribution (see McElreath for an overview [*17*]).
2. Earth mover's distance (EMD) (*18*): This is a true distance metric (bi-directional) that describes how much "work" it takes to move one distribution to another where "work" is the mass * distance.
3. Difference between the means.

Limitations

Our method has a few limitations. First, the video participants and the bicyclists were not the same people and thus we could not calculate an individual-level difference between ratings of comfort/safety. Instead, our analysis is at the aggregated group level and subject to ecological fallacy, meaning that findings at the aggregate level may not hold at the individual level (*19*); differences between imagined and first-person ratings for the groups do not necessarily mean that individuals would rate imagined and first-person comfort/safety with the same difference. Nonetheless, this is a first step toward estimating the magnitude of bias between bicycling comfort/safety surveys for imagined and first-person experiences. Second, our sample only covered female undergraduates and so should not be used to generalize to the entire population. Third, the sample size of the bicyclists was small, as is common for field studies, and much smaller than the sample of video participants. The small bicyclist sample size is likely to inflate the error of first-person comfort/safety. We partially addressed this by constructing triangular distributions of comfort/safety (as opposed to simple averages) for each bicyclist to provide a more representative view of comfort/safety for the small sample, and by using repeated subsampling of video participants to match sample size between groups (see above). Fourth, our sample of bicyclists was not random and may be biased toward women with more bicycling experience. We partially addressed this limitation by matching our video sample based on two common (across samples) attitude variables thought to influence comfort/safety responses (ability and general comfort).

Results and Discussion

Effect of Video Anchor

The effect of the anchor on video participants' ratings was minimal across individual video ratings. This can be seen by comparing the distributions of ratings between the three anchors (Figure 2). Each panel in the left four columns of Figure 2 shows the aggregate distribution of scores for a given video (road type by traffic combination). The three box/violin plots on each panel demonstrate similar distributions for responses across the three anchor types. For example, the top left panel shows responses for the video of a collector with light traffic. The box/violin plots show that responses had similar medians, interquartile, and full ranges for each of the three anchors (x axis). Furthermore, the effect of the anchor on the average rating for road type (column 5 of Figure 2) was so small that the distribution of ratings looks identical. There are a few cases where there appear to be subtle differences in response based on the anchor video. For example, the video participants who received Block 7 (collector with the major arterial anchor) appeared to rate the collector as slightly more comfortable/safe than those participants with the other anchors. However, this was only when traffic was moderate and heavy on the collector. Similar subtle differences can be observed for other experimental environments (Figure 2), but nothing is indicative of clear anchoring bias. This is especially true given that the direction of influence from the anchors seems to vary. For example, were anchoring occurring we would expect the scores for the major arterial to be ordered by anchor: greatest for the collector, less for the minor arterial, and least for the major arterial. However, Figure 2 suggests that the bias is sometimes positive and sometimes negative. The lack of a clear anchoring effect suggests that either the video participants treated the scale as absolute, or that if they employed a relative ranking, it is unlikely to have caused significant bias. This is somewhat surprising given the 1–10 rating scale is only vaguely defined by its end points and because we fused comfort and safety constructs. Although we had expected to see some evidence of systematic anchoring, the lack of anchoring allowed us to treat the anchoring video as a regular video sample with no special role in the ratings. Because of this, we included all video ratings and the average rated for each participant resulting in n*6 (7,218) ratings for further analysis.

Comparing Imagined and First-Person Ratings of Bicycling Comfort/Safety

Our results indicate that survey responses are slightly more negative for video participants compared with the bicyclists (mean rating difference of about 1–1.5 on a 10-point scale) (Table 2). That is, video participants reported a 10% to 15% lower rating for comfort/safety on average. Not only did the mean estimates differ, but the distributions of estimated comfort/safety differ, most substantially in the lower range of scores (see EMD from Tables 2 and 3, and Figure 3). These results are consistent when we compare random samples of 20 attitude-matched video participants with the bicyclists (see Table 3). The standard deviation (SD) of the random samples suggests that there is some amount of uncertainty in the estimated bias given the small sample size. However, even at 2 SD above the mean estimate, the video participant comfort/safety scores were more negative (Table 3). This indicates a clear negative bias for video participants, even after controlling for attitudes. Although there is a clear negative bias, Figure 3 shows that the first-person experienced and imagined ratings are consistent in terms of their relative frequencies of scores. For example, both measures agree that the most common score for the collector is in the 8–10 range, whereas the least common in the 1–4 range. This indicates that the imagined experiences resembled the real experiences.

We also took the video participants with the strongest attitudes (highest category for ability and general comfort) to simulate a biased video participant sample (Table 2). This comparison suggests that ability and general comfort have a strong influence on imagined comfort/safety ratings. The video participants who were most able and most comfortable bicycling had a positive bias for the collector and major arterial compared with the bicyclists in general. This result supports our decision to match the video sample based on the bicyclists' attitudes about ability and general comfort. It also parallels prior survey results that suggest people without bicycling experience report less comfort than their peers when bicycling on the same road types (*20*). This point is important for defining an appropriate subpopulation to survey as a basis for designing roads to make bicycling possible for the masses. These results indicate that the survey population should be defined not only based on demographics (e.g., age, gender) but also psychological measures (e.g., attitudes about ability and general comfort).

The differences between imagined and first-person ratings can probably be partially explained by the study design. We selected four levels of traffic conditions and presented

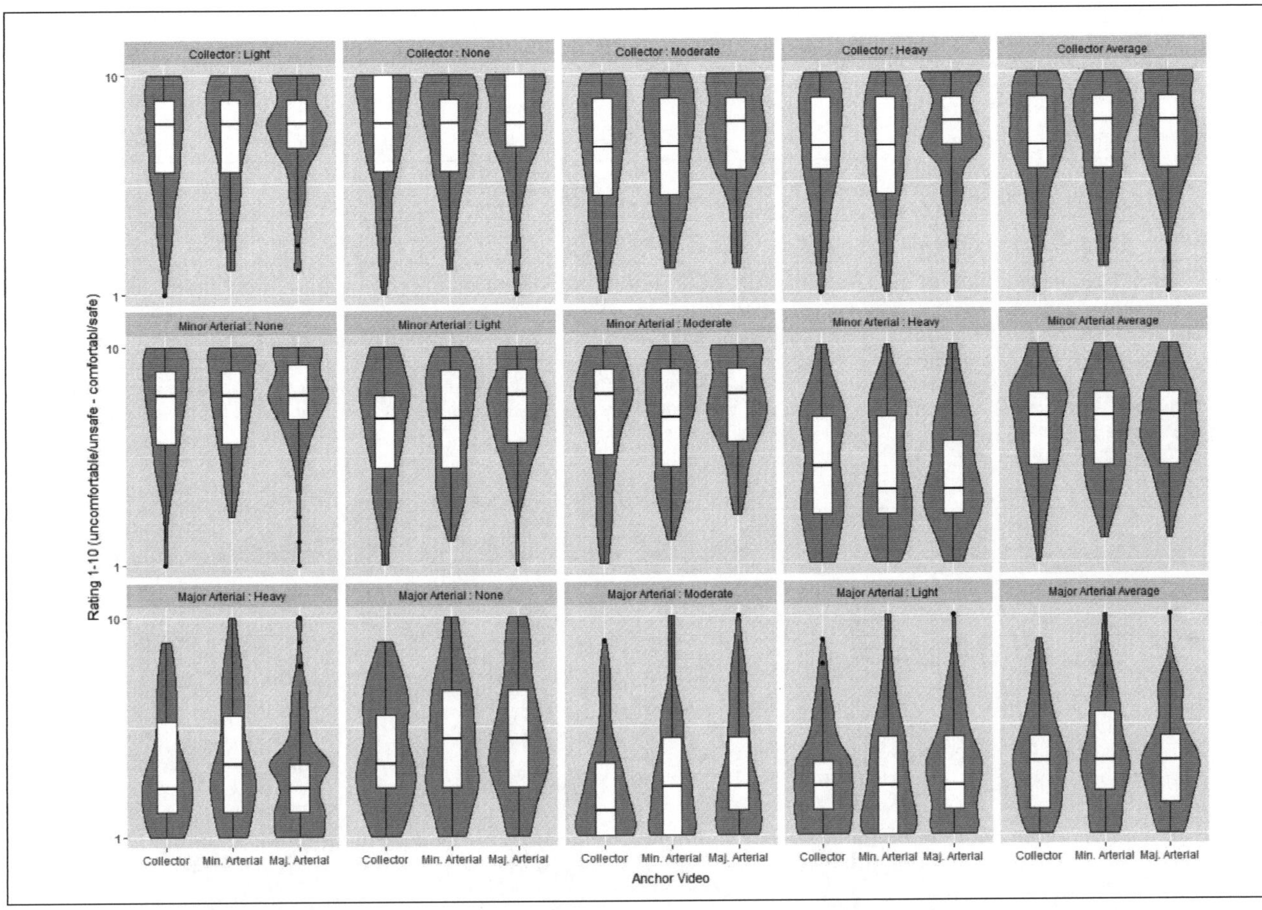

Figure 2. Box and violin plots showing the distribution of imagined ratings by anchor, grouped by experimental environments (road type by traffic condition). Panel rows represent road type, and columns represent the order the clips were presented. The last column is the user-provided average rating for all videos in their road type.

Table 2. Distance Statistics Between Video Participant and Bicyclist Comfort/Safety

Road type	KL[b]	EMD[c]	Δ means	*t*-test *p*-value
D(bicyclists||video participants)[a]				
Collector	0.35	1.14	−1.00	<0.01
Minor arterial	0.44	1.53	−1.45	<0.01
Major arterial	0.31	1.31	−1.29	<0.01
D(bicyclists||high ability and general comfort video participants)[a]				
Collector	0.43	0.78	+0.23	0.21
Minor arterial	0.29	0.74	− 0.15	0.15
Major arterial	0.24	1.07	+1.07	<0.01

[a]Where *D*(*P*||*Q*) is in general the distance between *Q* and *P*.
[b]Units of average difference in log probability (information lost).
[c]Units of probability mass * distance (e.g., 50% of probability mass moved 2 units on the 10-point scale would have an EMD = 1).

them as having equal weight on the ride on that road type, whereas the bicyclists only experienced pockets of moderate to high traffic (Figure 4). This might have caused the video participants to have a negative bias as their exposure was to proportionally more traffic. However, we might expect that negative experiences (those during heavy traffic)

Table 3. Distance Statistics Between Video Participant and Repeat Samples of Attitude-Matched Bicyclist Comfort/Safety

D(bicyclists||attitude matched video participants)[a]

Road type	mean KL [b]	sd	% ∞ KL[d]	mean EMD[c]	sd	mean Δ means	sd	*t*-test *p*-value 90% CI
Collector	0.40	(0.15)	2%	1.07	(0.31)	−0.89	(0.38)	$1e^{-8} - 0.1$
Minor arterial	0.48	(0.12)	0%	1.53	(0.28)	−1.44	(0.32)	$2e^{-11} - 1e^{-4}$
Major arterial	0.35	(0.11)	42%	1.31	(0.30)	−1.21	(0.35)	$1e^{-12} - 1e^{-3}$

[a]Where *D*(*P*||*Q*) is in general the distance between *Q* and *P*.
[b]Units of average difference in log probability (information lost).
[c]Units of probability mass * distance (e.g., 50% of probability mass moved 2 units on the 10-point scale would have an EMD = 1).
[d]KL divergence is directional (e.g., $D_{KL}(P||Q)$ is the divergence from *Q* to target *P*). This implies that when imagined comfort/safety has no responses for a given category while first-person comfort/safety has a positive count for that category, the KL divergence goes to infinity (indicating the method is unable to measure the difference between distributions). Because of this, the KL divergence for the sampled major arterial is not to be trusted given the large percentage of infinite divergence.

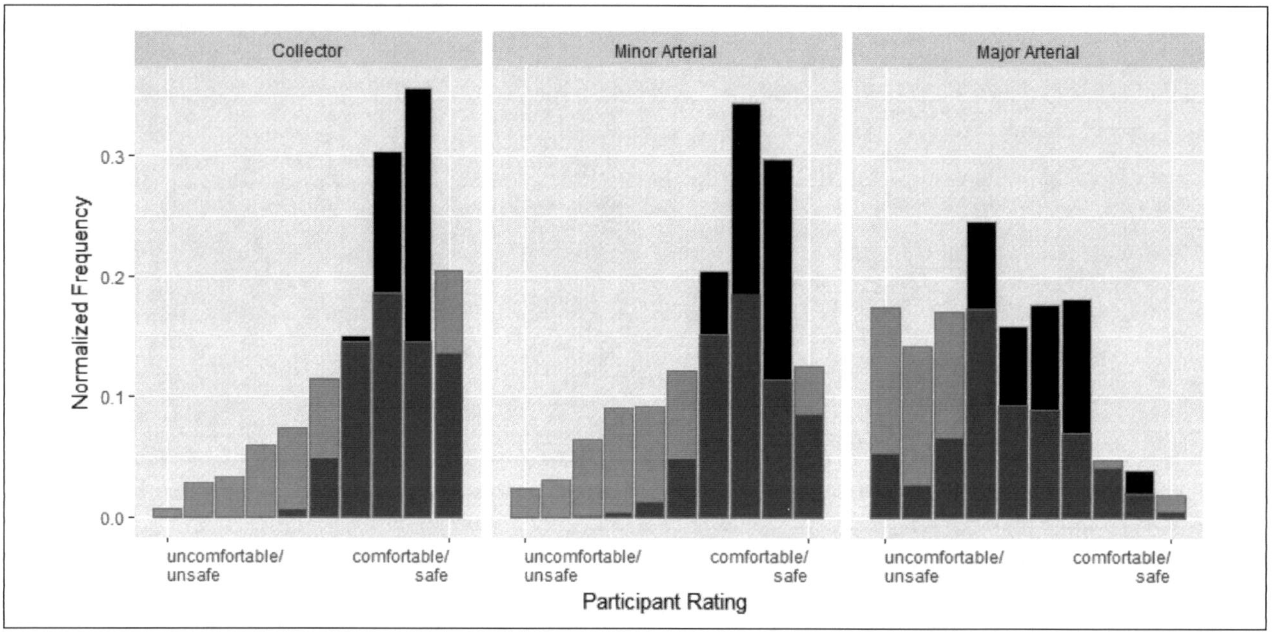

Figure 3. Aggregate normalized ratings of bicyclists' ratings (*black*) and video participants' ratings (*light red*), with overlap being *dark red.*

had a stronger influence on the bicyclists' ratings given that negative experience has a stronger psychological impact (*21*). Additional issues such as video wobble may have caused viewer discomfort, and lack of audio and peripheral views limited video participants' ability to anticipate passing vehicles. Simultaneously, there are reasons why we might have expected the video participants to have a positive bias. First, the video participants were not exposed to any real safety threat compared with the bicyclists, so we would not expect them to have experienced true fear for their safety. In addition, subtle features of the road that could not be adequately captured by video (e.g., side winds and rough pavements) could make a road seem less comfortable/safe for a bicyclist (because those features can cause unexpected interactions with traffic), but not to a video participant.

Another explanation is that bicyclists may be basing their ratings on a small subset of salient moments during their ride that are unknown to us and thus impossible to select for the video experiment. This possibility is suggested by behavioral theories of retrospective evaluations such as the so-called *peak end rule* that suggests retrospective evaluations are largely based on some "peak" (or most intense) point and the end of an experience (*22*). However, because negative experiences loom larger than positive ones (*21*), the peak end rule suggests the bicyclists would focus on the most stressful moments, causing the video participants to be biased in the opposite direction (i.e., more comfortable). Moreover, our use of minimum, mode, and maximum comfort/safety ratings by the bicyclists attempts to encourage broader evaluations of first-person experiences.

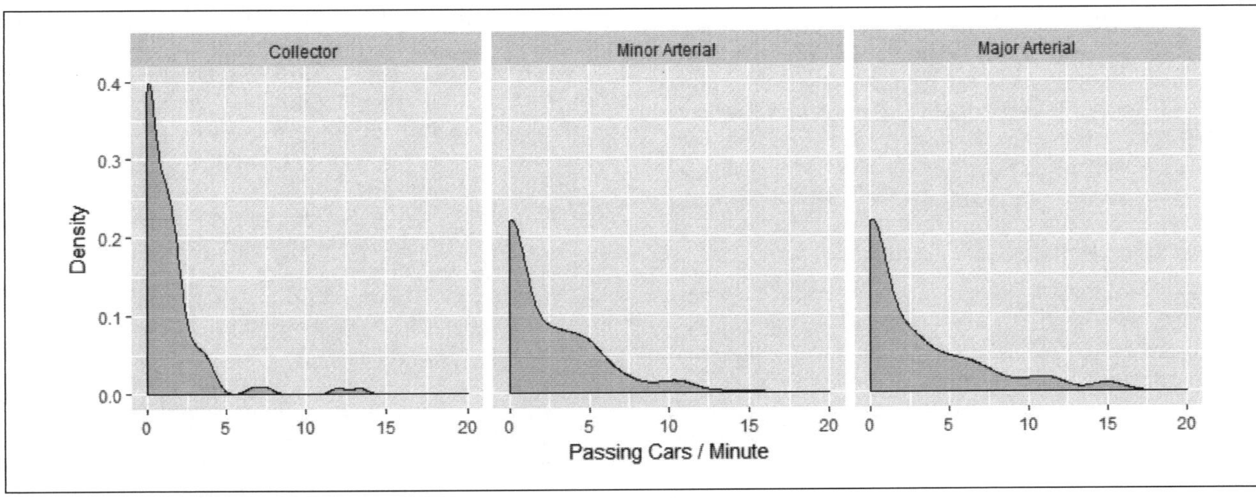

Figure 4. Bicyclists' traffic rate (of right-most travel lane) frequency by road type.

When comparing road types, our results suggest that imagined experiences may have more subtle biases. KL divergences indicate that imagined comfort/safety provide less information about first-person comfort/safety when judging the minor arterial compared with the collector and major arterial (tables 2 and 3). It could be that our choice of video clips for this road type is to blame. Trying to represent a bicycle ride of 10–15 min in four 10-s video clips was challenging and we may have exaggerated a negative feature of the environment for this road type. For example, we chose a video clip of a slow-passing bus on both the minor and major arterials, whereas only 8 and 12 of 20 bicyclists experienced a bus passing on those road types, respectively. In addition, we chose a video clip of an approach to a left-hand turn on the minor arterial. It is the only video clip that demonstrates a bicyclist crossing a vehicular lane to enter a pocket bike lane at an intersection. Although every bicyclist undertook this maneuver, it could be that emphasizing it in one of four video clips exaggerated its relative importance on the comfort/safety of the minor arterial as a whole. Further examination of a wider range of road environments would be needed to determine what it is about the environment that is difficult to imagine.

Implications and Conclusions

Our results show that imagined ratings resemble first-person experienced ratings but with a systematic negative bias. Some of the negative bias may be due to our study design; however, similar systematic negative biases have been reported when comparing other examples of imagined and first-person surveys. For example, Krach et al. (*23*) show that social embarrassment (a negative emotion) can be much stronger for vicarious observers than for first-person experiences. Although social embarrassment is a different experience than bicycling comfort/safety, this parallel between surveys of imagined and first-person experience suggests a broader psychological reason for these results that merits further exploration.

The implications of these results are two-fold. First, if road environments seem less comfortable and less safe when people imagine themselves bicycling than they do when they are really bicycling, non-bicyclists may need real bicycling experiences before they would consider bicycling for normal day-to-day travel. Our results thus provide indirect evidence in support of bicycling programs that explicitly create real bicycling experiences (e.g., bike to work days, ciclovias). Second, in terms of aiding road design, video surveys may provide a conservative estimate of bicycling comfort/safety *depending on the surveyed population*. To provide an estimate of comfort/safety that inspires road designs to make bicycling possible for the masses, important attitudes such as bicycling ability and comfort should be considered in selecting the survey population.

Although this negative bias from video surveys is intriguing, it is by no means conclusive. Future surveys of first-person versus imagined bicycling experience are warranted across other populations and geographies to determine its validity. As Davis is one of the most bicycling friendly places in the United States, perceptions of comfortable and safe bicycling as reported by Davis residents may not be representative of those of other populations. In addition, broadening the scope of this research to include different presentations of video clips (or other mediums including virtual and augmented reality) and more subtle differences in road environments (as is typically done in stated preference surveys) will improve our understanding of the relationship between imagined and first-person experienced bicycling comfort and safety.

Acknowledgments

This research was funded by the University of California at Davis National Center for Sustainable Transportation. The authors thank the participants, Drew Heckathorn of UC Davis for administering the 2016–2017 campus travel survey, and to UC Davis Transportation and Parking Services for their support. Additional thanks to Mike Boswell and Huyen Le for insightful comments about interpretation of the results in light of behavioral theories of retrospetive evaluations.

Author Contributions

The authors confirm contribution to the paper as follows: study conception and design: DF; data collection: DF; analysis and interpretation of results: DF, SH; draft manuscript preparation: DF, SH. All authors reviewed the results and approved the final version of the manuscript.

References

1. The New York City Bicycle Survey. Department of City Planning Transportation Division, NYC Department of City Planning, 2007.
2. Tilahun, N. Y., D. M. Levinson, and K. J. Krizek. Trails, Lanes, or Traffic: Valuing Bicycle Facilities with An Adaptive Stated Preference Survey. *Transportation Research Part A: Policy and Practice*, Vol. 41, 2007, pp. 287–301. doi:10.1016/j.tra.2006.09.007
3. Handy, S. L., Y. Xing, and T. Buehler. Factors Associated with Bicycle Ownership and Use: A Study of Six Small U.S. Cities. *Transportation (Amst)*, Vol. 37, pp. 967–985. doi:10.1007/s11116-010-9269-x.
4. Axhausen, K. W., and A. Erath. Between Abstraction and Hyperrealism: Is There a Right Point?, ABJ40 Workshop "Use of Visualization in Stated Surveys". Presented at 94th Annual Meeting of the Transportation Research Board. Washington, D.C., 2017.
5. Caulfield, B., E. Brick, and O. Thérèse. Determining Bicycle Infrastructure Preferences – A Case Study of Dublin. *Transportation Research Part D: Transport and Environment*, Vol. 17, 2012, pp. 413–417. doi:10.1016/j.trd.2012.04.001
6. Innocenti, A. Virtual Reality Experiments in Economics. *Journal of Behavioral and Experimental Economics*, Vol. 69, 2017, pp. 71–77. doi:10.1016/j.socec.2017.06.001
7. Chang, J. B., J. L. Lusk, and B. F. Norwood. How Closely do Hypothetical Surveys and Laboratory Experiments Predict Field Behavior? *American Journal of Agricultural Economics*, Vol. 91, 2009, pp. 518–534. doi:10.1111/j.1467-8276.2008.01242.x
8. FeldmanHall, O., D. Mobbs, D. Evans, L. Hiscox, L. Navrady, and T. Dalgleish. What We Say and What We Do: The Relationship Between Real and Hypothetical Moral Choices. *Cognition*, Vol. 123, 2012, pp. 434–441. doi:10.1016/j.cognition.2012.02.001
9. Fifer, S., J. Rose, and S. Greaves. Hypothetical Bias in Stated Choice Experiments: Is It a Problem? and If So, How Do We Deal With It? *Transportation Research Part A: Policy and Practice*, Vol. 61, 2014, pp. 164–177. doi:10.1016/j.tra.2013.12.010
10. Wardman, M. 1988. A Comparison of Revealed Preference and Stated Preference Models of Travel Behaviour. *Journal of Transport, Economics and Policy*, Vol. 22, 1988, pp. 71–91.
11. Yang, C., and M. Mesbah. *Route Choice Behaviour of Cyclists by Stated Preference and Revealed Preference*. Australasian Transport Research Forum, Brisbane, Australia, 2013.
12. Garrard, J., S. Handy, and J. Dill. City Cycling. *Women in Cycling* (J. Pucher, and R. Buehler, eds.), MIT Press, Cambridge, MA, 2012, pp. 211–234.
13. Beecham, R., and J. Wood. Exploring Gendered Cycling Behaviours Within a Large-scale Behavioural Data-set. *Transportation Planning and Technology*, Vol. 37, 2013, pp. 83–97. doi:10.1080/03081060.2013.844903
14. Emond, C. R., W. Tang, and S. L. Handy. 2014. Explaining Gender Difference in Bicycling Behavior. *Transportation Research Record: Journal of the Transportation Research Board*, 2014. 2125: 16–24.
15. Gudz, E., D. Heckathorn, and C. Thigpen. *Results of the 2015 - 16 UC Davis Campus Travel Survey*. UC Davis Institute of Transportation Studies, 2016.
16. Tversky, A., and D. Kahneman. Judgment under Uncertainty: Heuristics and Biases. *Science*, Vol. 80, No. 185, 1974, pp. 1124–1131. doi:10.1007/978-94-010-1834-0_8
17. McElreath, R. *Statistical Rethinking*. Chapman & Hall/CRC, 2015.
18. Levina, E., and P. Bickel. The Earth Mover's Distance is the Mallows Distance: Some Insights from Statistics. *Proceedings of the IEEE International Conference on Computer Vision*, Vol. 2, 2001, pp. 251–256. doi:10.1109/ICCV.2001.937632
19. Freedman, D. Ecological Inference and the Ecological Fallacy. *International Encyclopedia of the Social & Behavioral Sciences*, 2001, pp. 4027–4030.
20. Buehler, R., and J. Dill. Bikeway Networks: A Review of Effects on Cycling. *Transport Reviews*, Vol. 36, 2016, pp. 9–27. doi:10.1080/01441647.2015.1069908
21. Baumeister, R. F., E. Bratslavsky, C. Finkenauer, and K. D. Vohs. Bad is Stronger Than Good. *Review of General Psychology*, Vol. 5, 2001, pp. 323–370. doi:10.1037//1089-2680.5.4.323
22. Fredrickson, B. L., and D. Kahneman. Duration Neglect in Retrospective Evaluations of Affective Episodes. *Journal of Personality and Social Psychology*, Vol. 65, 1993, pp. 45–55. doi:10.1037/0022-3514.65.1.45
23. Krach, S., J. C. Cohrs, N. C. de Echeverría Loebell, T. Kircher, J. Sommer, A. Jansen, and F. M. Paulus. Your Flaws are My Pain: Linking Empathy to Vicarious Embarrassment. *PLoS One*, Vol. 6, 2011. doi:10.1371/journal.pone.0018675

The Standing Committee on Bicycle Transportation (ANF20) peer-reviewed this paper (18-02992).

Article

TRR
JOURNAL OF THE TRANSPORTATION RESEARCH BOARD

Transportation Research Record
2018, Vol. 2672(36) 125–135

DOI: 10.1177/0361198118788425
journals.sagepub.com/home/trr
SAGE

Analytical Model of Passing Events for One-Way Heterogeneous Bicycle Traffic Flows

Liang Xu[1], Meiqi Liu[2], Xiang Song[3], and Sheng Jin[1]

Abstract

Heterogeneous bicycle traffic flows, consisting of electric bicycles (e-bicycles) and regular bicycles (r-bicycles), have become the main traffic form on shared bicycle routes in China due to the increasing number of e-bicycles. As a result, overtaking occurs frequently among bicycles, which affects cyclists' safety and perception. This paper presents an analytical model to estimate the number of passing events in heterogeneous bicycle traffic flows. The relationships between passing events and the parameters of the heterogeneous bicycle traffic flow is established in the proposed model. The probability density functions of the speed of r-bicycles and e-bicycles are taken into consideration. The results of the model analysis show that the number of passing events increases with an increase in the flow rate and density. Both a difference in speed between different types of bicycle and the standard deviation of speed of each type of bicycle have positive correlations with the number of passing events. In addition, when the proportion of e-bicycles increases, the number of passing events first increases, and then decreases. The proposed model is calibrated against field data collected in Hangzhou. The results show that the model prediction is consistent with field observations. The model proposed in this paper provides an analytical approach to study the relationship between the characteristics of heterogeneous bicycle traffic flows and the number of passing events. This work can be considered a prerequisite for the development of the bicycle level of service criteria for heterogeneous bicycle flows.

Bicycles are one of the major travel modes of slow traffic and play an important role in urban road networks in many countries, especially China. In recent years, the number of electric bicycles (e-bicycles) has been growing rapidly as an alternative to regular bicycles (r-bicycles). Compared with r-bicycles, e-bicycles can provide higher speeds and longer range with the support of electric propulsion (*1*). In addition, both pollution and emission that e-bicycles emit are significantly lower than those emitted by motorcycles and cars (*2*). E-bicycles are energy efficient and environmentally friendly compared with cars and have longer range than r-bicycles. Given current traffic conditions in China, many provincial and city administrations have been promoting e-bicycles in recent years because of these advantages (*3*). Consequently, the non-motor traffic mode has gradually shifted from the "r-bicycle" to "r-bicycle and e-bicycle" mixed modes.

R-bicycles and e-bicycles always run on the same non-motorized lanes, which creates a heterogeneous bicycle traffic flow. There are significant differences between r-bicycles and e-bicycles in terms of speed, acceleration, and deceleration. According to field data from Kunming city in China, the average speed of e-bicycles is 21.86 km/h which is 7.05 km/h (47.6%) higher than that of r-bicycles (*4*). The flexibility of non-motor vehicles and the big difference in speed between r-bicycles and e-bicycles can result in particular overtaking and lane-changing behavior in heterogeneous bicycle traffic flows. Such particular behavior might result in great psychological pressure on riders and, to a certain extent, compromises with safety. Different types of bicycles are shown in Figure 1.

Overtaking is one of the most common types of behavior in heterogeneous bicycle traffic flows. However, only a limited number of existing studies focus on the effects of heterogeneous bicycle traffic. The Highway Capacity Manual (HCM) treats the number of passing events as a parameter when evaluating the level of service

[1]College of Civil Engineering and Architecture, Zhejiang University, Hangzhou, China
[2]Department of Transport & Planning, Delft University of Technology, Delft, The Netherlands
[3]Department of Civil and Environmental Engineering, Massachusetts Institute of Technology, Cambridge, MA

Corresponding Author:
Address correspondence to Sheng Jin: jinsheng@zju.edu.cn

of a bicycle lane (5), but this method cannot be directly applied in China due to heterogeneous bicycle traffic flows. Therefore, the main purpose of this paper is to study the relationship between overtaking behavior and the parameters of heterogeneous bicycle traffic flows, which can provide theoretical suggestions for the level of service of bicycles, as well as planning and management of non-motorized facilities.

The remainder of this study is organized as follows. The next section presents the overview of literature on passing events and bicycles. The following three sections: describe the field study, the basic features of the collected data, and the speed characteristics of bicycles; present the proposed analytical modeling approach; and discuss the calibration of the model. Finally, conclusions are presented.

Literature Review

Much research has been devoted to bicycle traffic, while the bicycle level of service (BLOS) has attracted much attention. Overtaking is considered an influencing factor for BLOS. Botma (1995) first proposed a method for determining the level of service for pedestrian paths and pedestrian-bicycle paths (*6*). In his study, the BLOS criteria were defined by simplifying the hindrance experienced by users as the frequency of passing and meeting maneuvers. Further, BLOS criteria were set up as a function of the pedestrian and cyclist volumes, path width, and speeds based on field studies conducted in the Netherlands. Allen et al. (1998) studied the number of events that occurred on an uninterrupted bicycle facility, and modified Botma's procedure (*7*). The frequency of events on an uninterrupted bicycle facility was related to the service volume of bicycles using or intending to use the facility, and did not have to be observed directly. HCM 2000 adopted Botma's procedure to determine the BLOS of exclusive bicycle paths (*8*). To estimate the BLOS on a shared-use path, Hummer et al. (2002) developed a new method based on field data collected in 10 cities in the U.S. (*9*). An adjustment for the number of delayed passes was added to the BLOS equation, which was applied to the procedure present in HCM 2010 (*5*). Khan (2001) studied the characteristics of passing and meeting maneuvers on the exclusive bicycle paths (*10*). Khan analyzed the speed, the lateral distance from the right edge of the bicycle path, the lateral spacing, and the length during passing. Minh et al. (2005) also proposed a method based on the characteristics of passing and paired riding maneuvers of motorcycles in Vietnam (*11*).

Since e-bicycles account for quite a small daily modal share in western countries, despite the fact that they are indeed practical for daily travel (*12*), most of the existing research has focused on bicycle traffic without e-bicycles. These methods are not appropriate for the heterogeneous bicycle traffic in China. Studies on heterogeneous bicycle traffic are limited. Li et al. (2010) analyzed the number of passing events, and introduced a new BLOS evaluation model for heterogeneous bicycle traffic (*13*). They (2013) also proposed a model to estimate the number of each type of passing maneuver on unidirectional two-, three- and four-lane bicycle paths (*14*). Zhao et al. (2013) used a cellular automata model to study passing events in heterogeneous bicycle traffic (*15*). The research revealed that e-bicycles contribute to most of the passing events in heterogeneous bicycle traffic. Lin et al. (2014) studied the phenomena and characteristics of moped-passing-bicycle events on shared lanes by studying the speeds of passing and passed vehicles, the lateral distance from the closer curb, the lateral spacing, and the longitudinal distance during passing (*16*).

The review of the literature suggests that, although many studies have analyzed the characteristics of passing maneuvers and established models to predict the number of passing events, most of them did not reveal deeper relationships between overtaking behavior and the characteristics of heterogeneous traffic flows. In this paper, an analytical model is proposed to estimate the number of passing events based on the relationships between passing events and the characteristics of heterogeneous bicycle traffic flows. This study can provide suggestions for the development of BLOS criteria and the construction of non-motor facilities.

Data Collection

Field Study

To obtain the actual parameters of heterogeneous bicycle traffic flows, several field studies were conducted on May 9 and June 4 in Hangzhou, Zhejiang, China. In Hangzhou, bicycles contribute to 34.3% of daily trips (of which e-bicycles account for more than two-thirds) according to a habitant trip survey in 2010. The weather was sunny and dry during the field study, which excluded the impact of adverse weather. Data was collected between 16:00 and 18:00 during evening peak hours in Hangzhou. The observed road sections were separated from motor lanes with physical separators and were far from intersections, bus stations and block accesses. Therefore, bicycles were not interrupted by pedestrians or motorized vehicles. In China, bicycle routes are mostly located on both sides of the road. Accordingly, every route used for data collection in this paper was one-way. The view of study sites is shown in Figure 2. Detailed information on the paths is shown in Table 1.

The equipment used in the field included a tape measure, cameras, tripods, and marking tapes. Four

Table 1. Detailed Information about Study Sites and Samples

Study site	Width (m)	Separation	Sample size	Bicycle type
Jiaogong Road	4.5	Barrier	2531	RB 29.16% EB 70.05% Others 0.79%
Shixiang Road	3.5	Barrier	2133	RB 7.83% EB 88.05% Others 4.12%

Note: RB = r-bicycle; EB = e-bicycle.

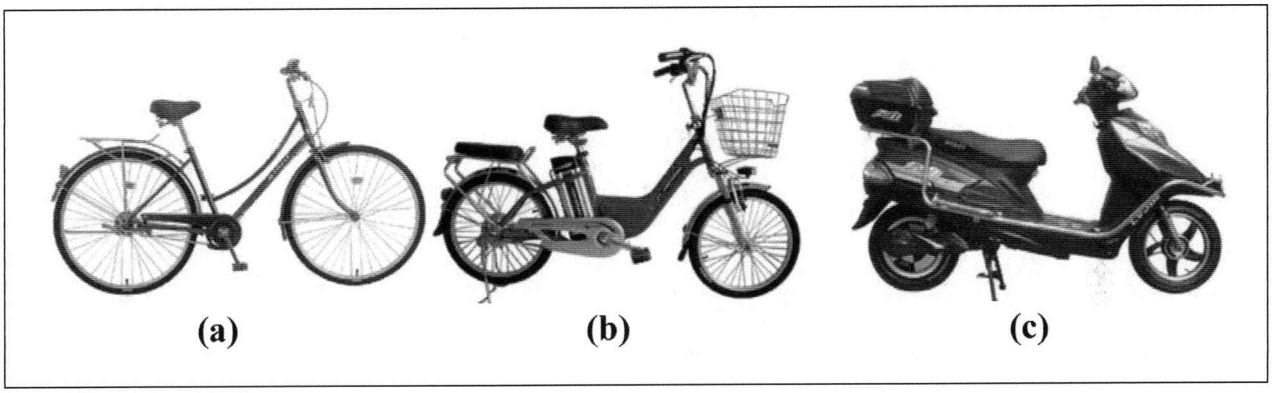

Figure 1. Different types of bicycles: (*a*) regular bicycle; (*b*) bicycle-style electric bicycle; (*c*) scooter-style electric bicycle.

reference lines were set on the path to clearly mark the sample area before recording. There was one access line, one egress line, and two central lines. The central lines were used to obtain the speed of bicycles. Three cameras were set up on the side of the bicycle route to record videos collecting data on r-bicycles' and e-bicycles' riding behavior. Cameras were carefully placed so that cyclists would be unlikely to be aware that they were being observed. The settings are shown in Figure 3. The data collection procedure was as follows. When a bicycle entered the sample area, its whole riding behavior was recorded. Each camera collected 25 frames per second, so the time accuracy was 0.04s. Later, the parameters of the traffic flow and the number of passing events were extracted manually. The features of bicycles and the timestamps of bicycles passing the access, central, and egress lines were recorded. Based on the recordings, the number of passing events in this section was obtained. The share of e-bicycles, flow, speed and density were obtained as well. In addition, when processing the videotapes, the abnormal behavior of bicyclists was marked, such as going in the wrong direction, stopping on the path for a phone call etc. The corresponding samples containing abnormal behaviors of cyclists were removed before model calibration.

Data Analysis

Two sections of two roads were selected for data collection. 2,531 observations were collected on Jiaogong Road for analysis and calibration. Among them, 1,773 observations were e-bicycles, 738 observations were r-bicycles, and the others were tricycles. Similarly, 2,133 observations were collected on Shixiang Road. Since tricycles are differ significantly from bicycles in size and other characteristics, we had to suppress their interference when processing raw data. The method of filtering of abnormal behavior and tricycles is discussed in below.

The characteristics of the speed are shown in Table 2 and Figure 4. E-bicycles make up more than half (70.05% and 88.05%) of the sample. The speed of e-bicycles is much higher than that of r-bicycles. And the standard deviation of the speeds of e-bicycles is larger. For two study sites, the values of the kurtosis and skewness of the empirical distribution of speed are under 1 except the for the value of r-bicycles for Shixiang Road. Since the sample size of r-bicycles for Shixiang Road observed is small, the value of the kurtosis and skewness is acceptable. The speeds of r-bicycles and e-bicycles can be assumed to be normally distributed. The parameters of the probability density function can be estimated using the maximum likelihood estimation method.

Figure 2. View of study sites: (*a*) Jiaogong Road; (*b*) Shixiang Road.

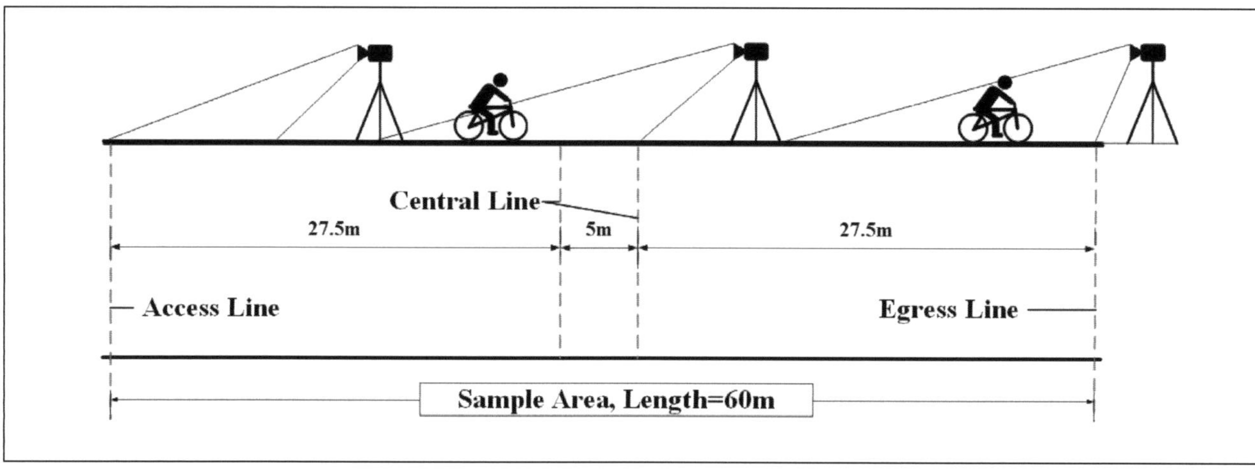

Figure 3. Setting details for each observation.

Table 2. Characteristics of the Speed

Study site	Bicycle type	Sample size	Mean (km/h)	SD (km/h)	Kurtosis	Skewness
Jiaogong Road	RB	738	15.35	3.25	0.336	0.304
	EB	1773	23.10	5.41	0.496	0.410
Shixiang Road	RB	167	14.49	3.40	1.342	4.053
	EB	1878	28.37	6.18	0.189	0.381

Note: SD = standard deviation.

Proposed Model

In this paper, heterogeneous bicycle traffic flows consisting of r-bicycles and e-bicycles are studied. In heterogeneous bicycle traffic flows, four types of passing maneuvers can be observed, including r-bicycles passing r-bicycles, r-bicycles passing e-bicycles, e-bicycles passing r-bicycles, as well as e-bicycles passing e-bicycles. All these situations are taken into consideration when obtaining the total number of passing events.

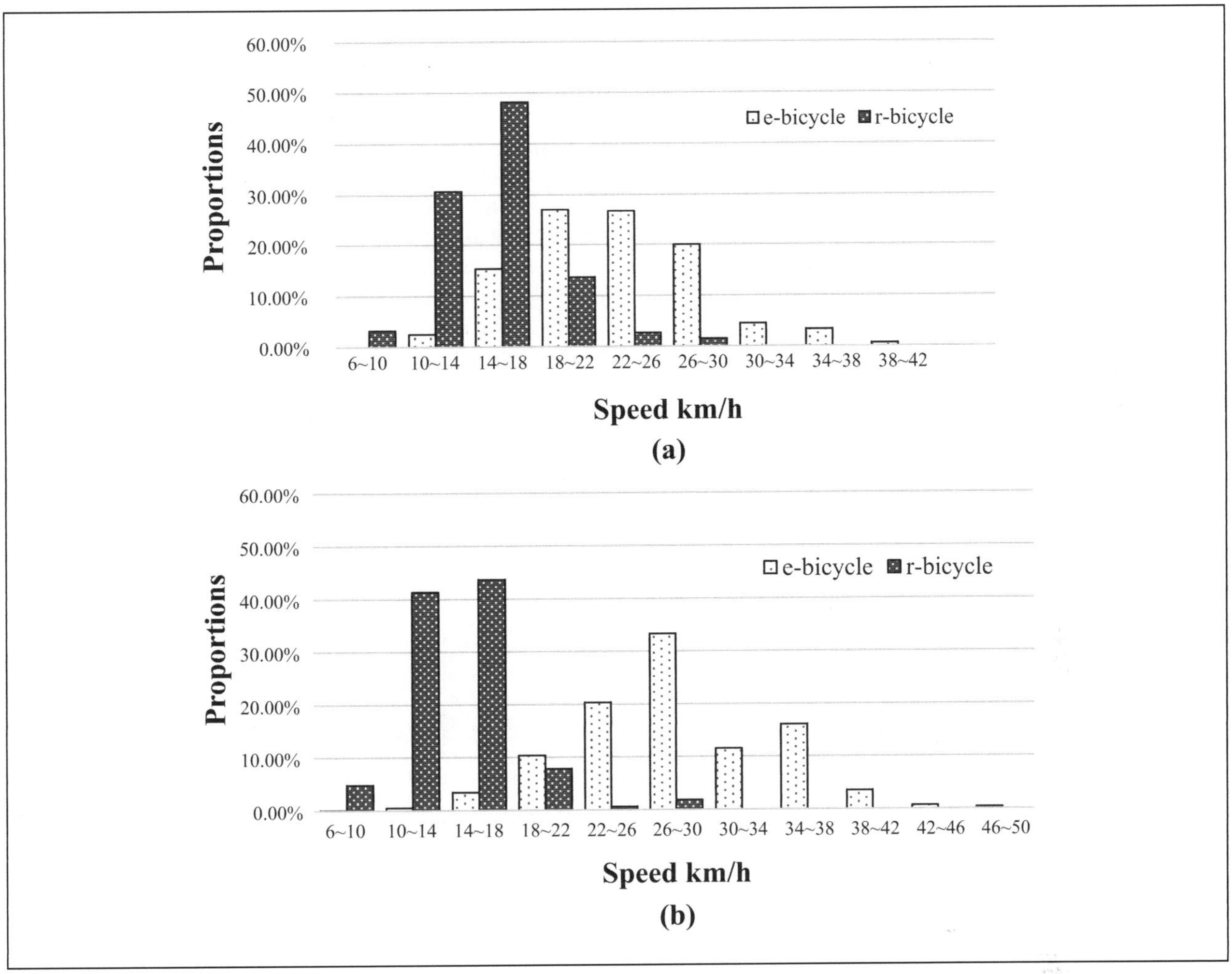

Figure 4. Distribution of field data on bicycle speed: (*a*) Jiaogong Road; (*b*) Shixiang Road.

Assumptions and Variables

Since heterogeneous bicycle traffic flows are quite complex, several assumptions are made to facilitate the model development:

1. R-bicycles and e-bicycles are distributed on the path uniformly.
2. The speeds of r-bicycles and e-bicycles are constant on the path.
3. The speeds of r-bicycles and e-bicycles are normally distributed and mutually independent.
4. The state of the bicycle traffic flow in the sample area does not change over a period of time.

Notation used in the following analysis is as follows:

L = the length of the path
Q = the flow rate of bicycles
Q_r = the flow rate of r-bicycles
Q_e = the flow rate of e-bicycles
k = the density of bicycles
v_r = the speed of r-bicycles
v_e = the speed of e-bicycles
p_r = the proportion of r-bicycles
p_e = the proportion of e-bicycles
$f_r(v)$ = the probability density function of speed of r-bicycles
$f_e(v)$ = the probability density function of speed of e-bicycles

Regular Bicycle Flow Model

First, a scenario with only r-bicycles on the path is modelled. There are several r-bicycles running on the path, which are defined as original bicycles. The parameters are assumed as follows. The speed of original bicycles is normally distributed. The number of original bicycles is

N. The length of the path is assumed to be L. When an r-bicycle enters the sample area, it makes several passing maneuvers. The speed of the new bicycle is assumed to be v_0. The time the bicycle spends to pass the sample area can be calculated as follows:

$$t_0 = \frac{L}{v_0} \tag{1}$$

Further, it is assumed that there is an original bicycle running on the path with the speed v. When the new bicycle enters the sample area, the distance between the original bicycle and the access line is assumed to be x. Therefore, the time the original bicycle takes to pass the sample area can be obtained by:

$$t = \frac{L - x}{v} \tag{2}$$

If $t_0 < t$, the new bicycle passes the original bicycle, and a passing event happens. Here $t_0 = t$ means the two bicycles reach the egress line at the same time. In this case:

$$t_0 = \frac{L}{v_0} = \frac{L - x}{v} = t \tag{3}$$

The critical value of x can be calculated as follows:

$$x = \left(1 - \frac{v}{v_0}\right)L \tag{4}$$

It is clear that the original bicycle can be passed by the new bicycle when the distance between the original bicycle and the access line of the sample area is less than the critical value x. The total number of the original bicycles passed by the new bicycle can be obtained by the following integral equation:

$$E = \frac{N}{L}\int_0^{v_0} f_r(v)x\,dv \tag{5}$$

where E is the number of passing events caused by the new bicycle.

We assume that the number of bicycles entering the sample area is n. The speeds of these bicycles follow the same normal distribution. The total number of passing events can be calculated as follows:

$$E_{\text{total}} = nN\int f_r(v_0)\int_0^{v_0} f_r(v)\left(1 - \frac{v}{v_0}\right)dv\,dv_0 \tag{6}$$

where E_{total} is the total number of passing events caused by the entering bicycles.

Heterogeneous Bicycle Flow Model

Heterogeneous bicycle traffic flows are much more complicated, but the method above can be extended to such traffic flows based on the same assumptions.

Since the four types of passing maneuvers coexist in heterogeneous bicycle traffic flows, their mutual impacts are difficult to quantify. To facilitate the development of the model, we assume that these four passing maneuvers are mutually independent. The total number of passing events can be calculated by taking the sum of the number of the four types of passing maneuvers.

Accordingly, the method above can be adopted to develop a model for heterogeneous bicycle traffic flows to estimate the number of passing events. For e-bicycles overtaking r-bicycles, the number of passing events E_{er} can be obtained as follows:

$$E_{\text{er}} = n_e p_r N\int f_e(v_0)\int_0^{v_0} f_r(v)\left(1 - \frac{v}{v_0}\right)dv\,dv_0 \tag{7}$$

where n_{e} is the number of e-bicycles going through the sample area.

Similarly, the number of passing events E_{ee} where e-bicycles overtake e-bicycles is given by:

$$E_{\text{ee}} = n_e p_e N\int f_e(v_0)\int_0^{v_0} f_e(v)\left(1 - \frac{v}{v_0}\right)dv\,dv_0 \tag{8}$$

Similarly, we have the following formulas when r-bicycles pass other bicycles:

$$E_{\text{rr}} = n_r p_r N\int f_r(v_0)\int_0^{v_0} f_r(v)\left(1 - \frac{v}{v_0}\right)dv\,dv_0 \tag{9}$$

$$E_{\text{re}} = n_r p_e N\int f_r(v_0)\int_0^{v_0} f_e(v)\left(1 - \frac{v}{v_0}\right)dv\,dv_0 \tag{10}$$

where n_{r} is the number of r-bicycles getting through the sample area, E_{rr} is the number of passing events where r-bicycles overtake r-bicycles, E_{re} is the number of passing events where r-bicycles overtake e-bicycles.

The flow rate, the density, and the speed are the essential parameters of the traffic flow, which can reveal the substantive characteristics of the traffic flow. However, the equations above do not depend on these parameters. Therefore, the relationships between passing events and heterogeneous bicycle traffic flow should be established to estimate the number of passing events more accurately. In the equations above, the numbers of r-bicycles and e-bicycles are used, and they can be divided by the time to obtain the flow rate of the type of bicycles. When the number of bicycles on the path is divided by the length of the path, the density of the type of bicycles is obtained:

$$Q_r = \frac{n_r}{t} \tag{11}$$

$$Q_e = \frac{n_e}{t} \tag{12}$$

$$k = \frac{N}{L} \tag{13}$$

Further, according to Greenshields' linear model, which has been studied by numerous researchers (*17*, *18*), the relationship between the flow rate and the density is given by:

$$Q = kv \tag{14}$$

The above formulas can be rewritten as follows to take into account the traffic flow parameters:

$$E_{er} = Q_e k p_r L \int f_e(v_0) \int_0^{v_0} f_r(v)\left(1 - \frac{v}{v_0}\right) dv dv_0 \tag{15}$$

$$E_{ee} = Q_e k p_e L \int f_e(v_0) \int_0^{v_0} f_e(v)\left(1 - \frac{v}{v_0}\right) dv dv_0 \tag{16}$$

$$E_{rr} = Q_r k p_r L \int f_r(v_0) \int_0^{v_0} f_r(v)\left(1 - \frac{v}{v_0}\right) dv dv_0 \tag{17}$$

$$E_{re} = Q_r k p_e L \int f_r(v_0) \int_0^{v_0} f_e(v)\left(1 - \frac{v}{v_0}\right) dv dv_0 \tag{18}$$

The total number of passing events is given by:

$$E_{total} = E_{er} + E_{ee} + E_{rr} + E_{re} \tag{19}$$

Model Analysis

This section discusses the impact of key parameters, such as the flow rate, the standard deviation of the speed and the proportion of e-bicycles, on the prediction of the number of passing events. Figure 5 shows the number of passing events based on different parameter values.

The flow rate of bicycles ranges from 1,500 bikes/h to 3,900 bikes/h. Figure 5*a* shows the effect of the flow rate of bicycles and the proportion of e-bicycles p_e on the number of passing events. As can be seen, the number of passing events increases in Q. And when the the proportion of e-bicycles increases, the number of passing events first increases, and then decreases.

Figure 5*b* illustrates the relationships among the proportion of e-bicycles p_e, the density of bicycles k and the number of passing events. With an increase in the density of bicycles, the number of passing events increases. In addition, when the proportion of e-bicycles p_e approaches 0 or 1, the influence of k on the number of passing events becomes less significant. This shows that passing events are more likely to occur in heterogeneous bicycle flow.

In Figure 5*c*, the standard deviations of the speeds of r-bicycles (S_r) and e-bicycles (S_e) are equal. By comparing different S_r and S_e, it is easy to see that the number of passing events increases with the standard deviation. Additionally, the larger the standard deviation, the lower the effect of the proportion of e-bicycles p_e on the number of passing events, because larger standard deviations lead to smaller differences between the speeds of the two types of bicycles.

In Figure 5*d*, the influence of the speed difference between r-bicycles and e-bicycles and the proportion of e-bicycles p_e on the number of passing events is shown. A positive correlation between the speed difference and the number of passing events is found. More passing events occur when the speed difference becomes larger. Besides, the influence of the speed difference on the number of passing events is small when p_e is near 0 or 1, but significant when p_e is around 0.6.

The figures show that higher mixing of r-bicycles and e-bicycles leads to a larger number of passing events, as the maximum of the number of passing events can be obtained when p_e is between 0 and 1.

Model Validation

To validate the model, field data on the density k, the flow rate Q, the proportion of e-bicycles p_e, and the average speed of r-bicycles and e-bicycles needs to be obtained. Before obtaining values of parameters, the time interval needs to be determined first, since it affects the result of data collection. Different time intervals also lead to different estimation results (*19*). The HCM 2010 suggests the number of passing events to be counted per minute (*5*). According to the assumptions of the proposed model, the method is suitable for studying steady flows. In China, the conditions of heterogeneous bicycle flows change frequently. It is difficult to find a flow that is steady for one minute. Furthermore, it is easy to obtain the number of passing events per minute by combining 15s intervals. Based on the experimental observations of the video of the bicycle traffic flow, we found that the state of the heterogeneous bicycle flow can be assumed stable for 15s. Therefore, for the sake of simplicity, we use 15s as the time interval to calculate the number of events in this paper. Further research will focus on

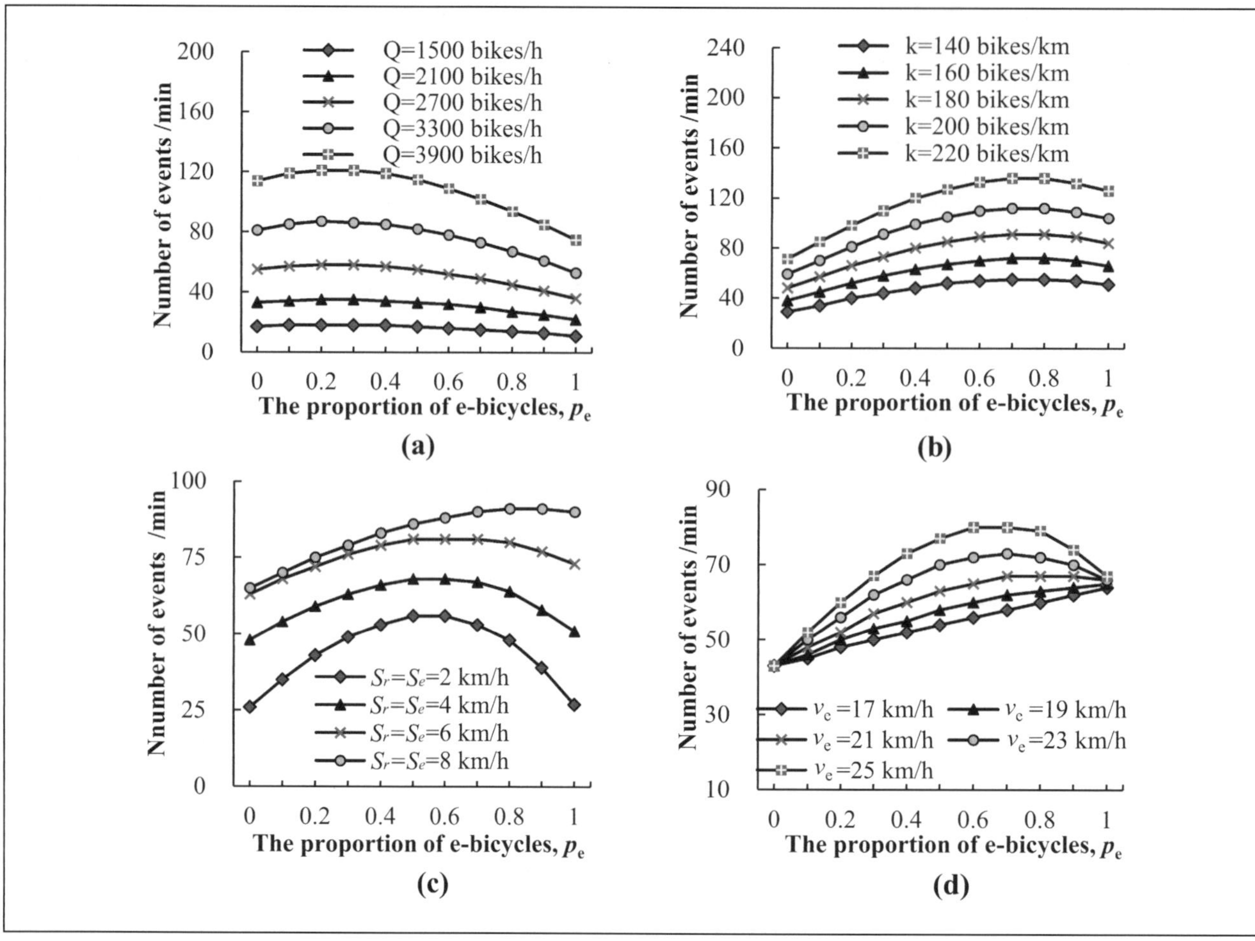

Figure 5. Number of passing events based on different parameters: (*a*) L = 60 m, v_r = 15 km/h, S_r = 3.4 km/h, v_e = 23 km/h, S_e = 5.3 km/h; (*b*) L = 60 m, v_r = 15 km/h, S_r = 3.4 km/h, v_e = 23 km/h, S_e = 5.3 km/h; (*c*) L = 60 m, v_r = 15 km/h, v_e = 23 km/h, k = 160 bikes/km; (*d*) L = 60 m, Sr = 3.4 km/h, S_e = 3.4 km/h, k = 160 bikes/km, v_r = 15 km/h.

how to determine the optimal sampling time interval for estimation. The raw data was processed according to the time interval determined. Further, samples containing the abnormal behavior of bicyclists and tricycles were removed from the dataset. Finally, 166 samples for Jiaogong Road and 195 samples for Shixiang Road were obtained for model validation. The results of the comparison between the estimated values and the field data are shown in Figure 6.

As can be seen in Figure 6,*a* and 6*b*, the fitted relationships between the values predicted by the model and the field data are quite good. The splattering spreads evenly on both sides of the line $y = x$. Figure 6,*c* and 6*d* show the residual plot of passing events between the prediction and field data. The vast majority of the residual values are less than 10. In addition, the residual increases with the number of passing events. In other words, prediction accuracy may decrease when the flow rate of bicycles becomes high enough, since disturbances among bicycles increase dramatically.

To quantify prediction accuracy, the root-mean-square error (RMSE) was employed. The index is given by the following equation:

$$RMSE = \sqrt{\frac{1}{N}\sum_{i=1}^{N}[\hat{n}(i) - n(i)]^2} \tag{20}$$

where

$\hat{n}$ = the predicted number of passing events

n = the measured number of passing events based on field data

N = the number of samples

The values of RMSE are 3.99 and 3.21 for Jiaogong Road and Shixiang Road, respectively. The results show that the model proposed in this paper can effectively predict the number of passing events.

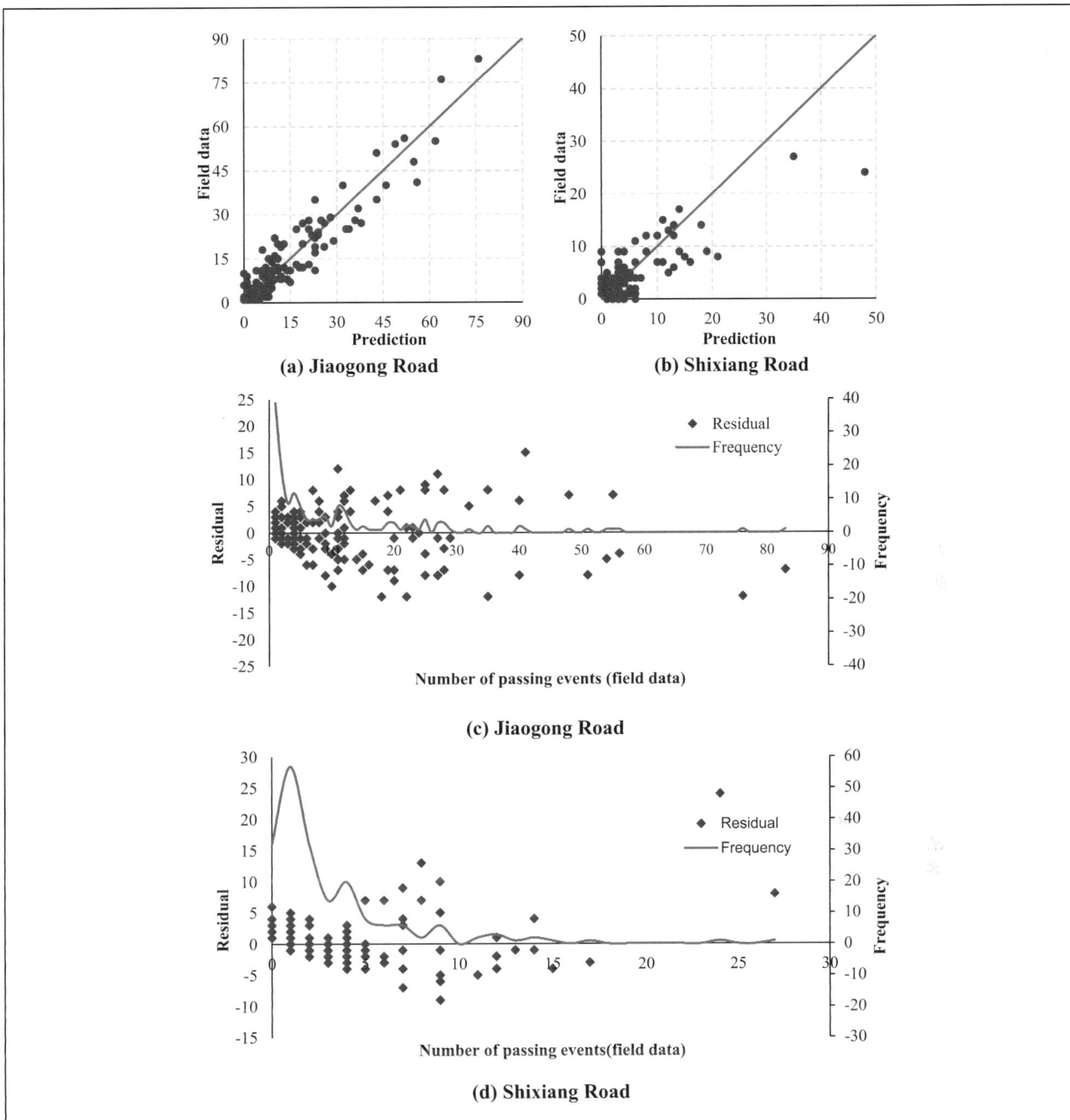

Figure 6. Relationships between prediction and field data: (*a*, *b*) fitted relationships between prediction and field data; (*c*, *d*) residual plot between prediction and field data.

Conclusions

In this paper, an analytical model for estimating the number of passing events under heterogeneous bicycle traffic conditions is presented. The probability density functions of the speeds of r-bicycles and e-bicycles are used in the model, which are not considered in many other existing models. Considering the spatial and temporal distribution of bicycles and the speed distributing functions, the probability of overtaking can be obtained. Finally, relationships between the characteristics of heterogeneous bicycle traffic flows and the number of passing events is established. By adopting the model proposed in this paper, the number of passing events can be obtained easily using the average speed of r-bicycles and e-bicycles, the standard deviation of the speed, the proportion of e-bicycles, and the values of the parameters of the heterogeneous bicycle traffic flow.

The proposed model proves to be accurate and reliable based on validation with field data. The values of RMSE between the prediction and field data are 3.99 and 3.21 for two study sites. The results show that the number of passing events increases with the total flow rate and density. The speed difference between the two types of bicycles and the standard deviation of the speed can also result in an increase in the number of passing events. In addition, when the proportion of e-bicycles increases, the number of passing events first increases, and then decreases. This shows that higher mixing of the two types of bicycles can increase overtaking behavior.

The model proposed in this paper provides an analytical approach to the study of relationships between the characteristics of heterogeneous bicycle traffic flows and the number of passing events. It can be seen as a prerequisite work for the development of BLOS criteria.

Admittedly, there are some limitations to our model, and some future studies are necessary:

1. The sample size is limited. Only two sites in Hangzhou were studied. Data should be collected on more bicycle paths or in other cities to calibrate and validate the model.
2. Bicycle routes in China are mostly one-way routes located on both sides of roads. Bidirectional bicycle routes with and without center-line separation are common in other countries and should be considered as well.
3. The speeds of e-bicycles may not follow a unimodal distribution in some cases due to the difference between bicycle-style and scooter-style electric bicycles. Bi-modal speed distributions should be taken into consideration.
4. The assumptions made in this study are idealized, and may not be true in reality. More disturbances and the joint probability density functions of the speeds of bicycles should be taken into consideration in the modeling process.
5. Passing events should be classified. Different types of passing events affect cyclists' perceptions differently. Therefore, different types of passing events should be considered in the future.

Acknowledgments

This work was supported in part by the National Natural Science Foundation of China (No. 91746105, 61773338, 61773337, and 51338008), and the Key Research and Development Program of Zhejiang (No. 2018C01007).

Author Contributions

The authors confirm contribution to the paper as follows: study conception and design: S. Jin, L. Xu; data collection: L. Xu; analysis and interpretation of results: L. Xu, M. Liu, X. Song; draft manuscript preparation: L. Xu, M. Liu, X. Song, S. Jin. All authors reviewed the results and approved the final version of the manuscript.

References

1. Weinert, J., C. Ma, X. Yang, and C. Cherry. Electric Two-wheelers in China: Effect on Travel Behavior, Mode Shift, and User Safety Perceptions in a Medium-sized City. *Transportation Research Record: Journal of the Transportation Research Board*, 2007. 2038: 62–68.
2. Cherry, C. R., J. X. Weinert, and X. Yang. Comparative Environmental Impacts of Electric Bikes in China. *Transportation Research Part D: Transport and Environment*, Vol. 14, No. 2, 2009, pp. 281–290.
3. Weinert., J., C. Ma, and C. Cherry. The transition to Electric Bikes in China: History and Key Reasons for Rapid Growth. *Transportation*, Vol. 34, No. 3, 2007, pp. 301–318.
4. Lin, S., M. He, Y. Tan, and M. He. Comparison Study on Operating Speeds of Electric Bicycles and Bicycles: Experience from Field Investigation in Kunming, China. *Transportation Research Record: Journal of the Transportation Research Board*, 2008. 2048: 52–59.
5. *Highway Capacity Manual 2010*. Transportation Research Board of the National Academies, Washington, D.C., 2010.
6. Botma, H. Method to Determine Level of Service for Bicycle Paths and Pedestrian-Bicycle Paths. *Transportation Research Record: Journal of the Transportation Research Board*, 1995. 1502: 38–44.
7. Allen, D., N. Rouphail, J. E. Hummer, and J. Milazzo. Operational Analysis of Uninterrupted Bicycle Facilities. *Transportation Research Record: Journal of the Transportation Research Board*, 1998. 1636: 30–36.
8. *Highway Capacity Manual*. TRB, National Research Council, Washington, D.C., 2000.
9. Hummer, J. E., N. M. Rouphail, J. L. Toole, R. S. Patten, R. J. Schneider, J. S. Green, R. G. Hughes, and S. J. Fain. *Evaluation of Safety, Design, and Operation of Shared-Use Paths*. Final Report. U.S. Department of Transportation, 2002.
10. Khan, S., and W. Raksuntorn. Characteristics of Passing and Meeting Maneuvers on Exclusive Bicycle Paths. *Transportation Research Record: Journal of the Transportation Research Board*, 2001. 1776: 220–228.
11. Minh, C. C., K. Sano, and S. Matsumoto. Characteristics of Passing and Paired Riding Maneuvers of Motorcycle. *Journal of the Eastern Asia Society for Transportation Studies*, Vol. 6, 2005, pp. 186–197.
12. Fyhri, A., and N. Fearnley. Effect of E-bikes on Bicycle Use and Mode Share. *Transportation Research Part D: Transport and Environment*, Vol. 36, 2015, pp. 45–52
13. Li, Z., W. Wang, X. Shan, J. Jin, J. Lu, and C. Yang. Analysis of Bicycle Passing Events for LOS Evaluation on

Physically Separated Bicycle Roadways in China. Presented at 89th Annual Meeting of the Transportation Research Board, Washington, D.C., 2010.

14. Li, Z., W. Wang, P. Liu, J. Bigham, and D. R. Ragland. Modeling Bicycle Passing Maneuvers on Multilane Separated Bicycle Paths. *Journal of Transportation Engineering*, Vol. 139, No. 1, 2013, pp. 57–64.
15. Zhao, D., W. Wang, C. Li C, Z. Li Z, P. Fu, and X. Hu. Modeling of Passing Events in Heterogeneous Bicycle Traffic with Cellular Automata. *Transportation Research Record: Journal of the Transportation Research Board*, 2013. 2387: 26–34.
16. Lin, D., X. Chen, B. Lin, and L. Li. Phenomena and Characteristics of Moped-passing-bicycle on Shared Lanes. Presented at 93rd Annual Meeting of the Transportation Research Board, Washington, D.C., 2014.
17. Qu, X., S. Wang, and J. Zhang. On the Fundamental Diagram for Freeway Traffic: A Novel Calibration Approach for Single-regime Models. *Transportation Research Part B*, Vol. 73, 2015, pp. 91–102.
18. Qu, X., J. Zhang, and S. Wang. On the Stochastic Fundamental Diagram for Freeway Traffic: Model Development, Analytical Properties, Validation, and Extensive Applications. *Transportation Research Part B*, Vol. 104, 2017, pp. 256–271.
19. Jin, S., X. Qu, D. Zhou, C. Xu, D. Ma, and D. Wang. Estimating Cycleway Capacity and Bicycle Equivalent Unit for Electric Bicycles. *Transportation Research Part A: Policy and Practice*, Vol. 77, 2015, pp. 225–248.

The Standing Committee on Bicycle Transportation (ANF20) peer-reviewed this paper (18-02651).

Article

TRR
JOURNAL OF THE TRANSPORTATION RESEARCH BOARD

Transportation Research Record
2018, Vol. 2672(36) 136–144

DOI: 10.1177/0361198118787092
journals.sagepub.com/home/trr
SAGE

Operational Evaluation of Central Sharrows and Dooring Zone Treatment on Road User Behavior in Ottawa, Canada

Ali Kassim[1], Karim Ismail[1], and Shawn McGuire[2]

Abstract

The purpose of this study is to determine how new pavement markings (central sharrows and indication of "dooring zone") influence cyclist and motorist interactions and positioning, especially with respect to the distance cyclists travel from the parking edge line. The markings are designed to discourage cyclists from riding in the dooring-zone area and to ride in the center of the travel lane to avoid serious safety hazards. A number of safety performance parameters were used to assess whether safer conditions existed after the new treatment was installed. Safe motorist behavior was defined as: (a) motor vehicle following a cyclist with or without visible oncoming traffic in the opposing lane, and (b) motor vehicle passing a cyclist at a safe lateral distance. Safe cyclist behavior was defined as: cyclist riding in the lane position indicated by the central sharrows and outside of the dooring-zone area. The findings indicate that the central sharrows and dooring-zone markings created more favorable conditions for cyclist safety and for motorist compliance. Cyclists rode further from the parking edge line and closer to the sharrows that were marked in the center of the travel lane, and motorists were shown to be less likely to pass and more likely to follow cyclists.

With active modes of transportation gaining popularity among users and planners, many cyclist-oriented traffic engineering treatments are employed to promote safe and efficient operation. Bicycle lanes are often promoted for their effectiveness in separating cyclists and vehicular traffic and potentially improving cyclist safety (*1*). Physical separation between these two traffic streams can also enhance perceived ride comfort by cyclists (*2*). Many urban streets lack sufficient roadway width to allow for proper bike lanes, yet they are expected to safely accommodate cyclists. This is particularly problematic when on-street parking is permitted. This is due to the risk of cyclists colliding with a suddenly opened door that blocks the cyclist's movement, commonly called "dooring" (*3*). Promoting responsible road sharing with cyclists can be a viable option to encourage cyclists to ride in the center of the lane and hence avoid the risk of dooring. However, this can be a daunting challenge which goes against a somewhat powerful driving culture in which cyclists normally ride on the side and motor vehicles dominate the lane. The effectiveness of different traffic control treatments to promote center-lane riding and responsible road sharing, as well as the precise response by road users to these measures, remains an open question worthy of focused investigation.

This study presents a before-and-after evaluation of two concurrent pavement markings: (a) a central shared-lane indication ("sharrows"), and (b) warning of dooring zone. This treatment was implemented in the summer of 2015 on a relatively narrow urban street in the City of Ottawa, Canada. The two-way street had permissible roadside parking on both sides, so that cyclists traveling on the side of the lane were helplessly exposed to the hazard of dooring. There was not enough space in the right of way to introduce bicycle lanes, with all the necessary traffic control countermeasures against dooring. In this study, central sharrows were painted on both lanes. The purpose of the study was to compare, before and after the treatment: (a) the average lateral distance that separates cyclists and the parking edge line, (b) the lateral distance between cyclists and moving motor vehicles, and (c) the propensity of motorists to overtake (or conversely to follow) a leading cyclist before and after the treatment. The goal of this treatment was to help reduce the number of potential dooring incidents and to bring awareness about dooring to motorists and cyclists.

The objectives of the treatment were:

[1]Department of Civil end Environmental Engineering, Carleton University, Ottawa, ON, Canada
[2]Coordinator, Cycling and Pedestrian Safety, Transportation Services, Ottawa, ON, Canada

Corresponding Author:
Address correspondence to Ali Kassim: ali.kassim@carleton.ca

1. Encourage cyclists to ride in the center of the travel lane, where the central sharrows are located.
2. Discourage cyclists from riding too close to parked vehicles, within the dooring-zone area, to avoid serious safety hazards.
3. Encourage motorists to be more patient and accommodating to slower moving cyclists riding in the center of the lane.
4. Encourage motorists to follow, rather than pass, a leading cyclist. Given that this street is only one lane in each direction, a patient response by motorists is likely to reinforce the message of the treatment and encourage cyclists to take the whole lane while maintaining a respectful response from motorists.
5. Educate users as to why cyclists are recommended to position themselves in the center of the lane. The signage reading "dooring zone" and "emportierage" is intended to help explain to both cyclists and motorists why it is recommended that cyclists take the lane.

In the next sections, this paper summarizes previous work in a review of the literature, presents the data collection and methodology, and offers a summary and discussion. A number of conclusions are listed at the end of this paper.

Literature Review

One of the traffic control options to promote efficient and safe sharing of roadway space is shared-lane markings (*4*, *5*). Sharrows are a type of lane marking intended to promote safe sharing of the road. They alert drivers to their obligations toward cyclists and may encourage cyclists to occupy the center of the lane and move away from roadside hazards (*6*, *7*). Some shared-lane treatments attempt to emphasize this message to cyclists by using a colored background (*8*, *9*) with results to some extent suggesting a favorable response from motorists and cyclists. When roadside parking is allowed, another hazard emerges from vehicle doors swinging suddenly in front of an unsuspecting cyclist. This dooring hazard is exacerbated by the limited space available on the roadway for the cyclist to perceive the impending risk and react by swerving and/or decelerating to avoid collision. A pavement indication of dooring may help in mitigating this risk. Dooring is a serious hazard for cyclists in an urban environment (*3*). There is no straightforward low-cost solution to this issue given all its contextual complexities.

Diverting cyclist traffic to pedestrian facilities may increase the risk of collision and injury for both cyclists and pedestrians (*10*). Constructing bicycle lanes to channelize cyclist traffic requires right of way, a challenging undertaking on many fronts, especially on narrow urban streets (*11*). Furthermore, bicycle lane safety is a complex topic (*1*) which relies primarily on spatial (and temporal in case of turning movements) separation of cyclists and vehicles to improve safety (*12*). While traffic control is primarily focused on separating moving vehicles and cyclists, the risk of dooring comes from parked vehicles. When right of way permits, adequate buffers can be provided to preclude dooring collisions. However, on narrow urban streets with permitted roadside parking, the situation creates more design challenges. Specially designed indications have been introduced and proved successful in deterring cyclists from riding too close to the lane edge (*8*, *13*). This is undoubtedly a favorable outcome, however, inviting cyclists to the lane center poses a challenge against the dominant mode of motor vehicles, where cyclists are much more physically vulnerable and risk more injury than drivers. The response from drivers should be monitored and promoted through proper signage and public education campaigns.

The implementation details of this treatment, as described in the next section, are novel. The evaluation was conducted by collecting data before the treatment and at two different times after it: shortly after the treatment and then one year after the treatment. Both motorist and cyclist behaviors were monitored in a before-and-after experimental setting. The study expands the current knowledge on pavement indications aimed at cyclist traffic and provides valuable decision-support information.

Data Collection and Methodology

Treatment Description

The treatment is composed of pavement markings in the form of sharrows, bordered with dashed white lines, placed in the center of the 3.65 m travel lane, spaced approximately 16.0 m apart. An extra text indication in both English and French was added in the 0.90 m area between the right-side dashed line and the parking edge line to indicate where cyclists are at risk of dooring, as shown in Figure 1. This treatment was implemented on both the eastbound and westbound directions of the 40 km/h section located on Wellington Street in the downtown of Ottawa, Ontario, shown in Figure 2 (*14*). The dooring-zone space was designed to be 0.90 m from the solid white parking edge line. The length of the treated road (study segment) is approximately 85.0 m.

Study Site and Equipment

For the purpose of examining if cyclist and/or motorist behavior changed with the introduction of central sharrows and dooring-zone indication, video camera installation was used to monitor cyclist and motor vehicle movements. In this paper, one segment of Wellington Street was selected to be studied in both eastbound and westbound directions. In this study, video data was collected using static high-definition video camera (1920 × 1080 pixels, approximately 30 frames/second). The video camera was installed on the side window of the instrumental parked vehicle, as shown in Figure 3.

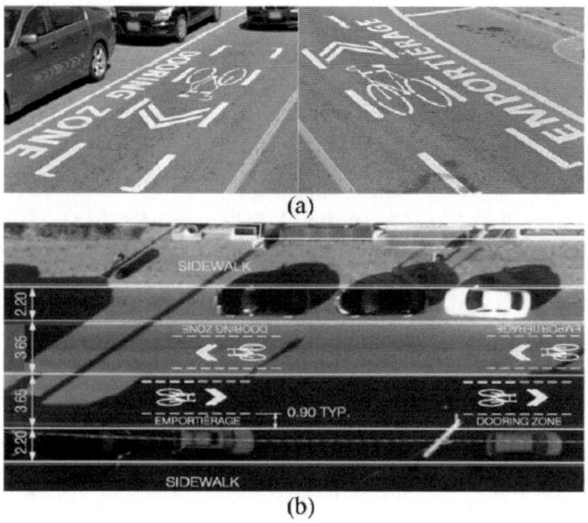

Figure 1. (*a*) Samples of video frame showing "post-treatment" marking indication with warning of dooring zone on the side in English and French. (*b*) Schematic diagram of treatment.

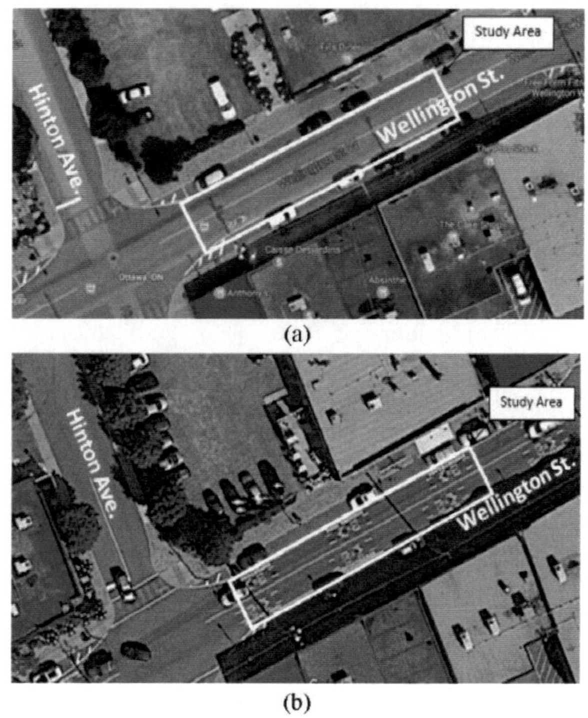

Figure 2. The study area (outlined) in two satellite images taken: (*a*) pre-treatment and (*b*) post-treatment (*14*).

Video Data Collection and Processing

Video data were collected in three phases where each phase consisted of two different days (a day on the weekend from approximately 10:00 a.m. to 2:00 p.m. and on a weekday from approximately 8:00 a.m. to 4:00 p.m.). Phase 1 (pre-treatment) video data was collected in May, 2015. Phase 2 (education phase) was conducted in August 2015 soon after the central sharrows were implemented. In this "education phase" the site was also visited by cycling "ambassadors" who performed outreach activities meant to inform cyclists of the purpose of the dooring zone. The campaign involved the creation and/or distribution of education materials through social media, media interviews, or personal interactions on the right of way with passers-by and road users. In the last phase, Phase 3, data was collected in July 2016 (one year after treatment). The volume of motor vehicle traffic was recorded within the study area by video data collection, every five minutes, for all phases and for each direction of travel: eastbound and westbound segments within the study area. The traffic count for all six days is summarized in Table 1.

The weather on all days of the video data collections was clear. The video data was labeled and converted to an appropriate format that enabled further processing. Video subsequences containing cyclists traveling in both directions while occupying the road were identified. The traffic volumes of the cyclists and the motor vehicles were counted for each direction separately along monitored sections. A total of 1,402 cyclists and 23,053 motor vehicles were counted. A subset of those that interacted with each other, for example, by passing or following, were examined in greater detail.

Likelihood of Vehicles to Pass/Follow a Leading Cyclist

When a motor vehicle was seen following a cyclist within the study area without attempting to pass, it was recorded as a *following* event. When the motor vehicle was seen to initiate passing the cyclist, the event was recorded as a *passing* event. When a cyclist was moving without being tailgated by a motor vehicle or there was no visible passing motor vehicle, no classification was made and the cyclist was deemed unexposed to the hazard of being passed by a tailgating motor vehicle.

For all phases, motor vehicles passing/following cyclists were also classified based on the opposing traffic conditions along the studied segment and were coded as follows: (a) the motor vehicle followed a cyclist with visible oncoming traffic in the opposing travel lane, (b) the motor vehicle followed a cyclist without visible opposing traffic, (c) the motor vehicle passed a cyclist with visible opposing traffic, (d) the motor vehicle passed a cyclist without visible opposing traffic. The rates of the motor vehicles that followed and passed cyclists were then calculated by using the following formulas:

$$R_f = \left(\frac{v_f}{v_f + v_p} \right) * 100$$

$$R_p = \left(\frac{v_p}{v_f + v_p} \right) * 100$$

Figure 3. Video camera set-up.

Table 1. Summary of the Motor Vehicle Traffic Volume Data Observation

Direction/day	Traffic parameter	Eastbound (EB) Phase 1	Eastbound (EB) Phase 2[a]	Eastbound (EB) Phase 3[b]	Westbound (WB) Phase 1	Westbound (WB) Phase 2[a]	Westbound (WB) Phase 3[b]
Weekday	Average hourly volume	353	355	377	352	312	384
Weekend		377	371	338	380	331	318
Weekday	Peak-hour volume	371	370	408	427	326	450
Weekend		432	406	388	415	371	357
Weekday	Peak-hour factor	0.84	0.81	0.83	0.82	0.82	0.85
Weekend		0.80	0.86	0.81	0.87	0.85	0.80

[a]Three months after the treatment was installed.
[b]One year after the treatment was installed.

where R_f and R_p; the rate of motor vehicles that followed and passed cyclists.

v_f and v_p; the number of motor vehicles that followed and passed cyclists.

Distance Cyclists Ride Away from the Parking Edge Line and from Motor Vehicle

Cyclists traversing the study area while on the roadway (neither on the sidewalk nor within empty parking spots) were detected and the lateral distance between a detected cyclist and the parking edge (continuous white line) was measured in both eastbound and westbound directions. Figure 4*a* shows a sample frame with an annotated distance d_1; which is the lateral distance between the shown cyclist and the parking edge at this moment for this cyclist. If the path of the cyclist was linear and parallel to the parking edge, then the lateral distance between the cyclist and the parking edge was measured anywhere within the study area. Not all cyclists moved in a parallel path to the parking edge. When the cyclist path exhibited non-linearity, the lateral distance between the cyclist and the parking edge was measured in two or sometimes three different locations within the study area along the parking edge, depending on the degree of non-linearity, and the average distance calculated for that cyclist.

Close-proximity interaction between a cyclist and a motor vehicle was also characterized by the measurement of a lateral distance between road users when they were visibly adjacent to each other. An exception to motor vehicle–cyclist adjacency condition is when the adjacent motor vehicle was not moving, in which case the lateral distance was not recorded. This is because the cyclist may navigate the available space within a context of static motor vehicles. Figure 4*b* shows a sample frame with distance d_2 laterally separating the cyclist and the motor vehicle. This lateral distance was measured along the perpendicular distance to the parking edge. All distances were measured in pixel coordinates and then converted to meters based on inferred geometric relationships between scale measured in meters/pixel perpendicular to the parking edge and location developed for each camera setting.

Figure 4. Lateral distance by camera video frame: (*a*) between a cyclist and the parking edge line, and (*b*) between a cyclist and the side of a motor vehicle.

Observed Performance Parameters

Different performance parameters were extracted from the video data collected. A complete list of performance parameters is given below:

1. When a cyclist and a motor vehicle were visibly adjacent to each other, both (a) the lateral distance between the side of the motor vehicle and the bicycle tires and (b) the lateral distance between the cyclist and the parking edge line were measured.
2. When the cyclist was moving with no tailgating motor vehicle or visible passing motor vehicle, no classification was made and the lateral distance between the cyclist and the parking edge line was measured.
3. Video events containing a motor vehicle following a cyclist within the study area, without attempting to pass, were recorded as *following* events. Based on the presence of an oncoming motor vehicle in the opposite lane, within the study area, two subsets of these events were identified, as mentioned in the introduction.
4. Video events containing a motor vehicle passing a cyclist within the study area were recorded as *passing* events. Similarly, and based on the presence of a motor vehicle in the opposite lane, within the study area, two subsets of these events were identified as, mentioned in the Data Collection section.

Analysis and Results

Impact of the Dooring Zone on Motor Vehicle Behavior

The dooring-zone indication in this treatment is intended to discourage cyclists from riding close to the parked vehicles. It is useful to compare counts of passing and following events in the pre-treatment and post-treatment periods. Table 2 shows a summary of these events. The effect of the central sharrows can be investigated through the comparison of pre and post-treatment exposure rates. There is an evident increase in the proportion of cyclists being followed, rather than passed, from 43% to 75% and to 76.6%. There is a corresponding reduction in the proportion of cyclists who were observed being passed by a motor vehicle, from 57% to 25% and to 23.4%. All changes in previous exposure rates when comparing Phase 1 to Phase 2 or to Phase 3 are statistically significant ($p < 0.01$). The changes occurred after the dooring-zone indication and central sharrows were implemented. These changes were beneficial in that cyclists were more likely take the middle of the travel lane and thus deter passing by tailgating motor vehicles. This can positively support the stated treatment objectives mentioned in the beginning of this paper.

It may be suspected that this change could be attributed to the opposing traffic conditions (which vary from case to case). Therefore, the passing/following events were classified based on the opposing traffic conditions: was there a visible opposing motor vehicle occupying the other traffic lane? The events were further disaggregated based on this condition and the results are summarized in Table 2. This is an indication that the passing behavior was primarily influenced by a cyclist's position in relation to the lane edge. Chi-square test (H_o: $p_1 = p_2$, where p is the proportion of count followed/passed out of total exposure events) was used and this change is statistically significant when comparing Phase 1 to Phase 2 or to Phase 3 ($p < 0.05$). These results positively support the treatment objectives.

Distance between Cyclists and Parking Edge Line

An important objective of this treatment is to encourage cyclists to ride in the middle of the travel lane (away from the dooring zone) where the central sharrows were implemented.

Table 2. Summary of the Motor Vehicle Passing and Following Events Involving Cyclists Outside the Dooring Zone Only

Phase/parameter	Phase 1	Phase 2[1]	Phase 2[2]	Phase 1 with/without opposing vehicle	Phase 2[1] with/without opposing vehicle	Phase 3[2] with/without opposing vehicle
Count followed	34 (43.0%)	93 (75.0%)	85 (76.6%)	25 (53.2%) 9 (28.1%)	67 (90.5%) 26 (52.0%)	54 (91.5%) 31 (59.6%)
Count passed	45 (57.0%)	31 (25.0%)	26 (23.4%)	22 (46.8%) 23 (71.9%)	7 (9.5%) 24 (48.0%)	5 (8.5%) 21 (40.4%)
Total exposure	79 –	124 –	111 –	47 32	74 50	59 52

Table 3. Summary of Measurements of Distance of Bicycles from Parking Edge Line

Percentile[a]	Phase 1	Phase 2[b]	Phase 3[c]
10th	0.24	0.43	0.45
25th	0.38	0.63	0.64
50th	0.57	0.92	0.90
75th	0.82	1.35	1.26
90th	1.10	1.63	1.57
Average distance (m)	0.64	1.01	0.96
Standard deviation (m)	0.37	0.48	0.42
Count > 0.9 m	87	193	202
Count ≤ 0.9 m	350	185	205
Total	437	378	407

[a]Percentile values represent proportion of observations less than given value in subsequent columns. For example, 10% of observations are less than 0.24 m for Phase 1.
[b]Three months after the treatment installed.
[c]One year after the treatment installed.

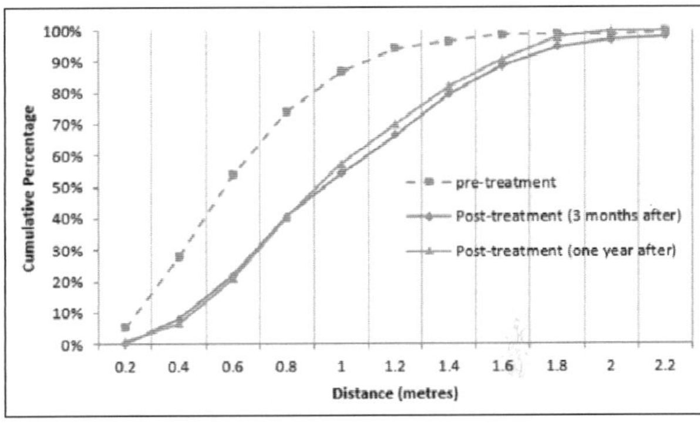

Figure 5. Cumulative distribution of distance of cyclists from parking edge line.

Lateral spacing data was obtained from frames extracted from the video data. The lateral distance was calculated manually by identifying and transforming screen pixel measurements to roadway metric measurements using a special script. These coordinate positions were extracted for two points along a straight line and identified visually the bicycle's front/or rear wheel contact point with (a) the road pavement surface, and (b) the parking edge line. If the path of the cyclist was linear and parallel to the parking edge line, the lateral distance was measured anywhere within the study area. If the cyclist's path exhibited non-linearity, the lateral distance for that event was measured in different locations within the study area along the parking edge, and the average distance was calculated for this cyclist.

Table 3 summarizes the lateral distance measurements between the parking edge line and cyclists for the two conditions, pre- and post-treatment, and for all phases. As shown in Table 3, there is a noticeable change in the average of the distance between the parking edge line and the cyclist. The average distance increased when comparing pre-treatment to post-treatment. The increase in the average is supportive of the treatment objective (which is to encourage cyclists to ride further from the dooring zone).

It is evident in all percentile values that the distances are greater after the treatment. This is also noticeable in Figure 5. The average lateral distance between cyclists and the parking edge line increased from 0.64 m to 1.01 m and to 0.96 m after the treatment. The difference is statistically significant (two-tail t-test, $p < 0.01$). In addition, the proportion of distances above 0.90 m increased from 87/437 ≈20.0% to 193/378 ≈51.0% and to 202/407 ≈50.0%. This change in proportion is statistically significant (Chi-square test, $p < 0.01$). This change can be explained by the cyclists' correct interpretation of the treatment and hence moving further from the dooring zone toward the center of the travel lane.

How Far are Cyclists and Vehicles from Each Other When Adjacent?

Table 4 shows the summary of the lateral distance between cyclists and traveling motor vehicles. As shown in Table 4, the average lateral distance in all phases between cyclist and motor vehicle when adjacent and visible within the study area was found to be 1.12 m, 1.31 m, and 1.10 m. Two findings are noted: first, approximately 48% of lateral spacing measurements were greater than 1.00 m before the treatment, while this percentage increased to 60% and 52% after the treatment. Both percentages reflect the limited space available for the two modes to travel side by side within the same lane; in particular when space available in the opposite lane

Table 4. Summary of Measurements of Distance between Bicycles and Motor Vehicles

Percentile[a]	Phase 1	Phase 2[b]	Phase 3[c]
10th	0.59	0.65	0.51
25th	0.80	0.77	0.66
50th	0.98	1.46	1.04
75th	1.34	1.81	1.48
90th	2.06	1.93	1.93
Average distance (m)	1.12	1.31	1.10
Standard deviation (m)	0.50	0.52	0.50
Count > 1.0 m	26	15	11
Count ≤ 1.0 m	28	10	10
Total	54	25	21

[a]Percentile values represent proportion of observations less than given value in subsequent columns. For example, 10% of observations are less than 0.59 m for Phase 1.
[b]Three months after the treatment installed.
[c]One year after the treatment installed.

is not usable or if a bypassing maneuver is not performed by the motorist. Second, the count of adjacency events declined by 39%, which is likely related to the fact that fewer motorists drove beside cyclists (or a cyclist was forced to ride to the side of a vehicle) in the post-treatment and instead chose to follow the cyclist. The standard deviation had a minor increase from 0.50 m to 0.52 m. While a positive improvement, it is important to examine how likely is a detectable true change. Given the sample size (54, 25 and 21 events) and standard deviation for all phases, this difference is not statistically significant (two-tail *t*-test, $\rho > 0.05$). This finding is likely due to the relatively low number of events and also to the limited space to accommodate cyclists and motor vehicles traveling side by side in all phases' conditions. In all phases (pre- and post- treatment) most motorists passed at safe distances from cyclists (*8*).

Intercept Survey

An intercept survey was conducted on August 18, 2016, where four city staff members were on site to communicate with residents about the new pavement markings. The survey was intended as a method of receiving feedback from the public and providing education to the public about the new treatment. It focused on two main questions: (a) where should cyclists ride with respect to the central -sharrow locations? and (b) what does the "dooring zone" refer to? In total, 83 cyclists and 9 pedestrians completed the survey, a total of 92 participants. The results found that 80.0% of the participants knew to ride in the center of the sharrow markings, and 67.8% of the participants understood that the dooring zone meant "the space in which a cyclist is in danger of getting hit by a motor vehicle door." The results can be found in Figure 6 below.

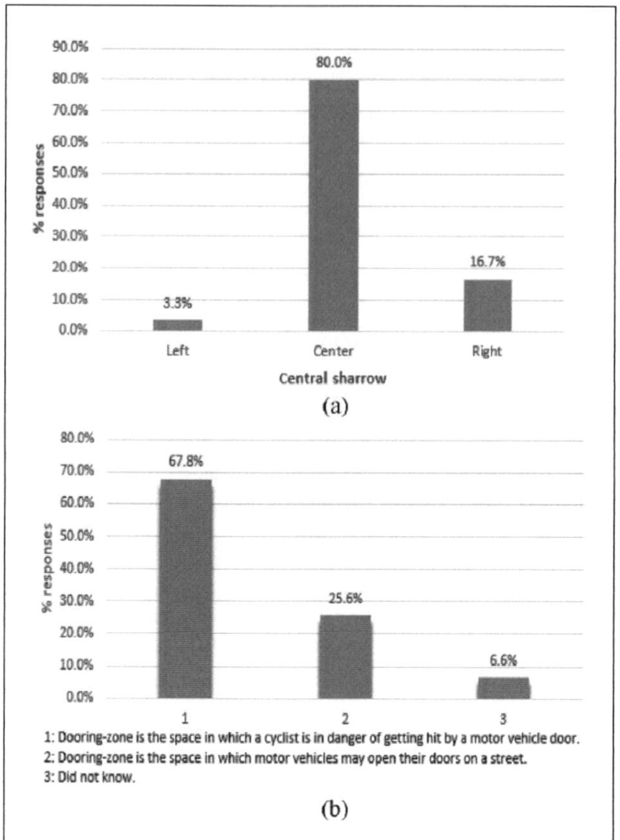

Figure 6. Survey responses distribution: (*a*) Q(a) where should cyclists ride with respect to the central sharrow locations? (*b*) Q(b) what does the "dooring zone" refer to?

Summary and Discussion

This study observes the change in behavior of motorists and cyclists associated with the painting of central sharrow markings along with written indication of dooring-zone boundaries. A before-and-after evaluation was conducted in order to evaluate the impact of dooring-zone and central sharrow markings. In order to assess the behaviors of cyclists and motorists for both conditions, different performance parameters were extracted from the video data collected: (a) lateral distance measurement, and (b) rates of vehicle lane change maneuvers (passing and following) with or without opposing vehicle presence.

A total of 1,402 cyclists and 23,053 motor vehicles were observed in both conditions for all phases. The observations were made within a total period of six days: two days in pre-treatment and four days in post-treatment conditions. The treatment was pavement markings in the form of sharrows in the center of the travel lane with dooring-zone boundaries marked beside the parking edge line on a two-way, two-lane, undivided urban street in Ottawa, Canada.

Two different lateral distance measurements were calculated for both conditions; the parking edge line to bicycle, and bicycle to moving motor vehicle. Most notable is the increase in average lateral distance between bicycle tires and the parking edge line from 0.64 m to 1.01 m and to 0.96 m. This difference was found to be statistically significant. This is a positive change as more cyclists rode outside of the dooring zone.

Passing and following events are the two different motorist behaviors that were observed in this study. The results, presented in Table 2, show that as the cyclists ride further from the parking edge line, there is an evident increase in the proportion of cyclists being followed, rather than passed, from 43% to 75% and to 76.6%. The results also show that there is a corresponding reduction in the proportion of cyclists being passed by vehicles, from 57% to 25% and to 23.4%. Results demonstrated that the treatment achieved the initial purpose of this experiment.

Conclusion

This paper examines the change in operational and safety measures for cyclists and motor vehicles associated with the painting of dooring-zone signage and central sharrows on a two-lane, two-way street in an urban area. The key findings are as follows:

- Dooring-zone markings had an effect on lateral distance between cyclists and the parking edge line, which increased in average distance between the bicycle tires and parking edge line. This is a positive change as more cyclists rode further from the parking edge line with the presence of the dooring-zone pavement markings.
- Fewer motorists chose to pass the leading cyclist in the post-treatment.
- Motorists were more likely to follow the cyclists along the same lane rather than pass them. The number of motor vehicles that followed the leading cyclist in the post-treatment increased.
- The average distance between the motor vehicle and the cyclist (passing distance) for all phases' conditions (pre- and post- treatment) was within a safe distance. This finding indicated that this average distance was unaffected by the treatment.
- Based on the intercept survey, the pavement markings were generally understood by cyclists with regard to where cyclists should position themselves on the roadway and what the term "dooring zone" refers to.

The findings of this study indicate positive trends in cyclist behavior with the presence of the dooring-zone markings. A positive change was also observed in motorist behavior when being positioned behind a cyclist. Fewer motorists chose to pass the leading cyclist in the post-treatment period compared with the pre-treatment period.

The behavioral change is favorable. It is noteworthy that there were still cyclists who rode within the dooring zone after the treatment. Similarly, some motorists also chose to narrowly pass cyclists after the treatment. Cyclists were not completely immune to the risk of collision in post-treatment conditions. More efforts are needed as to public awareness and treatment enhancement (e.g., colored background) to see further reduction in collision risk. Furthermore, in this treatment cyclists are effectively invited to join the mainstream of dominant vehicular traffic. They therefore face a heightened exposure to rear-end collisions. It is not very well understood if such exposure will undermine any of the treatment's evident success. Further investigation of the nature of near-misses or even analysis of collision trends may reveal any underlying hazards. A related question is whether the risk trade-off in exchanging dooring collisions for rear-end collisions is worthy of further investigation and whether there are any factors which may affect the manifestation of such a trade-off.

Acknowledgments

The study was supported by the Natural Science and Engineering Research Council of Canada and the City of Ottawa. The authors thank Rob Wilkinson and Kale Brown from the Transportation Service Ottawa for their valuable contribution during the intercept surveys.

Author Contributions

The authors confirm contribution to the paper as follows: study conception and design: Ali Kassim, Karim Ismail, Shawn McGuire; data collection: Ali Kassim; analysis and interpretation of results: Ali Kassim, Karim Ismail; draft manuscript preparation: Ali Kassim, Karim Ismail, Shawn McGuire. All authors reviewed the results and approved the final version of the manuscript.

References

1. DiGioia, J., K. Watkins, Y. Xu, M. Rodgers, and R. Guensler. Safety Impacts of Bicycle Infrastructure: A Critical Review. *Journal of Safety Research*, Vol. 61, 2017, pp. 105–119.
2. Bai, L., P. Liu, C. Y. Chan, and Z. Li. Estimating Level of Service of Mid-Block Bicycle Lanes Considering Mixed Traffic Flow. *Transportation Research Part A: Policy and Practice*, Vol. 101, 2017, pp. 203–217.
3. Schimek, P. Bicycle Facilities Adjacent to On-Street Parking: A Review of Crash Data, Design Standards, and Bicyclist Positioning. Presented at 96th Annual Meeting of the Transportation Research Board, Washington, D.C., 2017.
4. Brady, J., J. Loskorn, A. Mills, J. Duthie, and R. Machemehl. Effects of Shared Lane Markings on Bicyclist and Motorist Behavior. *ITE Journal*, Vol. 81, No. 8, 2011, p. 33.
5. Hunter, W., R. Srinivasan, L. Thomas, C. Martell, and C. Seiderman. Evaluation of Shared Lane Markings in Cambridge, Massachusetts. *Transportation Research Record: Journal of the Transportation Research Board*, 2011. 2247: 72–80.

6. Alta Planning + Design. *San Francisco's Shared Lane Pavement Markings: Improving Bicycle Safety*. Final Report. San Francisco Department of Parking and Traffic, 2004.
7. Furth, P. G., D. M. Dulaski, D. Bergenthal, and S. Brown. More Than Sharrows: Lane-Within-a Lane Bicycle Priority Treatments in Three U.S. Cities. Presented at 90th Annual Meeting of the Transportation Research Board, Washington, D.C., 2011.
8. Foletta, N., C. Nielson, J. Patton, J. Parks, and R. Rees. Shared Lane Markings on Urban Arterial in Oakland, California: Evaluation of Super Sharrows. *Transportation Research Record: Journal of the Transportation Research Board*, 2015. 2492: 61–68.
9. Kassim, A., K. Ismail, and S. Woo. Investigation of the Effect of Super-Sharrows on Cyclist and Vehicle Behavior. *Transportation Research Record, Journal of the Transportation Research Board*, 2017. 2659: 224–232.
10Aultman-Hall, L., and M. Kaltenecker. Toronto Bicycle Commuter Safety Rates. *Accident Analysis and Prevention*, Vol. 31, No. 6, 1999, pp. 675–686.
11. Burke, C., and D. Scott. Identifying Sensible Locations for Separated Bike Lanes on a Congested Urban Road Network: Case Study in Toronto, Canada. Presented at 96th Annual Meeting of the Transportation Research Board, Washington, D.C., 2017.
12. Duthie, J., J. Brady, A. Mills, and R. Machemehl. Effects of On-Street Bicycle Facility Configuration on Bicyclist and Motorist Behavior. *Transportation Research Record: Journal of the Transportation Research Board*, 2010. 2190: 37–44.
13. Van Houten, R., and C. Seiderman. How Pavement Markings Influence Bicycle and Motor Vehicle Positioning: Case Study in Cambridge, Massachusetts. *Transportation Research Record: Journal of the Transportation Research Board*, 2005. 1939: 3–14.
14Google [Online], 2016. https://www.google.ca/maps/@45.3997547,-75.7341252,17z/data=!3m1!4b1. Accessed June 5, 2016.

The Standing Committee on Bicycle Transportation (ANF20) peer-reviewed this paper (18-03262).

Article

TRR
JOURNAL OF THE TRANSPORTATION RESEARCH BOARD

Transportation Research Record
2018, Vol. 2672(36) 145–156

DOI: 10.1177/0361198118796350
journals.sagepub.com/home/trr
SAGE

Quality of Bicycle Traffic Management at Urban Road Links and Signalized Intersections Operating under Mixed Traffic Conditions

Sambit Kumar Beura[1] and Prasanta Kumar Bhuyan[1]

Abstract

This study proposes bicycle level of service (BLOS) models for the urban road links and signalized intersection approaches operating under heterogeneous traffic flow conditions prevailing in developing countries. Extensive data collection was carried out through field investigations, videography techniques, and perception surveys to inspect which parameters primarily affect the quality of bicycling in the mentioned facilities. As many as 74 road links and 70 intersection approaches of nine different Indian cities were included in the study locations. In addition, the perceived satisfaction ratings of 22,320 on-site bicyclists (i.e., 155 bicyclists from each site) were included in the database. These ratings ranged from 1 to 6 (excellent–worst bicycling conditions), and were used as the array of dependent variable in the model development process. Separate BLOS models were developed for road links and bicycle through movement at signalized intersections through the application of a novel evolutionary artificial intelligence technique, namely multi-gene genetic programming. The link model included eight significant variables in total whereas the intersection model included a total of seven. With compacted structures, both models represented excellent reliabilities in the present context and produced high coefficient of determination (R^2) values of above 0.86 with averaged observations. These models will help traffic planners and engineers to evaluate the service levels of urban bicycle facilities of developing countries. Statistical analysis of modeled parameters has reported that the traffic volume and crossing pedestrians have the highest influences on the quality of bicycle traffic management at urban road links and intersection approaches, respectively.

A thorough understanding of the operational characteristics of bicyclists facilitates the proper designing of bicycle-friendly transportation infrastructures. Although several attempts have been made in developed countries to explore this, the applicability of these findings is limited to homogeneous traffic conditions only. This kind of traffic consists of identical vehicles maneuvering in a lane-disciplined manner. However, about 80% of the world population resides in developing countries, where the road traffic is characterized by extreme vehicle heterogeneity and poor lane discipline. In these countries, the unavailability of proper bicycle facilities forces vulnerable bicyclists to mobilize on the main carriageways at their own risk. An in-depth operational analysis of on-street bicyclists under these complex situations is yet to be carried out. To fill this gap, the present study has aimed at the following: (1) to identify the variables which have significant influence on the quality of bicycling on urban road links and signalized intersection approaches; and (2) to develop separate bicycle level of service (BLOS) models for these facilities.

Links are the portions of roadways between successive intersections (or ramp terminals), whereas intersections are the at-grade junctions where two or more roads meet or cross. Major intersections are operated with traffic signals to safeguard traffic and pedestrian movements; these are called signalized intersections. In this study, the quality of bicycle through movement at these two facilities is analyzed and modeled. Geometric, traffic, and built-environmental attributes are collected from 74 links and 70 signalized intersection approaches located in nine Indian cities. Variables which have significant influence

[1]Department of Civil Engineering, National Institute of Technology Rourkela, India

Corresponding Author:
Address correspondence to Sambit Kumar Beura: sambit.beura@gmail.com

on the BLOS offered at the respective facilities are identified by correlating various facility attributes and the personal characteristics of bicyclists (socio-demographic and travel-related) with perceived satisfaction scores of 22,320 on-site bicyclists. Subsequently, reliable BLOS models are developed for both facilities by establishing empirical relationships between significant variables and perceived satisfaction scores. These models can greatly assist administrative personnel to judge the ability of existing road facilities in managing bicycle traffic. Subsequently, judicious decisions could be taken to achieve better service levels.

Regression analysis is widely used for the analysis various properties of the transportation system, including the BLOS. However, this study has implemented a novel evolutionary approach, namely multi-gene genetic programming (MGGP) for the better analysis of BLOS. The main motivation behind using this technique is that it preserves the modeling capabilities of both genetic programming (GP) (*1*) and statistical regression methods. This property has made it highly advantageous over many other prediction tools available at present. MGGP is also capable of developing compacted model equations (regression-like), which are largely favorable for field implementations. Another advantage of this technique is that it does not require detailed information of the process phenomena, and develops an empirical model automatically from the data at hand, called data-driven approach (*2*).

Background Studies

The Bicycle Safety Index Rating model (*3*) is the first ever service prediction model developed for urban streets carrying homogeneous traffic. It comprises two submodels, namely the Roadway Segment Index (RSI) model and the Intersection Evaluation Index (IEI) model. RSI model estimates the service classes of road links by utilizing information on traffic volume, number of lanes, speed limit, width of outside traffic lane, pavement conditions, and location factors. IEI model estimates the BLOS of road intersections by utilizing information on main-street and cross-street traffic volumes, type of signalization, presence of left- and right-turn lanes, number of through lanes, curb radii, and sight distance. Discussed below, several other attempts were made in the subsequent years to further explore the BLOS criteria.

Studies on Road Links

Epperson modified the location and pavement factors of the RSI model, and renamed it as the modified Roadway Condition Index model (*4*). The lane width was also multiplied with speed limit to place greater weightage on narrow roads with high traffic speeds. The Interaction Hazard Score model revealed the importance of roadside land-use intensity and curb cut frequency (*5*). The Bicycle Stress Level model used the curb-lane parameters only (width, traffic volume, and traffic speed in the curb-lane) to estimate the link BLOS (*6*). The Bicycle Suitability Rating model signified the importance of traffic volume and traffic speed in the facility BLOS (*7*). Afterwards, the Bicycle Compatibility Index model highlighted the importance of bicycle lane parameters and right-turning vehicles (*8*). Some studies also revealed the importance of pavement surface quality and bicycle lane parameters (*8–10*). The *Highway Capacity Manual* (HCM)-2000 considered average travel speed, average delay, and hindrance to define the link BLOS (*11*). Further, HCM-2010 considered number of through lanes, effective width of the outside lane, pavement conditions, mid-segment demand flow rate, traffic speed, and percentage of heavy vehicles to define the link BLOS (*12*). Kang and Lee revealed that the BLOS is largely determined by number of lanes, width of the bicycle path, pedestrian volume, and number of encounters (*13*).

The above studies were carried out under heterogeneous traffic flow conditions. An investigation of the bicycle activity on multi-lane urban road segments carrying heterogeneous traffic has revealed that traffic volume has the highest influence on the facility BLOS (*14*). Recently, two BLOS models have been proposed for the assessment of urban road links carrying heterogeneous traffic. Of these models, one is ordered-probit-based (*15*) and the other one is multivariate adaptive regression splines (MARS)-based (*16*). Various quantitative variables such as outside lane width, traffic volume, traffic speed, pavement condition index, land-use pattern, interruptions by unauthorized stoppages of intermittent public transits, on-street parking activities, and frequency of driveways carrying a high volume of traffic were observed to have significant influences on the BLOS. However, these two models are very complex in structure and need high computational efforts for their field implementations. Thus, there remain huge opportunities to develop less complex BLOS models for the present context.

Studies on Signalized Intersections

HCM-2000 (*11*) used the controlled delay (s/bicycle) parameter only to define signalized intersection BLOS. However, average bicycle delay, conflicts, and exposure also have substantial influences on this (*17*). Landis et al. found that the outermost lane width, number of lanes and traffic volume on the subject approach, and intersection crossing distance have significant influences on the quality of bicycle through movement at signalized intersections (*18*). HCM-2010 considered these three parameters in conjunction with left- and right-turning traffic volumes to model the intersection BLOS (*12*). The Bicyclist Intersection Safety Index model revealed that the operational conditions of through bicyclists are mainly influenced by main-street and cross-street traffic volumes, main-street speed limit, number of right-turn lanes, and the presence of bicycle lane, signalization, turning vehicular traffic, and on-street parking activities on the main street (*19*).

Jensen modeled the operational conditions of through bicyclists at signalized intersections, and exposed the importance of bicycle crossing facilities, width of bicycle facility at stop lines, and the type of bicycle facilities before intersections (*20*). Strauss et al. revealed that the presence of wider crosswalks and bus stops increases bicyclists' injury occurrence, while the provision of raised medians decreases the same (*21*). However, no attempt has been made yet to model the BLOS of signalized intersections carrying heterogeneous traffic.

Method of MGGP Modeling

MGGP is a variant of GP (*1*) that mimics the biological evolution of living organisms and develops non-linear models using the Darwinian principle of natural selection and survival of the fittest. The GP formalism starts with the formation of an initial population of numerous individuals (non-linear transformations of input variables) generated at random. Each individual resembles a tree structure whose nodes are the elements either from a functional set or from a terminal set. As illustrated in Parent-1 of Figure 1*a*, the root node of the GP tree carries one or more branches extending from a functional node and ending at a terminal node. The functional set includes arithmetic operators, mathematical functions, Boolean operators, logical expressions, or any other user-defined functions. Conversely, the terminal set includes variables, constant, or both. The elements functional and terminal sets are chosen at random and constructed together to build an individual.

In this study, the root mean square error (RMSE) of observed and predicted outputs is used as the objective function. It assesses the prediction ability an individual in terms of how well it performs in the problem environment. Unless the problem is so small and simple, most of the individuals of the initial population exhibit extremely poor fitness. Thus, the genetic operations such as reproduction, crossover, and mutation are performed with their specified frequencies (p_r, p_c, and p_m, respectively) to create an offspring population of better-fitted individuals. During the "reproduction" process, high-fitness parental individuals are simply duplicated to the offspring population. The "crossover" operation produces better-fitted individuals by exchanging the randomly chosen parts of two parental individuals selected in proportion to their fitness value. Figure 1*a* illustrates a typical crossover of two parental individuals $\ln\{a^2/(b+2)\}$ and $\exp(a/b^{0.5})$ to produce two offspring individuals $\ln(a^2/b^{0.5})$ and $\exp\{a/(b+2)\}$. During the "mutation" process, better-fitted individuals are produced by randomly changing a node of the parent. Figure 1*b* illustrates a typical mutation operation on $\ln(a/b)$ to produce $\ln(a \times b)$.

After the completion of genetic operations, the newly produced offspring population replaces the initial population. The offspring population again undergoes the genetic operations to create another population of better-fitted offspring individuals. This process continues over many generations and terminates when either the maximum number of generations or a threshold fitness value is achieved. The best-fit individual appeared in any generation defines the output of the GP formalism. The MGGP formalism is also carried out through the traditional GP procedure, except that each individual in MGGP is a weighted linear combination of two or more GP trees (referred as "genes"). Thus, the mathematical expression of any MGGP-based model can be represented as:

$$y = \left(\sum_{i=1}^{n} w_i G_i \right) + w_0 \tag{1}$$

where

y = model output,
n = number of genes in the evolved model,
w_i = weight of ith gene,
G_i = mathematical expression of ith gene, and
w_0 = bias parameter.

w_i and w_0 parameters are estimated by using the ordinary least squares method. Koza (*1*), Poli et al. (*2*), and Searson et al. (*22*) can be consulted for further details on the GP and MGGP formalisms.

Optimum Parameters of the MGGP Algorithm

The execution time, complexity, and accuracy of MGGP-based models are regulated by controlling the values of various algorithm parameters such as population size, number of generations, maximum allowable number of genes (G_{max}), maximum tree depth (d_{max}), p_r, p_c, and p_m, and so forth. Lower values of these parameters lead to the formation of a less complex but inferior model in less time. Thus, the optimum values of these parameters exist for a given data set for which a less complex but reliable model is obtained (*22*). However, no concrete guideline is available at present to estimate the optimum values these

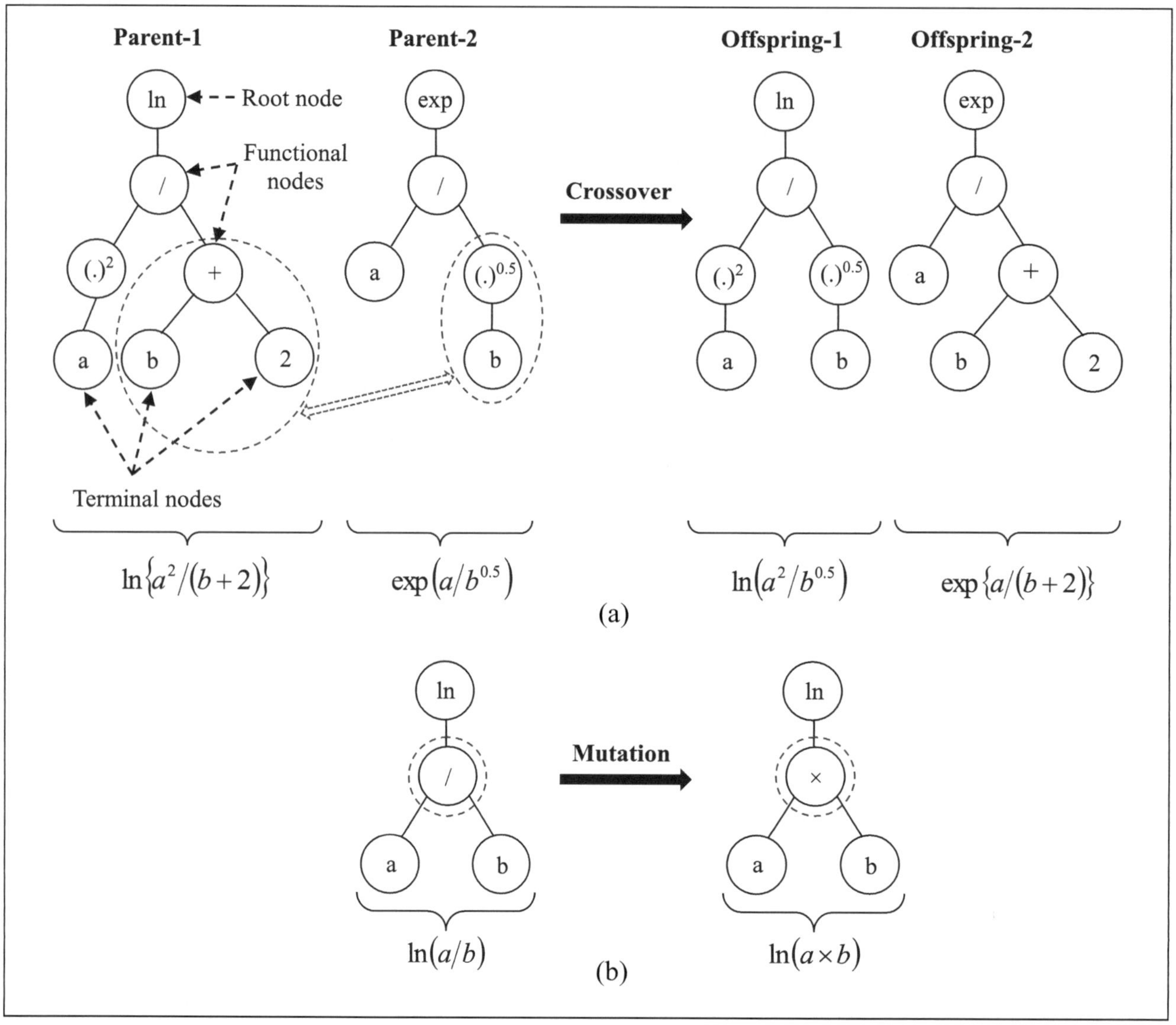

Figure 1. Typical evolutionary operations in GP: (*a*) crossover and (*b*) mutation.

parameters accurately. Hence, to attain this, a step-wise selection procedure is adopted in this study. The parameter to be optimized was scanned thoroughly in a wide range while keeping the nominal values of all other parameters unchanged. The model performance with each configuration was tested and the value at which the concerned parameter produced the best result was recorded. Subsequently, the optimal values of the remaining parameters were estimated in a similar manner. During this process, the MGGP algorithm parameters were varied in the following ranges.

- population size: 200–1500 individuals;
- number of generations: 100–500;
- G_{max} and d_{max}: 2–5; and
- p_r, p_c, and p_m: 0.01–0.07, 0.75–0.9, and 0.05–0.15, respectively.

A tournament selection strategy with tournament size of seven was adopted for the selection of parental genes from the pool of available solutions. To keep the final models as simple as possible, the instructions defined in the functional set were limited to $+$, $-$, $\times$, $\div$, $(.)^2$, $(.)^3$, sqrt(.), exp(.) and ln(.). Criteria used to terminate the program execution were 500 generations or a fitness value less than 0.0001, whichever is earlier. A pool of BLOS models was developed for each facility, and the best one was selected by making a trade-off between the complexities and prediction precisions of evolved models.

Database Preparation

To carry out a comparative analysis of BLOS, a diverse sample of 74 road links and 70 signalized intersection approaches was identified from various parts of India. Wide diversities in site characteristics were considered, including traffic volume (high and low), traffic speed (high and low), number of traffic lanes (one or more), bicycle facilities (bicycle lanes, paved shoulders, and wide curb lanes), and roadside developments (residential, office, commercial, and industrial). Another criterion was that the selected sites have adequate amounts of bicycle traffic, which would allow the productive collections of bicyclists' characteristics and perceived BLOS ratings. Based on the windows of opportunity for videotaping and the willingness of local contacts to participate, the following cities were selected as the study locations.

1. Bhubaneswar, Odisha state (29 links, 25 intersection approaches);
2. Lucknow, Uttar Pradesh state (14 intersection approaches);
3. Rajahmundry, Andhra Pradesh state (11 links);
4. Nagpur, Maharashtra state (7 intersection approaches);
5. Tirupati, Andhra Pradesh state (12 intersection approaches);
6. Kottayam, Kerala state (12 links);
7. Rourkela, Odisha state (19 links, 6 intersection approaches);
8. Kurnool, Andhra Pradesh state (3 intersection approaches); and
9. Anantapur, Andhra Pradesh state (3 intersection approaches).

The above sites are parts of 2-, 4- and 8-lane divided or undivided roads located in urban, sub-urban, or residential areas of the respective cities. Facilities such as wide curb lanes, sidewalks, medians, and parking lanes are unevenly available in these locations. Considered road links represent wide diversities in the geometric (3–14 m carriageway, 0.5–5 m medians, and 2–3.5 m shoulders) and traffic flow conditions (250–5100 Passenger Car Units per hour or PCUs/h of peak traffic flowing with an average speed of 23–46 km/h). Similarly, considered intersection approaches are the parts of T-type or 4-legged isolated intersections. Here, the main-street and cross-street traffic volumes vary in the ranges of 395–4085 PCUs/h and 208–2110 PCUs/h, respectively, during the peak hours. In addition, the volume of crossing pedestrians varies in the range of 33–1700 ped/h. Detailed statistics of various other important parameters are given in the "Statistical analyses and results" section.

Collection of Quantitative Data

Various quantitative attributes, listed in Table 1, were identified by the team members as having potential impacts on the quality of bicycling. These variables were collected through visual observations, geometrical measurements, and the videography technique. Visually observable and field-measurable variables were collected during the inventory survey. Video cameras were mounted on nearby buildings, foot-over bridges, or any other rigid structures to record other parameters. Video recordings were carried out during the expected peak hours of Indian traffic, that is, 8.30–10.30 a.m. or 4.30–6.30 p.m. Subsequently, the peak one hour was determined by the running average method. This hour was chosen as the analysis period because bicyclists encounter the worst operating conditions during this time.

The mid-segment traffic flow was recorded over a longitudinal trap of 30 m length. Such a long trap was chosen mainly for the accurate measurement of the traffic speed. Speed measures such as spot speed or space mean speed, which are normally measured for homogeneous traffic, should not be considered for the heterogeneous traffic because there exists a large variation in the speeds of individual vehicles. The average crossing time of motorists on the longitudinal trap was extracted with an accuracy of 0.1 s, and the average traffic speed (km/h) was estimated. The peak-hour volume was expressed in PCUs/h by using the PCU conversion factors given in *Indian Road Congress* code of practice-106 (*25*). Other necessary attributes listed in Table 1 were also extracted from the peak 1-hour videos.

Similar procedures were also followed to collect various attributes of the intersection approaches. The camera was positioned in such a way to record the unobstructed view of subject and opposing approaches simultaneously with traffic signal timings. Subsequently, all required parameters (Table 1) were extracted. As per the HCM-2010 (*12*), on-street parking activities should be observed within at least 250 ft. (76.2 m) upstream of the stop line at intersection approaches. As the situations under heterogeneous traffic conditions are more complex, this activity was observed over a slightly elongated stretch of 100 m length. The surrounding developmental patterns were also observed within the same length.

Opinion Survey of On-Site Bicyclists

The opinion survey was intended to collect perceived BLOS scores and various information on the socio-demographic and travel characteristics of bicyclists. Bicyclists who immediately traveled over the full length of desired links or crossed through desired intersections were recruited for interview. In total 24,480 bicyclists

Table 1. Attributes Included in the Database

Attribute (unit or scale)	Facility	Collection procedure or data source
Geometric design		
Number of traffic lanes	All[a]	Visual observation
Total length of road link (km)	Link	Google Earth
Carriageway width (m)	All[a]	Field measurement
Effective width of the outside lane[b] (m)	Link, subject approach	Field measurement
Presence (1 = yes, 0 = no) and width (m) of the shared-use path, bicycle lane, sidewalk, parking lane, median, curb and gutter facilities	All[a]	Visual observation, field measurement
Total number of driveways	Link, subject approach	Visual observation
Number of driveways carrying high volumes of traffic[c]	Link, subject approach	Visual observation
One-way or two-way (1 for one-way, 2 for two-way)	Link, subject approach	Visual observation
Number of intersection legs	Intersection	Visual observation
Intersection crossing distance (m)	In subject direction	Field measurement
Presence of crosswalk (1 = yes, 0 = no)	Subject approach	Visual observation
Left-turn curb radius (m)	Subject approach	Google Earth
Presence of exclusive left-turn lane (1 = yes, 0 = no)	Subject approach	Visual observation
Intersection sight distance (ISD) availability (1 = good, 2 = moderate, 3 = poor)[d]	Subject approach	Visual observation
Traffic characteristic and control		
Total traffic volume (PCUs/h)	All[a]	Videography technique
Volumes of the through, left-turning and right-turning traffic (PCUs/h)	All[a] except link	Videography technique
Traffic speed (km/h)	Link, subject approach	Videography technique
Percentage of heavy vehicles (%)	All[a]	Videography technique
Interruptions by roadside stoppages of intermittent public transits (1 = high, 0.5 = medium, 0 = minimal)	Link	Visual observation
Approximate vehicular ingress–egress to the on-street parking area (veh/km)	Link	Videography technique
On-street pedestrian volume (ped/h)	Link	Videography technique
Traffic control (1 = signal, 2 = manual)	Subject approach	Visual observation
Green period '*g*' (sec), cycle length '*C*' (sec), and the green time over cycle length '*g/C*'	Subject approach	Videography technique
Right-turn on red allowance (1 = yes, 0 = no)	Subject approach	Visual observation
Average bicycle delay (s/bicycle)	Subject approach	Videography technique
Platoon ratio	Subject approach	Videography technique
Volume of crossing pedestrians across the path of through bicyclists (ped/h)	Subject approach	Videography technique
On-street parking turn-over (1 = high, 0.5 = moderate, 0 = minimal)	Subject approach	Visual observation
Built-environment and other		
Surrounding developments (1 = highly commercial, 0.5 = moderately commercial, 0 = minimal commercial)	Link, subject approach	Visual observation
Pavement Condition Index (varies from 5 = excellent through 1 = worst)	Link, subject approach	Visual observation
Street-lighting condition (1 = good, 2 = moderate, 3 = poor)	Link, subject approach	Visual observation
Visibility and maintenance of signaling devices (1 = good, 2 = moderate, 3 = poor)	Subject approach	Visual observation

Note: Directions of turning lanes and turning movements (left or right) corresponds to the left-hand drive rule followed in India.
[a]All facilities include the link, subject approach, opposing approach and conflicting approaches.
[b]Effective width of outside lane is the sum total width of outermost lane, paved shoulder, and the width of paving between outermost lane stripe and outer edge of pavement minus the average width reduction owing to various encroachments in the outermost lane (*9*);
[c]High traffic on a driveway corresponds to a flow rate of at least 4000 vehicles/day (i.e., peak-hour traffic of 400 vehicles/h or above) (*23*); and
[d]Signalized intersection approaches considered in this study allow left-turns on red. The ISD at these locations requires that, (1) the first vehicle on the subject approach should be visible to the drivers of the first vehicles on all other approaches of the intersection, and (2) the departure sight triangle for left-turning vehicles should be provided (*24*). Based on these requirements, the ISD availability is classified as: 1 = good (if both of the ISD requirements are fulfilled), 2 = moderate (if only one requirement is fulfilled, but not the other), and 3 = bad (if none of the requirements is fulfilled).

(approximately 170 bicyclists from each site) comprising employees, homemakers, students, and retirees participated in the survey. Children under the age of 14 were not interviewed, as they may provide inaccurate responses. The minimum bicycling experience of participants was 1 year, which ensured that the interviewees have good knowledge of the urban bicycling environment.

Survey participants represented a diverse mix of socio-demographics and travel-related characteristics. The age groups of participants were classified as: young (14–25 years old), middle-aged (26–60 years old), and elderly (above 60 years old). This classification was based on the critical thresholds that the age of 25 is usually used to differentiate the young and middle-aged persons (*26*), and the retirement age (60 years in India) is usually used for the elder age identification. The female participation rate at individual sites varied in the range of 41–46%. Similarly, 26–32% of the total participants at individual sites were young, 51–69% were middle-aged, and 9–17% were elderly. However, as per Census 2011, 48% of the total Indian population are females and the remaining 52% are males. Further, the total Indian population of 14 years old and over includes 33% young, 57% middle-aged, and 10% elderly persons. Thus, the participation rates of females and young bicyclists were lower in the survey sample as compared with the national statistics. In this regard, the database was weighted using the gender and age proportions from the national demographics. The responses of about 15 middle-aged and five elderly males were excluded at random. Following this weighting, the participants at individual sites were characterized as follows:

- sample size: 22,320 bicyclists (155 from each site);
- gender: 45–50% females, and 50–55% males;
- age group: 28–35% young, 51–63% middle-aged, and 8–14% elderly;
- bicycling experience: 19–28% have 1–5 years of experience, 42–56% have 5–10 years of experience, and 18–27% have over 10 years of experience;
- daily average bicycling distance: 24–32% ride 1–5 km/day, 46–59% ride 5–10 km/day, 11–17% ride 11–20 km/day and 1–3% ride more than 20 km/day;
- user type: 39–44% regular bicyclists and 55–61% occasional bicyclists; and
- trip purpose: recreation, exercise, education/work, personal errand, and visit a friend/relative.

Survey participants were asked to rate the respective sites on a scale of 1–6 (extremely satisfied–extremely dissatisfied). For instance, if the conditions on road links were such that they felt extremely satisfied as a bicyclist, they were instructed to give a rating of 'one.' Similarly, if the conditions at intersection approaches were such that they felt very much satisfied while crossing it, they were instructed to give a rating of 'one.' The final database included 11,470 observations for road links, and 10,850 observations for intersection approaches. The application of Cochran's sample size formula (*27*) on these data samples revealed that the allowed errors in the estimation of mean perceived BLOS scores for both facilities were limited to 1% only (as estimated at 95% confidence level). This indicated that the collected data sets were sufficient for the further analyses.

Statistical Analyses and Results

This section presents a detailed discussion on selection of variables, development of BLOS models, interpretation of obtained results, and the ranges of BLOS classes (A–F).

Variable Selection

Spearman's correlation analysis was carried out to identify the significant variables for road links and intersection approaches. The array of the dependent variable included perceived BLOS scores obtained at respective sites, and the array of independent variables included various quantitative attributes, and socio-demographic details and travel characteristics of bicyclists. Attributes related to socio-demographic and travel characteristics were inputted by using suitable categorical or ordinal scales. For instance, the gender of bicyclists was inputted as 0 for female and 1 for male, and the age group was inputted as 1 for young, 2 for middle-aged and 3 for elderly bicyclists.

Listed in Table 2, eight quantitative attributes were observed to have significant correlations (p-values < 0.001) with perceived BLOS scores at road links, and that for intersection approaches included seven quantitative variables. Although the socio-demographic and travel characteristics of bicyclists altered their perceived satisfaction levels, these alterations were found to be insignificant. In the case of multi-collinear variables, the one having the highest correlation with perceived BLOS scores was selected as the model input. For instance, average delay, g/C and opposing traffic volume were highly correlated with each other with Spearman's rho (ρ)-values of above 0.8. Of these variables, average delay was selected as the only model input as it had the highest correlation with perceived BLOS score ($\rho = 0.569$).

Table 2 shows the descriptive statistics (range, mean, and standard deviation) of significant variables alongside their correlations (ρ) and significances (p-value) with the perceived BLOS score. Descriptive statistics ensured that the attributes were collected from widely diversified

Table 2. Descriptive Statistics and Significance of Variables Used for Model Developments

Variable	Unit or value	Range	Mean	Standard deviation	Spearman's ρ	p-value
Link attributes						
Effective width of the outermost lane ($WOTL_{Eff}$)	m	2–7	3.68	0.90	−0.532	<0.001
Peak-hour traffic volume per lane (PHV/L_{Link})	PCUs/h/lane	148–2586	1069.6	489.10	0.550	<0.001
Average traffic speed (S)	km/h	23–46	33.94	4.88	0.410	<0.001
Pavement Condition Index (PCI)	1–5 (scale)	2.5–4.5	3.85	0.37	−0.465	<0.001
Commercial activities on the roadside area (CA_{Link})	1 (high), 0.5 (medium), 0 (minimal)	0–1	0.49	0.44	0.520	<0.001
Interruptions by roadside stoppages of intermittent public transits ($IIPT$)	1 (high), 0.5 (medium), 0 (minimal)	0–1	0.45	0.39	0.409	<0.001
Vehicular ingress-egress to the on-street parking area (P_{Link})	Veh/h/km	0–6000	745.78	1442.45	0.530	<0.001
Frequency of driveways carrying a high volume of traffic (DW_{Freq})	Driveways/km	0–3	0.83	0.95	0.439	<0.001
Overall perceived BLOS score ($BLOS_{Perc,\ Link}$)	Scale of 1–6	1.55–5.40	3.49	0.71	–	–
Intersection approach attributes						
Approach width (AW)	m	3–14	8.84	2.73	−0.483	<0.001
Peak-hour traffic volume on the approach ($PHV_{Approach}$)	PCUs/h	395–4086	1609.7	831.74	0.466	<0.001
Crossing pedestrian volume (PV)	ped/h	33–1700	409.15	370.81	0.676	<0.001
Turning vehicular volume across the path of through bicyclists (V_{Turn})	PCUs/h	69–703	288.8	140.05	0.462	<0.001
Average bicycle delay (D)	s/bicycle	15–52.2	27.29	8.79	0.569	<0.001
On-street parking turn-over ($P_{Approach}$)	1 (high), 0.5 (medium), 0 (minimal)	0–1	0.46	0.34	0.625	<0.001
Commercial activities ($CA_{Approach}$)	1 (high), 0.5 (medium), 0 (minimal)	0–1	0.51	0.43	0.563	<0.001
Overall perceived BLOS score ($BLOS_{Perc,\ Approach}$)	Scale of 1–6	1.3–5.7	3.64	1.02	–	–

bicycling environments, and the database can develop well-generalized BLOS models. In addition, p-values of ρ concluded that the selected variables are significantly correlated with the perceived BLOS score. A positive ρ-value indicates that the perceived BLOS score increases with increasing values of the concerned variable (i.e., service quality decreases), and a negative value indicates the reverse. A higher ρ-value means that the variable concerned has a higher influence on the BLOS criteria. With these theories, it can be concluded from Table 2 that the peak-hour traffic volume per lane and effective width of the outermost lane have the highest (adverse) and second highest (positive) influences on the link BLOS. Similarly, crossing pedestrian volume, vehicular ingress–egress to the on-street parking area, and average bicycle delay have by far the most significant influences (adverse) on the intersection BLOS.

Model Development

In the present context, bicyclists irrespective of their socio-demographic and travel characteristics were observed to perceive similar kinds of services under a given set of road conditions. Thus, the overall perceived scores obtained at individual sites were used as the array of output variable. Therefore, the final database of road links included 74 observations, and that for signalized intersections included 70 observations. In each case, 70% of the total data were used for model training, and the remaining 30% were used for model testing. Splitting of the total data into these two groups was done in such a way that both groups covered the wide variations in the contained variables. The MGGP algorithm was executed in MATLAB to develop the following models.

Link Model. The best BLOS model for urban road links was obtained with a population size of 800 individuals at 400 generations, G_{max} of 3, d_{max} of 4, and p_r, p_c, and p_m values of 0.02, 0.84, and 0.14, respectively. In total three genes were retained in the link model as the G_{max} value was three. Tree structures and mathematical expressions of individual genes are presented in Figure 2a. This figure also demonstrates the process followed to derive the final expression of the link model by incorporating the expressions of individual genes, respective gene weights (w_1, w_2, and w_3), and the value of bias parameter (w_0). The estimated values of w_1, w_2, and w_3 are 0.0512, −0.486, and 0.401, respectively, which all are highly significant at the 95% significance level (i.e., p-values $<$ 0.05). In addition, the estimated value of w_0 is 0.476, which is also significant with p-value < 0.1. Values of these coefficients along with mathematical expressions of three genes [$CA_{Link}{}^4 \times (WOTL_{Eff} - CA_{Link})^2$, $\ln(WOTL_{Eff} - 1.55) - \ln(DW_{Freq} + S)$, and $\ln(PHV/L_{Link} + P_{Link}) - (PCI - IIPT^{0.5})$] were inputted in Equation 1 to derive the final link model as follow:

$$\begin{aligned} BLOS_{Link} = {} & 0.476 + 0.041 \times \ln\left(PHV/L_{Link} + P_{Link}\right) \\ & - 0.401 \times PCI + 0.486 \times \ln\left(S + DW_{Freq}\right) \\ & - 0.486 \times \ln(WOTL_{Eff} - 1.55) + 0.0512 \\ & \times WOTL_{Eff}{}^2 \times CA_{Link}{}^4 \\ & - 0.102 \times WOTL_{Eff} \times CA_{Link}{}^5 \\ & + 0.401 \times IIPT^{0.5} + 0.0512 \times CA_{Link}{}^6 \end{aligned} \tag{2}$$

where

$BLOS_{Link}$ = Overall predicted BLOS score for a road link.

The signs of coefficients obtained in Equation 2 mean that the service quality of road links is positively influenced (i.e., $BLOS_{Link}$ value is negatively influenced) by $WOTL_{Eff}$ and PCI, whereas all other parameters negatively influence the same.

Intersection Model. The best BLOS model for signalized intersection approaches was obtained with a population size of 1200 individual at 400 generations, G_{max} of 2, d_{max} of 4, and p_r, p_c, and p_m values of 0.02, 0.84, and 0.14, respectively. Figure 2*b* presents the tree structures and the mathematical expressions of two genes retained in the final model.

The estimated values of gene weights (w_1 and w_2) and the bias parameter (w_0) of intersection model are 0.000837, 0.702, and 0.4 respectively, which all are highly significant with p-values $<$ 0.05. The final intersection model is derived as follow by inputting the mathematical expressions of two genes [$PHV_{Approach}{}^{0.5} \times (D - P_{Approach}) + V_{Turn} \times CA_{Approach} + V_{Turn} \times P_{Approach}$, and $\ln(PV + D) - \ln(AW + 7.79)$], and the values of model coefficients (w_1, w_2, and w_0) in Equation 1:

$$\begin{aligned} BLOS_{Int} = {} & 0.4 + 0.000837 \\ & \times \left(V_{Turn} \times P_{Approach} + V_{Turn} \times CA_{Approach} + PHV_{Approach}{}^{0.5} \times D\right) \\ & + 0.000837 \times PHV_{Approach}{}^{0.5} \times P_{Approach} + 0.702 \times \ln(PV + D) - 0.702 \times \ln(AW + 7.79) \end{aligned} \tag{3}$$

where

$BLOS_{Int}$ = Overall predicted BLOS score for an intersection approach.

The signs of coefficients obtained in Equation 3 mean that all input parameters except AW have negative influences on the performance of signalized intersection approaches.

Reliability of Developed Models

Prediction performances of developed models were assessed in terms of R^2, RMSE and the mean absolute percentage error (MAPE) of overall perceived and predicted BLOS scores. Obtained results are summarized in Table 3. As observed, the link model has produced a high R^2 value of 0.872, and minimal RMSE and MAPE values of 6.18% and 0.254, respectively, with the training data. Thus, the model has a reliable goodness-of-fit. It has also produced a high R^2 value of 0.868, and minimal MAPE and RMSE values of 7.05% and 0.290, respectively, with the testing data sets. Thus, the model is well validated in the present context. Further, the overfitting ratio (OR) has been estimated to witness the generalization ability of the model. 'OR' is defined as the ratio of RMSE in model testing to the same in model training. A value of 'one' represents the ideal generalization ability. The link model has shown an excellent generalization ability in the present context with an OR value of 1.142. As observed in Table 3, the intersection model has also produced very high R^2 values and minimal MAPE and RMSE values with both training and testing data sets. Thus, the model has excellent goodness-of-fit and validity in the present context. The model has also shown an excellent generalization ability with an OR value of 1.082.

The statistical performance of the link model was compared with the performances of an ordered probit-based (*15*) and a MARS-based (*16*) model previously proposed in the present context. The former model estimates the BLOS score of an urban road link in two steps. First, the probabilities of attaining individual service scores (*1–6*) on the desired link are estimated by the ordered probit modeling approach. Then, the overall BLOS score is estimated by taking the weighted average of six probability values as obtained. Alternatively, the MARS-based model utilizes the normalized values of independent variables and returns the BLOS score normalized within the range of 0–1. The same is further denormalized to find out the real BLOS score. As depicted in Table 3, the newly developed and existing link models

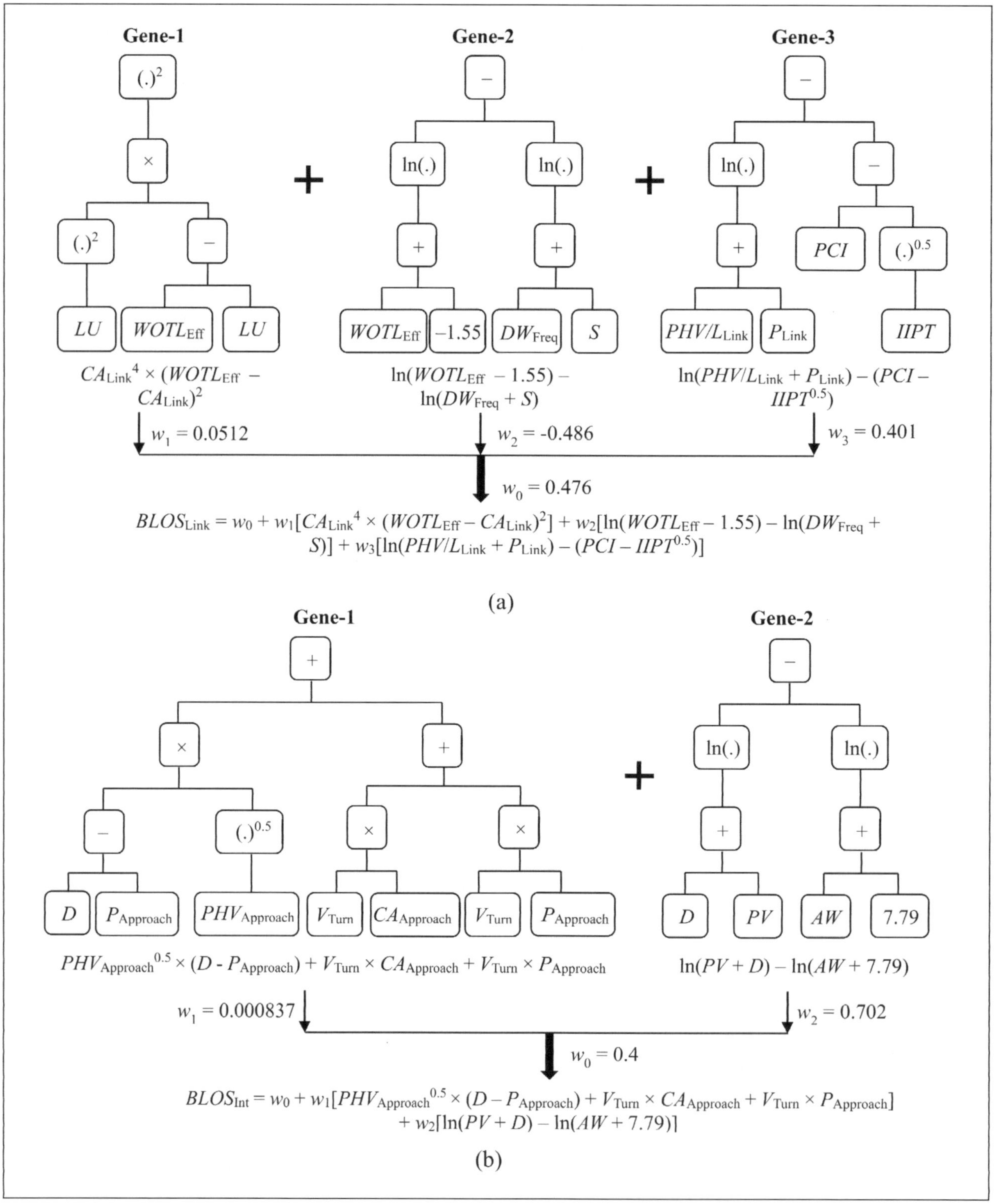

Figure 2. MGGP model formulations for (*a*) road links and (*b*) signalized intersection approaches.

Table 3. Prediction Precisions of BLOS Models

BLOS model	Data	R^2	MAPE (%)	RMSE	OR
Newly developed models					
MGGP-based link model	Training	0.872	6.281	0.254	1.142
	Testing	0.868	7.051	0.290	
MGGP-based intersection model	Training	0.905	7.522	0.303	1.082
	Testing	0.918	6.987	0.328	
Existing models					
Ordered probit-based link model (*15*)	Training	0.841	6.914	0.286	0.905
	Testing	0.873	6.235	0.259	
MARS-based link model (*16*)	Training	0.908	4.770	0.217	1.129
	Testing	0.932	6.819	0.245	

have comparable performances, and the MARS-based model is marginally better than the others are. However, the MGGP-based model is the most advantageous one from the point of view of its simplicity and manageability. The sensitivity of this model could also be easily understood by non-modelers. Performances of existing intersection models are not presented as all of them are developed for homogeneous traffic conditions and are highly likely to show inferior performances in the present context.

Ranges of BLOS Classes (A–F)

A service scale has been defined to convert the model predicted BLOS scores into letter-graded service classes (A–F). Here, the letters 'A' through 'F' designate excellent, good, average, below average, poor, and very poor quality of services, respectively. The scale used to collect perceived BLOS scores varied from one to six. Thus, the middle value of this scale (3.5) corresponds to the boundary between service classes 'A–C' and 'D–F.' Considering the symmetry of this boundary point, the service scale is defined as:

- BLOS A: BLOS score $\leq$ 1.5
- BLOS B: 1.5 $<$ BLOS score $\leq$ 2.5
- BLOS C: 2.5 $<$ BLOS score $\leq$ 3.5
- BLOS D: 3.5 $<$ BLOS score $\leq$ 4.5
- BLOS E: 4.5 $<$ BLOS score $\leq$ 5.5
- BLOS F: BLOS score $>$ 5.5

Conclusion

The quality of bicycling on transportation infrastructures is usually quantified by BLOS models. However, most of the existing BLOS models are developed for homogeneous traffic conditions only, which are not transferable to the heterogeneous traffic conditions prevailing in developing countries. Although ordered probit-based and MARS-based models are available in these countries for the assessment of urban road links, those are very complex and somewhat unfavorable for field implementation. Besides, there exists no standard model for the assessment of road intersections carrying heterogeneous traffic. To fill this gap, the bicycling environments persisting on 74 urban road links and 70 intersection approaches (for through movement only) of nine Indian cities are thoroughly analyzed in this study. As observed from the Spearman's correlation analysis, the BLOS of road links is decided by eight quantitative variables, and that for intersection approaches include seven variables (Table 2). Incorporating these variables, highly reliable and less complex BLOS models are developed for these facilities with the help of the MGGP technique. Data sets used for the training and testing of these models are collected from widely diversified road conditions (Table 2). Thus, the developed models would provide reliable performances in various developing cities across the world. Traffic planners and engineers could use these models to quantify the efficiency of existing bicycle facilities and recognize the extent of needs for further improvements.

The field implementation of the proposed service scale has revealed that above 95% of the total investigated sites are offering average to very poor service levels (BLOS C–F). To improve the quality of these facilities, the most important variables should be prioritized. As observed in this study, traffic volume and outermost lane width are the most important variables for road links, and crossing pedestrian volume, on-street parking activities, and average bicycle delay are the most important variables for signalized intersection approaches. Among the limitations of this study, the proposed models do not address the influence of bicycle lane parameters. This is because bicycle lanes were not available at the study locations. Thus, this could be investigated in future studies. The developed models may not provide desirable performances in highly populated metropolitan cities, where road conditions are more complex than the locations considered in this study. In such cases, either the proposed models could be used with some calibrations, or similar new models could be developed if the choice of significant variables alters.

Author Contributions

The authors confirm contribution to the paper as follows: study conception and design: Sambit Kumar Beura, Prasanta Kumar Bhuyan; data collection: Sambit Kumar Beura; analysis and interpretation of results: Sambit Kumar Beura, Prasanta Kumar Bhuyan; draft manuscript preparation: Sambit Kumar Beura, Prasanta Kumar Bhuyan. Both authors reviewed the results and approved the final version of the manuscript.

References

1. Koza, J. R. *Genetic Programming: On the Programming of Computers by Means of Natural Selection*. MIT Press, Cambridge, MA, 1992.
2. Poli, R., W. B. Langdon, and N. F. McPhee. *A Field Guide to Genetic Programming*. Lulu Enterprises, UK Ltd., Egham, United Kingdom, 2008.
3. Davis, J. *Bicycle Safety Evaluation*. Auburn University, City of Chattanooga and Chattanooga-Hamilton County Regional Planning Commission, TN, 1987.
4. Epperson, B. Evaluating Suitability of Roadways for Bicycle Use: Toward a Cycling Level-of-Service Standard. *Transportation Research Record: Journal of the Transportation Research Board*, 1994. 1438: 9–16.
5. Landis, B. W. Bicycle Interaction Hazard Score: A Theoretical Model. *Transportation Research Record: Journal of the Transportation Research Board*, 1994. 1438: 3–8.
6. Sorton, A., and T. Walsh. Bicycle Stress Level as a Tool to Evaluate Urban and Suburban Bicycle Compatibility. *Transportation Research Record: Journal of the Transportation Research Board*, 1994. 1438: 17–24.
7. Davis, J. Bicycle Test Route Evaluation for Urban Road Conditions. In *Transportation Congress: Civil Engineers—Key to the World Infrastructure*, ASCE, San Diego, CA, 1995, pp. 1063–1076.
8. Harkey, D. L., D. W. Reinfurt, and M. Knuiman. Development of the Bicycle Compatibility Index. *Transportation Research Record: Journal of the Transportation Research Board*, 1998. 1636: 13–20.
9. Landis, B. W., V. R. Vattikuti, and M. T. Brannick. Real-Time Human Perceptions: Toward a Bicycle Level of Service. *Transportation Research Record: Journal of the Transportation Research Board*, 1997. 1578: 119–126.
10. Jensen, S. U. Pedestrian and Bicycle Level of Service on Roadway Segments. *Transportation Research Record: Journal of the Transportation Research Board*, 2007. 2031: 43–51.
11. *Highway CapacityManual*. TRB, National Research Council, Washington, D.C., 2000.
12. *Highway Capacity Manual 2010*. Transportation Research Board of the National Academies, Washington, D.C., 2010.
13. Kang, K., and K. Lee. Development of a Bicycle Level of Service Model from the User's Perspective. *KSCE Journal of Civil Engineering*, Vol. 16, No. 6, 2012, pp. 1032–1039. https://doi.org/10.1007/s12205-012-1146-z.
14. Chellapilla, H., S. K. Beura, and P. K. Bhuyan. Modeling Bicycle Activity on Multi-Lane Urban Road Segments in Indian Context and Prioritizing Bicycle Lane to Enhance the Operational Efficiency. *Proc., 12th Transportation Planning and Implementation Methodologies for Developing Countries (TPMDC) Conference*, IIT Bombay, Mumbai, Maharashtra, India, 2016, pp. 1–11.
15. Beura, S. K., and P. K. Bhuyan. Modeling Quality of Bicycle Accommodations on Urban Road Segments Using Functional Networks and Multivariate Adaptive Regression Spline Techniques. Presented at 96th Annual Meeting of the Transportation Research Board, Washington, D.C., 2017.
16. Beura, S. K., H. Chellapilla, and P. K. Bhuyan. Urban Road Segment Level of Service Based on Bicycle Users' Perception under Mixed Traffic conditions. *Journal of Modern Transportation*, Vol. 25, No. 2, 2017, pp. 90–105. https://doi.org/10.1007/s40534-017-0127-9.
17. Crider, L. B., J. Burden, and F. Han. *Multimodal LOS-Point Level of Service Project*. Final Report. Department of Urban and Regional Planning, University of Florida, Gainesville, FL, 2001.
18. Landis, B., V. Vattikuti, R. Ottenberg, T. Petritsch, M. Guttenplan, and L. Crider. Intersection Level of Service for the Bicycle through Movement. *Transportation Research Record: Journal of the Transportation Research Board*, 2003. 1828: 101–106.
19. Carter, D., W. Hunter, C. Zegeer, J. Stewart, and H. Huang. Bicyclist Intersection Safety Index. *Transportation Research Record: Journal of the Transportation Research Board*, 2007. 2031: 18–24.
20. Jensen, S. U. Pedestrian and Bicycle Level of Service at Intersections, Roundabouts and Other Crossings. Presented at 92nd Annual Meeting of the Transportation Research Board, Washington, D.C., 2013.
21. Strauss, J., L. F. Miranda-Moreno, and P. Morency. Cyclist Activity and Injury Risk Analysis at Signalized Intersections: A Bayesian Modelling Approach. *Accident Analysis & Prevention*, Vol. 59, 2013, pp. 9–17. https://doi.org/10.1016/j.aap.2013.04.037.
22. Searson, D. P., D. E. Leahy, and M. J. Willis. GPTIPS: An Open Source Genetic Programming Toolbox from Multi-Gene Symbolic Regression. *Proc., International Multi-Conference of Engineers and Computer Scientists*, Vol. I, Hong Kong, Newswood Ltd., 2010, pp. 77–80.
23. *Driveway Geometrics*. Engineering Policy Guide, EPG 940.16. Missouri Department of Transportation, 2010. http://epg.modot.org/.
24. American Association of State Highway and Transportation Officials. *A Policy on Geometric Design for Streets and Highways*. American Association of State Highway and Transportation Officials, Washington, D.C., 2011.
25. IRC-106-1990. *Guidelines for Capacity of Urban Roads in Plain Areas*. Indian Road Congress, New Delhi, India, 1990.
26. Bai, L., P. Liu, C. Y. Chan, and Z. Li. Estimating Level of Service of Mid-Block Bicycle Lanes Considering Mixed Traffic Flow. *Transportation Research Part A: Policy and Practice*, Vol. 101, 2017, pp. 203–217. https://doi.org/10.1016/j.tra.2017.04.031.
27. Cochran, W. G. *Sampling Techniques*. John Wiley & Sons, Inc., New York, NY, 1997.

The Standing Committee on Bicycle Transportation (ANF20) peer-reviewed this paper (18-02132).

Article

TRR
JOURNAL OF THE TRANSPORTATION RESEARCH BOARD

Transportation Research Record
2018, Vol. 2672(36) 157–166

DOI: 10.1177/0361198118783171
journals.sagepub.com/home/trr
SAGE

Design and Evaluation of K-Pass: A Bicycle-Friendly Modification of Speed Bumps

Vinod Vasudevan[1], Aniruddha Rajurkar[1], Rahul Soni[1], and Akhil Tiwari[1]

Abstract

Traffic safety is a major concern across the world. Since speed is a major cause of traffic crashes, various traffic calming measures are used to enhance the safety of the road users in urban areas. Speed humps and bumps are the most widely used traffic calming devices. As a considerable number of bicyclists use the road along with motorized users, it is important to consider them while deploying any traffic calming devices. Studies have shown that bicyclists experience significantly larger discomfort as compared to motorized vehicles while passing over speed humps. In this study, a modification has been proposed to enhance riding comfort of bicyclists without compromising the traffic calming attribute of a speed bump. This modification is named "K-pass." Analysis of data collected before and after the modification justifies the effectiveness in reducing the discomfort of bicyclists. Over a period of eight months, 70% of the bicyclists chose to use K-pass. Data also revealed that a high proportion of motorized two-wheeler users also used the K-pass at reduced speeds. It is expected that the K-pass will be a useful modification to the existing speed humps and speed bumps in areas where bicycle usage is expected.

In urban areas, traffic calming measures have become a necessity as different types of road users and vehicles interact with each other continuously. Traffic calming measures play an important role in traffic safety. Various kinds of traffic calming measures used are speed bumps, speed humps, speed cushions, speed tables, and chicanes. Among all, speed humps (and variations such as speed bump, speed table) are the most commonly used. It is mainly used to achieve lower speeds in speed sensitive areas such as schools, parking lots, toll plazas etc. Speed humps have the advantage of being largely self-enforcing and of creating a visual impression, real or imaginary, that speeding is not permitted on a given stretch of the road (*1*). They are cost-efficient and restrict the number of over-speeding drivers. Assembling and dismantling speed humps is very quick and they are easy to maintain. On the other hand, speed humps pose various limitations such as the inconvenience caused to emergency vehicles and discomfort to local residents who frequently use the road. One major problem created by speed bumps is that non-motorized vehicles such as bicyclists suffer severe vibrations while crossing even at slow speeds. To overcome this shortcoming, various designs of speed humps with bicycle pass lanes, speed tables with smooth channels, and speed kidney have been introduced (*2*, *3*). The efficacy of these designs is yet to be validated in traffic conditions of developing nations, which have a huge proportion of two-wheelers and do not have proper road discipline. Thus, this study aims to modify a speed bump to make it bicycle friendly and validate its effects through before–after traffic analyses.

Literature Review

Traffic calming is the combination of mainly physical measures that reduce the negative effects of motor vehicle use, alter driver behavior and improve conditions for non-motorized street users (*4*). Various measures under traffic calming are meant for controlling speed as well as managing traffic flows. In urban areas, the driving speed plays a decisive role in the severity of the crashes. The chances of fatal injury to pedestrians increase from 10% to 75% if the speed of the vehicle causing the crash is increased from 30 to 50 km/h (*5*). Therefore, the use of traffic calming (speed controlling) measures is unavoidable in urban and residential areas. Variety of measures used include speed humps, raised crosswalks, raised intersections, center islands and chicanes, to name a few (*6*). Amongst these, speed humps/bumps are the most popular owing to their simplicity as well

[1]Indian Institute of Technology Kanpur, Kanpur, India

Corresponding Author:
Address correspondence to Vinod Vasudevan: vinodv@iitk.ac.in

as effectiveness. These work on the principle of causing a certain degree of vibrations/jerks to a vehicle negotiating the hump at speeds higher than the design speed. The design details of such devices are reported in various guidelines and manuals (*7*).

The driver senses vibrations while crossing speed humps in vertical direction. As per Kassem and Nasser (*8*), vertical acceleration higher than 0.6 g resulted in discomfort. Fwa and Liaw (*9*) recommended designing a speed hump geometry resulting into vertical acceleration between 0.6 and 0.9 g. Zainuddin et al. (*10*) recommended selecting the optimum dimensions of a hump based on limiting the vertical acceleration to an acceptable range.

Apart from their efficacy, speed bumps also create a few undesirable issues. In urban/residential areas, where speed humps are predominantly used, the road users consist of pedestrians, bicyclists, emergency response vehicles and public transportation vehicles, in addition to the motorized personal vehicles. These road users have peculiar characteristics which are not complemented by the speed humps. Some of the challenges faced by these road user groups while passing hump/bump are discussed as follows:

- In the absence of sidewalk/footpath (which is the case in most of the developing and under-developed countries), pedestrians moving with baby strollers face difficulties.
- Persons with disability, who use wheelchair due to their special requirements, face severe difficulty in passing.
- Bicyclists suffer whole-body vibrations and loss of momentum when they negotiate a speed hump (*11*).
- Repeated exposure to uncharacteristic impact, when a bicyclist crosses hump, may lead to various health issues like lower back problems (*12–14*).

Some of these issues have been acknowledged and remedies have been designed across the world (e.g., *15, 16*). Provision of separate footpath for pedestrians is the most obvious solution. As far as timings of emergency response vehicles are concerned, researchers have designed a speed hump with channels positioned wide enough so as to match with the wheel base of the emergency response vehicles; thus preventing over speeding by the common four wheeled vehicles (*2*). This paper focuses specifically on non-motorists. A previous study (*11*) showed that even at speeds lower than the design speed limit, a high degree of discomfort was caused to the bicyclists. The commonly used commuter bicycles in developing countries are very rigid in structure and hardly have any shock absorbing accessories. This may act as a discouraging factor for popularizing cycling as a major mode in an urban area. The study of the policy being followed by the city of Copenhagen, Denmark, highlights the design of by-pass lanes for bicyclists in harmony with traffic calming measures (*17*). With the objective of excluding the bicyclists from the discomfort caused by a speed hump, several design modifications are being used globally, some of which are referred to below. Bicyclists do not have to ride over the bump, instead a narrow channel is left at the edge of the carriageway without the bump extending into it. The narrow channelizing island/strip also contributes in separating the bicyclists' lane from the main driveway (*18*). At these places, the bicycle lane is separated by use of yellow colored bollards on both sides of the speed hump location (*19*). This prevents a four-wheeler from using the two strips cut out in the hump for passage of bicyclists.

Designs mentioned above have been proving successful in most developed cities. The effectiveness of these measures hinges on the responsible behavior of a road user who is aware of and inclined to obey traffic rules. However, from developing nations' perspective, there are other factors which merit consideration. Firstly, the types of vehicles (both motorized and non-motorized) are highly heterogeneous. Secondly, road users do not follow proper road discipline. Thirdly, motorized two-wheelers maneuver around the speed hump at high speeds to avoid vibrations. Such maneuvers may prove hazardous for the bicyclists/pedestrians. Therefore, any modified passage created for bicyclists, such as by-pass lanes, has a high chance of being abused by riders of motorized two-wheelers, making the speed hump/bump ineffective in reducing speed or enhancing safety of the area.

In summary, although a few design modifications for speed humps are proposed, they are mainly aimed at reducing discomfort for specific types of motorized vehicles. Of the ones aimed at bicyclists, chances of abuse by motorized two-wheelers are high. Since motorized two-wheelers constitute approximately 60% to 70% of road users in most of the developing countries, these enhancements could negate the effectiveness of speed humps/bumps (*20, 21*). Hence, a newer design is required to address this issue.

Objective

The objective of this study is to propose a design modification to the speed hump/bump in developing countries so that the discomfort caused to the bicyclist and other vulnerable road users is reduced without compromising the primary function of reducing speeds of motorized vehicles and enhancing safety of the location.

Design

In this section, although bicyclists are addressed directly, they represent other vulnerable road users as well, since any design improvement to accommodate bicyclists will make it welcoming to other vulnerable road users.

One way to reduce discomfort to bicyclists is to reduce the vertical movement of bicyclists while going through the hump location. In this study, the working principle of chicane is explored for both ends of a speed hump/bump.

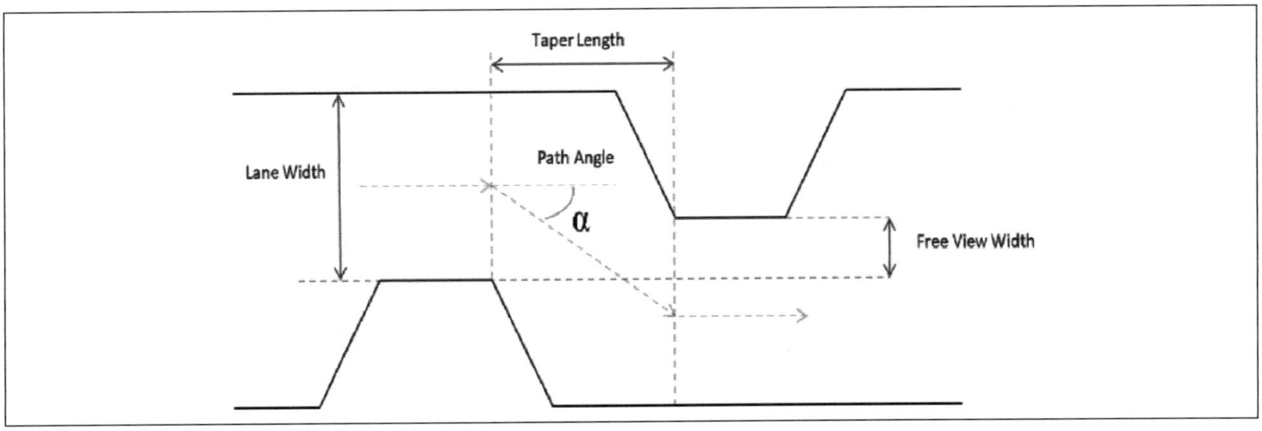

Figure 1. Design of chicane.

Chicane is a speed calming device which works on the principle of sharp change in horizontal alignment (without any change in vertical alignment) over a small distance forcing the motorists to reduce their speed. Figure 1 shows a schematic top view of a chicane. The path angle "α" depicts the change in the driving direction of a driver. As per Leonardi (*22*), a shift of one lane width or deflection angle of at least 45° is recommended to prevent the drivers from maintaining their driving speed and alignment. For a given length of the chicane segment, as the deflection angle "α" increases further, the drivers are forced to reduce the driving speed that much more.

In a standard chicane, a portion exists which provides free view of the road. This may affect the effectiveness of the speed hump as along with providing smoother ride for bicyclists, it may also allow two-wheelers to pass through this portion at faster speeds, hence nullifying the main functionality of the speed hump. Hence the major challenge is to adapt the concept of chicane to achieve the goal of giving a comfortable ride to bicyclists without allowing fast passage of motorized two-wheelers. Although it can be achieved using different methods, the way adapted in this study is discussed in this section.

Since bicyclists are expected to use the sides of the roads, the modification needs to be made at both ends of the speed bump. To avoid vibrations, both end portions of the speed bump need to be flattened out. Special provisions need to be introduced to force riders of motorized two-wheelers to slow down, in case they choose to use the modified portion of the bump. One way is to utilize the design philosophy of chicane and introduce a couple of mini bumps, one each at upstream and downstream of the modified portion of the bump. Plan view of this concept is shown in Figure 2. The modified portion, at both ends, of the bumps is named K-pass henceforth. Here, "K" stands for the city Kanpur, India, where this design was conceptualized and evaluated.

As shown in Figure 2, the distance from the main bump (*X*), width of removed portion of the existing bump (*Y*), and

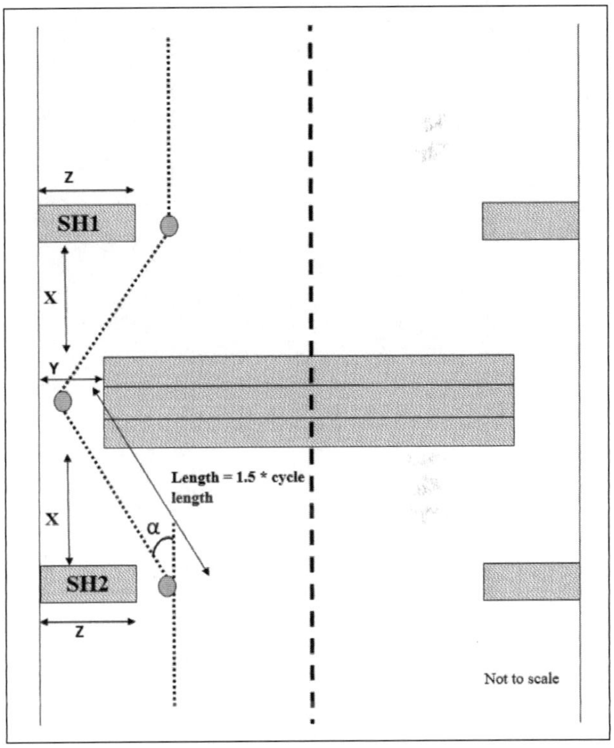

Figure 2. Design elements of K-pass.

width of each mini bump (*Z*) will become the governing dimensions of K-pass. To achieve the goal of smoother ride for bicyclists and disallow high-speed motorized two-wheelers, appropriate values of these need to be determined. Although speed humps/bumps with any acceptable height/breadth (*h/b*) ratio can be considered for the two mini bumps (labeled SH1 and SH2 in Figure 2), to maintain aesthetics of the location, it is recommended that the *h/b* ratios of mini bumps are kept the same as that of the original hump/bump. *Y* is the length of portion of original speed bump which was

Figure 3. Experimental setting.

cut from both edges of speed hump/bump. *Y* can be assumed to be the same as that of the bicycle lane. In case bicycle lanes are not present, at least 60 cm need to be provided to enable easier passage. The path expected to be traversed by the cyclist is shown by dotted line in Figure 2. The change in alignment is depicted by angle α. Estimation of the remaining design parameters are discussed in the following two sub-sections.

Locations of Mini Bumps with Reference to Main Hump/Bump

As shown in Figure 2, the spacing between the mini bumps and the main bump will influence the speed and ease of maneuvering through K-pass. It is assumed that the bicyclist/rider of motorized two-wheeler traverses a path inclined at some angle to the straight alignment. Based on the recommendations of the European manual for design of chicane (*22*), the deflection angle of 45° is fixed. When a bicycle passes through the cut portion of the original speed hump/bump, along the expected path as shown using dotted line in Figure 2, half length of the bicycle will be behind the hump and half will be in front of the hump. Hence, the length of the path that a bicycle traces from mini bump to the cut portion is assumed to be 1.5 times the length of the bicycle. Hence a length of 1.5 times the length of the vehicle will provide minimum distance required for a vehicle to maneuver around the original bump. Assuming the length of average bicycle to be 1.75 m, the length of path traced by bicyclist can be estimated to be 2.62 m ($1.5 \times 1.75 = 2.62$). From the geometry of the diagram, the minimum value of *X* for a bicycle comes out to be 1.86 m ($1.5 \times 1.75 \times \text{Sin } 45$). If value of α is increased beyond 45°, the *X* value would reduce, thus making it difficult to maneuver vehicles through the K-pass.

In this case, the value of *Y* has been fixed to 60 cm. The maximum value of *Z* is governed by the deflection angle α as mentioned earlier. For α value to be 45°, it can be found that value of *Z* is about 2.16 m ($X + Y/2$). Providing any value of *Z* higher than 1.5 times the value of *Y* would increase the difficulty of maneuver for the bicyclists riding in the edge strip of the carriageway since they have to go into the main carriage way and share road with motorized vehicles, resulting in chances of vehicle–bicycle conflicts, before using K-pass. Hence for design purpose, it is recommended to use a value *Z* to be at least 1.5 *Y* so that the possibilities of free view width are avoided.

The values estimated previously provide a starting point for experiment. To find the reasonable value of *X*, satisfying the objective of the problem, the value obtained in previous step (1.86 m) is taken as a reference. Trials were carried out by placing the safety cones in a way to represent alignment of proposed K-pass as shown in Figure 3. The values of *Y* and *Z* were taken as 60 and 90 cm respectively to give the chicane effect without providing the free view width shown in Figure 1. The first and third cones represent the inner end points of the proposed mini bumps placed at *Z* distance from edge of the carriageway (one upstream and one downstream of original bump), whereas the central cone represents the outer end point of the original bump placed at *Y* from the end of carriageway. Now, the value of *X* was varied over a range of values from 1 to 3.5 m with 0.5 m increments. For various values of *X*, using three volunteers each riding motorized two-wheelers (motor cycle and scooter) and bicycles, trial runs were conducted. The subjects were asked to ride at the maximum comfortable speed through the set-up. The speed at which a driver could pass through was chosen to be an indicator of the ease of negotiating this arrangement at the corresponding value of *X*. The behavior of a road user towards any modification of infrastructure would evolve gradually as the user gets familiar with it over passage of time (*23*). Therefore, initial ten runs by the riders were not considered. Thereafter, each rider was asked to go through the arrangement 15 times. The speeds of the vehicles were recorded 20 m before the set-up and within the set-up. The observed speeds at the set-up during trial runs are presented in Figure 4.

From Figure 4, it can be noted that for initial spacing of 1 and 1.5 m, it was impossible for any vehicle to maneuver through the set-up without disturbing the cone or without

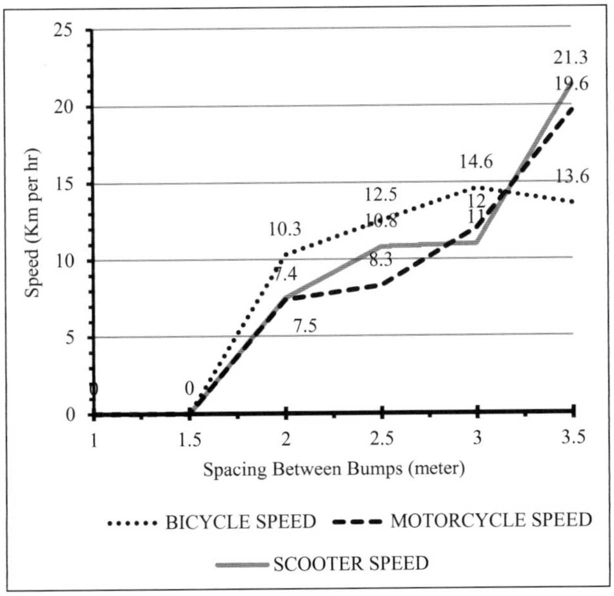

Figure 4. Variation in speeds in response to variation in spacing between mini bumps and main bump (*X*).

placing foot on ground while passing. As the lengths of typical bicycles and motorcycles are 1.75 and 2 m respectively, entering the portion depicted by the first two cones and turning immediately was impossible for riders of the different vehicles considered. Hence, the graph shows zero speed for spacing 1 and 1.5 m. As the spacing was increased to 2 m, the bicyclists found it easier to maneuver. However, the riders of motorcycle and scooter had to slow down significantly. At a spacing of 2.5 and 3 m, similar trends in speeds were observed for these three users, with bicyclists being able to ride at increased speeds. Although further increase in spacing to 3.5 m resulted in increase in comfortable speeds of bicyclists, the riders of motorized two-wheelers also found it easier to maneuver through the set-up at speeds over 20 km/h. These trials provided vital information that the spacing has to be some value between 2.0 and 3.0 m. Any value higher than this will nullify the effect of speed bumps for motorized two-wheelers.

In summary, the preferred dimension of K-pass is as follows:

- *X* should be between 2.0 and 3.0 m;
- Y should be same as the width of the bicycle lane (if one exists); if the bicycle lane does not exist, a minimum value of 60 cm is required for two-lane roads; a larger value can be chosen for wider roads; and
- Z should be at least 1.5 times the value of Y.

It is worth mentioning that if wider bicycle paths (with width over 1.0 m) are present, the values of *Y* need to be adjusted such that the corresponding values of *Z* do not become very large. A large value of *Z* will force bicyclists into the direct path of motorized vehicles, compromising their safety.

Assessment of the Effectiveness of the K-Pass

To understand the effectiveness of the design, it is important to evaluate K-pass. Effectiveness of K-pass was measured by conducting a before and after study. The following measures of effectiveness (MOEs) were used for evaluation.

1. Percent of bicyclists using K-pass: As the K-pass is designed for bicyclists' comfort, the percentage of total bicyclists getting channelized to make use of this facility will be one of the key indicators.
2. Percent of motorized two-wheelers using K-pass: The K-pass is to be designed in such a way as to discourage motorized two-wheelers from maneuvering through the smoothened section at higher speed. Therefore, low percentage of motorized two-wheelers passing through K-pass is preferred.
3. Speed of bicyclists: The speed bumps were forcing bicyclists to go at very low speeds. It is expected that the bicyclists will be able to negotiate K-pass at higher speeds.
4. Speed of motorized two-wheelers: Although it is preferable that motorized two-wheelers do not use the K-pass, if the speeds of these vehicles are low enough so as not to harm bicyclists, then it may be considered acceptable for them to use the K-pass. Thus, an acceptable low average speed of motorized two-wheelers passing through the K-pass will be an indicator of its effectiveness.

Data Collection

The site having two-wheeler predominant traffic flow was selected inside the campus of the Indian Institute of Technology Kanpur, popularly known as IIT Kanpur. The road considered at the site was a 6 meter-wide two-lane road. This road is used extensively by students, who rely heavily on bicycles as their major mode, and other campus residents, who use motorized two-wheelers for commute. The site had three speed bumps of average height 4 cm and average breadth 64 cm placed back-to-back. This set-up is very common in India and the discomfort it causes to users is even higher than that of standard speed humps. In order to evaluate the MoEs, the width of existing speed bump was divided into six segments, five (*A, B, C, D* and *E*) of which were of width 60 cm (as shown in Figure 5). Segment *F* was the remaining width of the road which is supposed to be used by traffic flowing in the other direction. The video recording camera was set up adjacent to the road, 15 m away from the bump site to record the user behavior. The number of users passing the speed bump was divided based on the segment through which

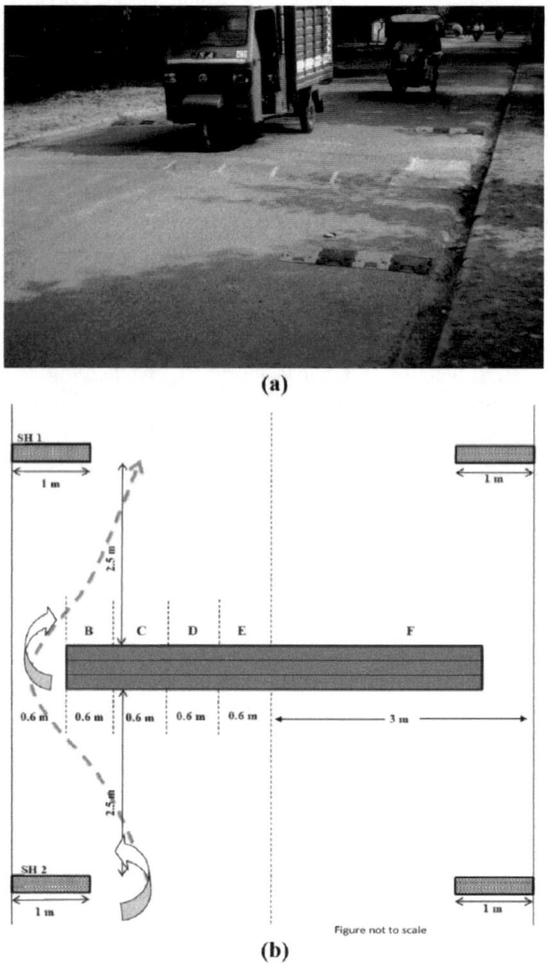

Figure 5. K-pass and sections of existing bump depicted in (*a*) site photograph and (*b*) schematic diagram.

they crossed the bump. Bicyclists and motorized two-wheelers were segregated into different categories. During the same time duration, the spot speeds were randomly measured for 75 bicycles and 75 motorized two-wheelers, at two locations: (i) 20 m prior to K-pass site; and (ii) at the K-pass.

The details of the K-pass installed, and the segments considered, are shown in this Figure 5.

Data were collected in three stages:

(a) Pre-modification: 6 October, 2016 for three hours
(b) Post-modification: After 48 hours' installation. the data were collected for 3 hours.
(c) Follow-up: A follow-up data collection was carried out eight months after installation (19 and 20 June, 2017 for 3 hours each).

Each of these stages are referred to as "before," "after," and "follow-up" respectively in the remaining part of this paper.

Results

Descriptive Analysis

The comparison of the spatial distribution of traffic over the width of the speed bump, before and after modification can be made from the Figure 6*a* and *b*.

Prior to modifying the bump, the percentages of bicyclists passing through *A, B, C, D, E* and *F* sections were found to be 9.9%, 51.1%, 24%, 11.4%, 3.1%, and 0.4% respectively (with sample size of 675). These values changed to 19.9%, 41.9%, 28%, 7.8%, 2.4%, and 0% respectively (with sample size of 372) during after-study data collection. Further, after passage of eight months, the values changed to 70.4%, 2%, 9.3%, 10.8%, 0.6%, and 7% respectively (with sample size of 1,063). At the same time, the percentages of motorized two-wheelers passing through the same sections were observed to be 2.1%, 17.8%, 30.7%, 26.5%, 15.4%, and 7.5% respectively (with sample size of 332) before modification; 10.1%, 14%, 26.8%, 25.1%, 15.1%, and 8.9% respectively (with sample size of 179) immediately after modification and 50.9%, 0.7%, 7%, 12%, 11%, and 18.3% respectively (with sample size of 837) after passage of eight months.

The data collected on speeds for the three time periods showed a notable improvement in speeds of bicyclists. The average speed of bicyclists was 9.89 km/h with a standard deviation of 2.86 km/h before implementation of K-pass. The average speed was improved to 10.68 km/h with a standard deviation of 2.56 km/h after implementation and finally to 11.97 km/h with a standard deviation of 1.06 km/h during follow-up period. For motorized two-wheelers, the average speeds (with standard deviation in parenthesis) for before, after, and follow-up periods are: 19.80 km/h (3.85 km/h), 13.50 km/h (2.17 km/h), and 12.87 km/h (2.83 km/h) respectively. It is worth noting that for before cases, since K-pass did not exist, the speeds of all vehicles (within the defined category) passing over the bumps were considered, whereas for after and follow-up cases, data of vehicles using the K-pass are reported.

Statistical Analysis

Data presented in the previous section showed that the K-pass has been able to attract bicyclists with higher speeds. Although motorized two-wheelers were attracted to the K-pass the passing speed was reduced noticeably. To make these results valid, it is important to show that the changes observed are statistically significant. The following hypothesis testing were conducted to achieve the results:

1. *Hypothesis testing for spatial distribution*

To validate the desired impact of modification in speed bump on the spatial distribution of bicyclists, the Z-statistic-based hypothesis testing is carried out as follows:

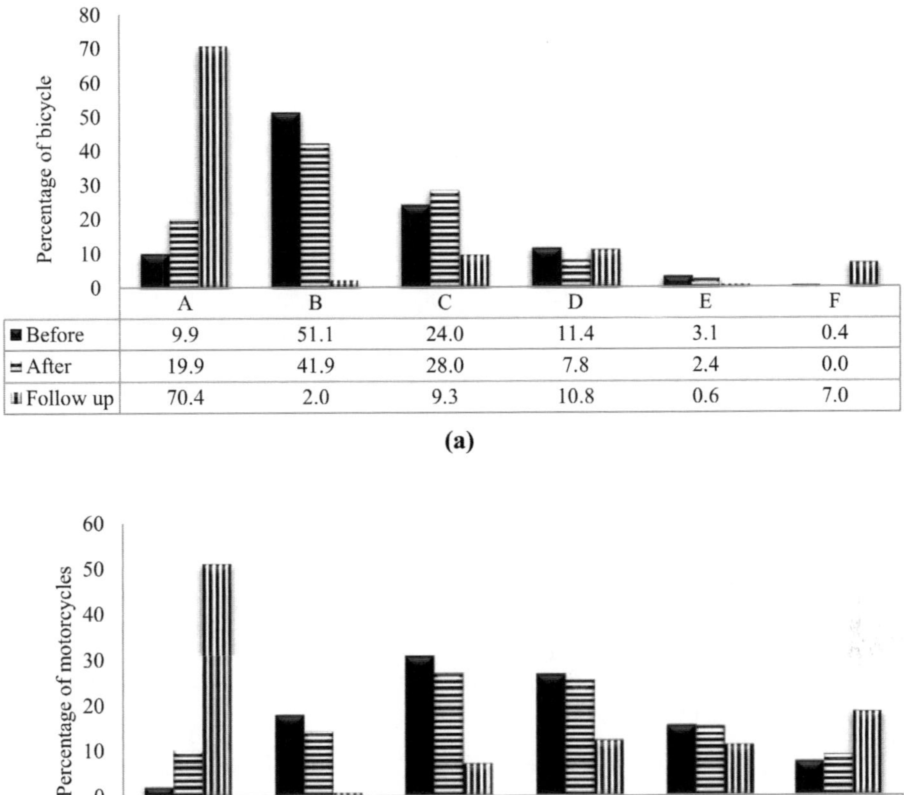

	A	B	C	D	E	F
Before	9.9	51.1	24.0	11.4	3.1	0.4
After	19.9	41.9	28.0	7.8	2.4	0.0
Follow up	70.4	2.0	9.3	10.8	0.6	7.0

(a)

	A	B	C	D	E	F
Before	2.1	17.8	30.7	26.5	15.4	7.5
After	10.1	14.0	26.8	25.1	15.1	8.9
Follow up	50.9	0.7	7.0	12.1	11.0	18.3

(b)

Figure 6. Spatial distributions across speed bump for (*a*) bicycles and (*b*) motorized two-wheelers.

Null hypothesis (H_0) suggests that there is no increment in the proportion of bicyclists/motorized two-wheelers using K-pass (section *A*). The alternate hypothesis (H_A) suggests an increment in proportion of users using K-pass. One-tailed hypothesis testing is used with confidence level of 99%.

$$H_o : p_2 = p_1$$
$$H_A : p_2 > p_1$$

Test statistics:

$$z_{cal} = \frac{\left(\widehat{p_1} - \widehat{p_2}\right) - \left(p_1 - p_2\right)}{\sqrt{\left(\widehat{p}\right)\left(1-\widehat{p}\right)\left(\frac{1}{n1} + \frac{1}{n2}\right)}} \quad (1)$$

$$\widehat{p} = \frac{x_1 + x_2}{n_1 + n_2} \quad (2)$$

$$z_{cr(0.01)} = -2.33$$

where,

p_1 and p_2 are the proportions of bicycles/motorized two-wheelers passing through the K-pass prior to modification and post modification (after and follow-up) respectively.

x_1 and x_2 are the number of bicycles/motorized two-wheelers passing through section *A* prior to and post modification respectively.

n_1 and n_2 are the total number of bicycles/motorized two-wheelers passing through the road section prior to and post modification respectively.

The results are shown in Table 1. The null hypothesis is rejected at 99% confidence level for both bicyclists and motorized two-wheelers, for both the stages of studies after installation of K-pass. The results prove that K-pass attracts both bicyclists and motorized two-wheelers significantly. Although attracting motorized two-wheelers towards it is undesirable, it was expected.

Table 1. Hypothesis Testing: Spatial Distribution of Motorized Two-Wheelers and Bicyclists

	Before			After			Follow-up			Before–after		Before–follow-up	
	P_1	x_1	n_1	P_{2A}	x_{2A}	n_{2A}	P_{2F}	x_{2F}	n_{2F}	Z_{cal}	H_o	Z_{cal}	H_o
Bicyclists	0.099	67	675	0.199	74	372	0.703	748	1063	−4.54	Reject*	−24.63	Reject*
Motorized two-wheelers	0.021	7	332	0.101	18	179	0.509	426	837	−4.00	Reject*	−21.88	Reject*

*At confidence level of 99%.

Table 2. Hypothesis Testing: Speed Distribution of Motorized Two-Wheelers and Bicyclists

	Before		After		Follow-up		Before–after		Before–follow-up	
	X_1	s_1	X_{2A}	s_{2A}	X_{2F}	s_{2F}	$t_{cal.}$	H_o	$t_{cal.}$	H_o
Bicyclists	9.89	2.86	10.68	2.56	11.97	1.06	−1.78	Reject**	−5.91	Reject*
Motorized two-wheelers	19.80	3.85	13.50	2.17	12.87	2.83	7.15	Reject*	12.57	Reject*

*At confidence level of 99%, **at confidence level of 95%.

2. *Hypothesis testing for speed variations*

To validate the impact of K-pass on the speeds of both bicyclists and motorized two-wheelers, a T-statistic-based hypothesis testing has been carried out (as population variance is not known) for each user type for data recorded immediately after the modification as well as during follow-up study.

Bicyclists hypothesis testing: Null hypothesis (H_o) is assumed such that there is no change in average speed of bicyclists passing through K-pass. Hence, alternate hypothesis (H_A) supports that there is an increase in the speed after modification. The confidence levels at which testing is carried out are 90%, 95% and 99%. The results are shown in Table 2.

$$H_o : X_2 - X_1 = 0$$
$$H_A : X_2 - X_1 > 0$$

Test statistics:

$t_{cr(0.01,\ 148)} = 2.35$, $t_{cr(0.05,\ 148)} = 1.65$, $t_{cr(0.10,\ 148)} = 1.287$ for one tailed test.

$$t_{cal} = \frac{(X_1 - X_2) - (\mu_1 - \mu_2)}{s_p\sqrt{\left(\frac{1}{n_1} + \frac{1}{n_2}\right)}} \quad (3)$$

$$s_p = \sqrt{\frac{(n_1 - 1)s_1^2 + (n_2 - 1)s_2^2}{n_1 + n_2 - 2}} \quad (4)$$

where,

X_1 and X_2 are mean speed of bicyclists passing through K-pass prior to modification and post modification respectively (after and follow-up).

n_1 and n_2 are the number of bicycle passing through section *A* prior to and post modification respectively (n_1, $n_2 = 75$).

s_1 and s_2 are the standard deviations observed in speeds for sample n_1 and n_2 respectively.

s_p is the pooled standard deviation.

The null hypothesis is rejected only at 95% confidence level for bicyclists when before data was compared with after data. However, when before data is compared with the follow-up data, the null hypothesis is rejected at 99% confidence level, showing that the improvements observed in bicycle speeds after introduction of K-pass are statistically significant.

Motorized two-wheelers hypothesis testing: Null hypothesis (H_o) is assumed such that there is no change in average speed of motorized two-wheelers; and alternate hypothesis (H_A) supports that there is decrease in the speed after modification. The results are shown in Table 2.

$$H_o : X_2 - X_1 = 0$$
$$H_A : X_2 - X_1 < 0$$

where,

X_1 = mean speed of motorized two-wheelers passing through entire width prior to modification.

X_2 = mean speed of motorized two-wheelers passing through entire width post modification (after and follow-up).

$n_1 = 75$ and $n_2 = 18$ for before–after study (since number of motorists passing through section A was observed to be low) and $n_1 = 75$ and $n_2 = 75$ for before–follow-up study.

For before–after study, $t_{cr(0.01,\ 91)} = 2.368$, $t_{cr(0.05,91)} = 1.66$, $t_{cr(0.10,\ 91)} = 1.29$ for one-tailed test.

For before–follow-up study, $t_{cr(0.01,\ 148)} = 2.35$, $t_{cr(0.05,91)} = 1.65$, $t_{cr(0.10,\ 91)} = 1.287$ for one-tailed test.

As shown in Table 2, in this case, null hypothesis is rejected at 99% confidence levels during after study as well as follow-up study. This shows that the speeds of the motorized two-wheelers who are using the K-pass have reduced significantly.

Discussion of Results

This research is aimed to design and evaluate a mechanism to allow comfortable passage for non-motorized users on speed humps/bumps. A newly designed modification named K-pass was deployed at a site and data were collected for comparative analysis. The analysis of MOEs prove that K-pass is effective in improving rider comfort without compromising safety of the area. Some of the highlights of the results are discussed as follows:

- The comparison of spatial distribution for bicyclists shows that the percentage of bicyclists passing through the K-pass increased to 70.4% after passage of eight months. This can be considered as an indicator of the bicyclists becoming aware of the benefits offered by the K-pass. These changes are statistically significant.
- The comparison of spatial distribution for motorized two-wheelers shows that the percentage of motorized two-wheelers passing through K-pass increased to 50.9% (after eight months). This change is statistically significant as well. Although it is not a desirable outcome, this result is expected, as motorized two-wheelers also will change their paths for a comfortable ride quality.
- For bicyclists, the average speed while negotiating the original bump was 9.8 km/hr. This was significantly improved, showing that while K-pass attracted majority of bicyclists to use this facility, it also enabled them to ride at higher speeds.
- For motorized two-wheelers, the average speed was significantly reduced to values closer to those of bicyclists. This shows that although K-pass attracted motorized two-wheelers because of its comfortable ride quality, they were willing to reduce speeds significantly. Thus, the main function of the speed bump is not compromised by the introduction of K-pass.

Conclusions

The objective of this study was to design a bicycle-friendly speed bump suitable for heterogeneous traffic flow conditions observed in most of the developing countries. Though various designs of speed bumps which enable smooth passage of bicyclists and emergency response vehicles exist, most of them are likely to be misused by road users in heterogeneous traffic conditions in absence of active enforcement. Thus, using the basic concept from the design of chicane, a modification, named as K-pass, was designed to allow comfortable passage of bicyclists. The design was validated by conducting before and after data collection. The results clearly indicate that the bicyclists are attracted to the K-pass, which allows them to pass the bump at their natural riding speeds without experiencing much discomfort. The average speed of bicyclists has been found to have increased by 21%. The improvement is statistically significant. At the same time, the chicane effect incorporated in the design forced the motorized two-wheelers to reduce their approach speeds significantly while passing through the K-pass. This geometry of modification has prevented misuse of the facility by motorized two-wheelers, which otherwise would have created more hazard for bicyclists. The motorized two-wheelers have been observed to negotiate the pass at speeds nearly equal to those observed for bicyclists. This is a desired outcome of the modification. This modification can be extensively used not only at other speed hump/bump locations but also with other traffic calming measures such as speed tables, speed cushions, rumble strips etc. to enhance the cycling experience in a city/region. The dimensions in the design (width of the mini bumps and the cut portion in main bump) can be tweaked further to cater for any peculiar site challenge/requirement to achieve favorable traffic calming effect. The results also point to an interesting finding that motorized two-wheelers are willing to accept lower speeds, provided they are given option of comfortable passage across speed bumps.

Acknowledgments

The authors would like to acknowledge the support from the Institute Works Department at Indian Institute of Technology, Kanpur for installation of the modified design. We also would like to thank Mr. Prince and Mr. Atul Kumar, graduate students at IIT Kanpur for their help during the experiments and data collection.

Author Contributions

The authors confirm contribution to the paper as follows: study conception and design: Vinod Vasudevan, Aniruddha Rajurkar; data collection: Aniruddha Rajurkar, Rahul Soni, Akhil Tiwari; analysis and interpretation of results: Rahul Soni, Akhil Tiwari, Vinod Vasudevan; draft manuscript preparation: Aniruddha Rajurkar. All authors reviewed the results and approved the final version of the manuscript.

References

1. Barber, G. L. *Speed Bump*. U.S. Patent 4,362,424. U.S. Patent and Trademark Office, Washington, D.C., December 7, 1982.
2. Batson, S. M. Offset Speed Tables for Reduced Emergency Response Delay. In *Intersection Safety: Achieving Solutions Through Partnerships*, Institute of Transportation Engineers and Federal Highway Administration, 2004.
3. García, A., A. Moreno, and M. Romero. Development and Validation of Speed Kidney, a New Traffic-Calming Device. *Transportation Research Record: Journal of the Transportation Research Board*, 2011. 2223: 43–63.
4. Transportation Association of Canada. *Canadian Guide to Neighbourhood Traffic Calming*, 1998, ISBN 1551871033,

9781551871035. http://www.tac-atc.ca/en/projects/progress/traffic-calming-guide. Accessed July 25, 2017.

5. Ashton, S. J. Pedestrian Injuries: The Influence of Vehicle Design. In *Road Safety Research and Practice* (H. C. Foot et al., eds.), Praeger, Santa Barbara, 1981.
6. Ahmed, A., and R. Bagchi. Traffic Calming in Indian Perspective. *International Journal of Chemical, Environmental & Biological Sciences (IJCEBS)*, Vol. 1, No. 3, 2013, ISSN 2320–4079; EISSN 2320–4087.
7. Johnson, L., and A. J. Nedzesky. A Comparative Study of Speed Humps, Speed Slots and Speed Cushions. Presented at the ITE Annual Meeting, August 1–4, 2004; Lake Buena Vista, Florida 2004.
8. Kassem, E., and Y. Al-Nassar. Dynamic Considerations of Speed Control Humps. *Transportation Research Part B: Methodological*, Vol. 16, No. 4, 1982, pp. 291–302.
9. Fwa, T. F., and C. Y. Liaw. Rational Approach for Geometric Design of Speed-Control Road Humps. *Transportation Research Record: Journal of the Transportation Research Board*, 1992. 1356: 66–72.
10. Zainuddin, N. I., M. A. Adnan, and J. MdDiah. Optimization of Speed Hump Geometric Design: Case Study on Residential Streets in Malaysia. *Journal of Transportation Engineering*, Vol. 140, No. 3, 2013, p. 05013002.
11. Patel, T., and V. Vasudevan. Impact of Speed Humps on Bicyclists. *Safety Science*, Vol. 89, 2016, pp. 138–146.
12. Adams, M. A., and W. C. Hutton. The Effect of Posture on the Lumbar Spine. *Bone & Joint Journal*, Vol. 67, No. 4, 1985, pp. 625–629.
13. Griffin, M. J. *Handbook of Human Vibration*. Academic Press, London, 2012.
14. Hostens, I., Y. Papaioannou, A. Spaepen, and H. Ramon. A Study of Vibration Characteristics on a Luxury Wheelchair and a New Prototype Wheelchair. *Journal of Sound and Vibration*, Vol. 266, No. 3, 2003, pp. 443–452.
15. Okunribido, O. O., M. Magnusson, and M. Pope. Delivery Drivers and Low-Back Pain: A Study of the Exposures to Posture Demands, Manual Materials Handling and Whole-Body Vibration. *International Journal of Industrial Ergonomics*, Vol. 36, No. 3, 2006, pp. 265–273.
16. Blood, R. P., J. D. Ploger, M. G. Yost, R. P. Ching, and P. W. Johnson. Whole Body Vibration Exposures in Metropolitan Bus Drivers: A Comparison of Three Seats. *Journal of Sound and Vibration*, Vol. 329, No. 1, 2010, pp. 109–120.
17. *Good, Better, Best: The City of Copenhagen's Bicycle Strategy 2011–2025.* http://www.eltis.org/sites/eltis/files/casestudies/documents/copenhagens_cycling_strategy.pdf. Accessed June 20, 2017.
18. Dunhill, L. *Cllrs Agree to Spend £5k on 'Smart' Speed Bump Study*. Bucks Free Press. http://www.bucksfreepress.co.uk/news/9396470.VIDEO5k_to_be_spent_on_smart_speed_bump_study/?ref=rss. Accessed May 12, 2017.
19. *Traffic Calming Report.* Lynnwood Public Works Department, Lynnwood, WA, 2012. http://www.lynnwoodwa.gov/Assets/Departments/Public+Works/Streets/-Documents/Traffic+Calming+Report+10-2012.pdf. Accessed June 20, 2017.
20. Ministry of Statistics and Programme Implementation, Government of India. *Statistical Year Book 2016*. Accessed October 20, 2017 http://www.mospi.gov.in/statistical-year-book-india/2016/189. Accessed October 20, 2017.
21. Poushter, J. *Car, Bike or Motorcycle? Depends on Where You Live*, Palermo, Italy, 2015. http://www.pewresearch.org/fact-tank/2015/04/16/car-bike-or-motorcycle-depends-on-where-you-live/.
22. Leonardi, S. Experimental Survey to Test Traffic Calming Measures Effectiveness in Urban Areas. *Proc., 4th International SIIV Congress*, Palermo, Italy, 2007, pp. 12–14.
23. Caroline, A., E. M. M. Jongen, K. Brijs, T. Brijs, S. Daniels, and G. Wets. A Simulator Study on the Impact of Traffic Calming Measures in Urban Areas on Driving Behavior and Workload. *Accident Analysis & Prevention*, Vol. 61, 2013, pp. 43–53.

The Standing Committee on Bicycle Transportation (ANF20) peer-reviewed this paper (18-06554).

Sage Publications USA
sagepub.com

ISBN 978-1-5108-8244-7

CURRAN ASSOCIATES INC.
proceedings.com

3rd Shared Task on Discourse Relation Parsing and Treebanking (DISRPT 2023)

Toronto, Canada
14 July 2023